Connectionist Models of
Memory and Language

Connectionist Models of Memory and Language

Edited by

Joseph P. Levy, Dimitrios Bairaktaris,
John A. Bullinaria, Paul Cairns

UCL
PRESS

First published in 1995 by UCL Press

UCL Press Limited
University College London
Gower Street
London WC1E 6BT

and
1900 Frost Road, Suite 101
Bristol
Pennsylvania 19007-1598

The name of University College London (UCL) is a registered trade mark used by UCL Press with the consent of the owner.

British Library Cataloguing in Publication Data
A catalogue record for this book is available from the British Library.

Library of Congress Cataloging-in-Publication Data are available.

ISBN: 1-85728-368-6 HB

Printed and bound by
Biddles Ltd., Guildford and King's Lynn, England.

Contents

Preface

There are many different perspectives on the nature of cognitive science. Some see it as a loose conglomeration of diverse viewpoints and approaches to the study of cognition; not really a science at all but a collection of interfacing disciplines. Others view it as a much more convergent enterprise, in which the problem of characterizing intelligence can only be addressed by bringing computational, mathematical, behavioural and neurobiological constraints to bear simultaneously. Alan Newell's work exemplified this latter approach. His efforts to understand cognition always combined a keen awareness of the depth of the problem; a strong commitment to finding real solutions; and an abiding respect for the importance of using evidence from human behaviour as a guide to the evaluation of the validity of theory.

The cognitive scientists who converged in Edinburgh in September of 1993 clearly shared Newell's commitment to computational sufficiency and behavioural adequacy. Some, like Newell, had primarily computational interests, while others – more like Herb Simon, I suppose – had interests that originated in the nature of human cognition. And some of the participants reflect the synthesis of these approaches, combining mathematical analysis and computational modelling with a vital interest in experimental evidence based on experimental analysis of human behaviour. Though the meeting – and this book arising from it – have been billed as explorations of connectionist models, both reflect a broad theoretical perspective, one that situates the models in the broader fabric of efforts to rationalize intelligence in terms of models of optimal (i.e. Bayesian) inference in the presence of uncertainty.

The topics covered at the meeting include a number of the issues that have been raised in recent years as cognitive science has sought to come to terms with the real meaning and significance of connectionist approaches to cognition. Several of the papers in the first section deal with issues raised by McCloskey and Cohen's observations about interference in connectionist networks. It is clear from Sharkey & Sharkey's chapter that the presence of interference in some connectionist models is reasonably well understood, and techniques have been developed to minimize and even eliminate it. Equally importantly, Murre makes clear in his chapter how the same class of models that exhibit negative transfer in some cases of sequential learning can often exhibit positive transfer as well. Thus it would be a bad idea to suggest abandoning the broad class of connectionist models that can under some training regimes exhibit interference. The other chapters in this section deal with a range of aspects of learning and memory and indicate the diversity of learning and memory phenomena that are susceptible to a connectionist analysis.

The section on reading addresses the debate about the role of rules vs. connections in capturing language structure. Several of the chapters here suggest that at least some of the concerns that have dampened enthusiasm about connectionist models can finally be laid to rest. Early connectionist models – ones I had some hand in, as a matter of fact (Rumelhart & McClelland 1986, Seidenberg & McClelland 1989) – did not really achieve an adequate level of lawfulness in their behaviour. And now we think we understand why: These models used representations that dispersed the regularities inherent in the systems they tried to learn (see the chapter by Plaut, McClelland and Seidenberg for full discussion). Models that condense the regularities in different ways are much more systematic: When trained to read a corpus of 3000 monosyllables, the very different models of Plaut et al. and Bullinaria learn not only to read the corpus but also to generalize, as well as humans do, to pronounceable nonwords. Bullinaria addresses some additional issues, including the question of how the learner might discover which letters in a word correspond to which of the word's phonemes. Damper's chapter situates connectionist models in the broader context of other computational approaches to learning the relation between spelling and sound.

The relation between statistical inference and connectionist models is addressed in section III. This section includes a valuable chapter by Kentridge, relating connectionist architectures to the theory of automata, and another by Chater, illuminating the strong links between connectionist models and models of statistical inference. Finch, Chater and Redington offer a model of syntax acquisition based on distributional statistics that illustrates the increasing convergence of connectionist approaches and statistical approaches to learning. The section on speech and audition may seem to deal with rather specialized topics, yet in fact some of the issues addressed are of quite general significance. Chapters 15 and 16 join the debate on whether cognitive processing is interactive or feedforward. The insight that arises from the juxtaposition of these papers is that what is crucial is the integration of constraints arising from several of the levels linguists and other theorists have classically identified. These constraints may be captured by classically interactive systems, of the sort considered by Harley and MacAndrew, in which processing units on different levels represent cognitive units at different levels of linguistic analysis (e.g. features, phonemes, words);

or they may be captured by recurrent models, in which featural, phonological and lexical constraints are inextricably intertwined, as in the chapter by Cairns et al. These more psychologically oriented papers are nicely complemented by the computational considerations that are the focus of the Smith chapter and the chapter by Abu-Bakar and Chater.

Overall the book reflects the diversity of influences and modelling techniques that have arisen within the connectionist framework, as it interacts with other frameworks and with current issues in specific research areas. The appreciation for the importance of combining computational and empirical approaches shines through all the contributions and bodes well for the future of cognitive science – connectionist and otherwise – among the new generation of researchers who are the main contributors to this volume.

Jay McClelland
February 1995

References

Rumelhart, D. E. & J. L. McClelland 1989. On learning the past tenses of English verbs. In *Parallel distributed processing: explorations in the microstructure of cognition*, vol. 2, *Psychological and biological models*, J. L. McClelland & D. E. Rumelhart (eds), 216–71. Cambridge, Mass.: MIT Press.

Seidenberg, M. S. & J. L. McClelland 1989. A distributed, developmental model of word recognition and naming. *Psychological Review* **96**, 523–68.

Acknowledgements

The task of organizing a research workshop is an example of distributed cognition *par excellence*. The enjoyable and productive meeting of the Neural Computing and Psychology Workshop in Edinburgh in September 1993 would never have been achieved without hard work and good will from a large number of people. The resulting book too was a product of large-scale co-operation. We would like to acknowledge the help we had from all those who contributed to these endeavours and we apologise for any omissions we make.

The academic side of the workshop was organized by a committee consisting of the editors of this book. The practical details were efficiently dealt with by Margaret Rennex, Fiona-Anne Malcolm and the staff of Pollock Halls at the University of Edinburgh. The workshop itself was made so enjoyable by the depth and variety of contributions from its attendees:

Mukhlis Abu-Bakar, Carmen Paz Suarez Araujo, Dimitrios Bairaktaris, Gordon Brown, Antony Browne, John Bullinaria, David Cairns, Paul Cairns, Nick Chater, Bob Damper, Kris Doing Harris, David Elkan, Jonny Farringdon, Steve Finch, David Glasspool, John Gray, Ulrike Hahn, Peter Hancock, Amy Hand, Trevor Harley, Costa Hondros, Masja Kempen, Bob Kentridge, Joe Levy, Richard Loosemore, Siobhan McAndrew, Jay McClelland, Peter Menell, Mauricio Iza Miqueleiz, Jaap Murre, Mike Oaksford, Mike Pake, Elena Perez-Minan, Nam Seog Park, Noel Sharkey, Richard Shillcock, Julian Smith, Leslie Smith, Keith Stenning, Martin Redington, Hirohide Ushida and David Willshaw.

Support for the workshop was received from the Human Communication Research Centre, Department of Psychology and Centre for Cognitive Science at the University of Edinburgh. We also acknowledge support from the Economic and Social Research Council, Science and Engineering Research Council (as it then was) and Medical Research Council, chiefly under their combined guise of the Joint Council Initiative for Cognitive Science and Human Computer Interaction, for funding our research.

ACKNOWLEDGEMENTS

The volume you see here owes much to Andrew Carrick and Sheila Knight of UCL Press for their patient forbearance of editing by committee. We were glad of the expert copy editing skills of Rich Cutler. We are particularly appreciative of the efforts of those colleagues who helped us by refereeing chapters including: Nick Braisby, Neil Burgess, Zoltan Dienes, Kris Doing Harris, Jonny Farringdon, Gareth Gaskell, Geoff Goodhill, Peter Hancock, Graham Hitch, George Houghton, Glyn Humphries, Dennis Norris, Malti Patel, Hinrich Schütze, Julian Smith, Keith Stenning and Tim Shallice.

We'd like to thank Betty Hughes for administrative assistance, the computing support team at HCRC and Cognitive Science for, well, computing support, and Art Blokland for last minute advice on the intricacies of IPA. Naturally, we owe enormous debts of gratitude to all the contributing authors – even those who delivered their manuscripts a little late! A special thank you to Jay McClelland for not only giving two talks at the workshop, chairing a session and preparing a chapter for this volume but also writing the informative Preface.

Joe Levy, Stirling
Dimitrios Bairaktaris, Stirling
John Bullinaria, London
Paul Cairns, London
March 1995

Contributors

Mukhlis Abu-Bakar, Department of English Language and Literature, National University of Singapore, Singapore

Dimitrios Bairaktaris, Department of Computing Science, University of Stirling, Stirling FK9 4LA, UK

Gordon D. A. Brown, Department of Psychology, University of Warwick, Coventry CV4 7AL, UK

John A. Bullinaria, Department of Psychology, Birkbeck College Main Building, Malet Street, London WC1H 7HX, UK

Paul Cairns, Centre for Cognitive Science, 2 Buccleuch Place, Edinburgh EH8 9LW, UK

Nick Chater, Department of Experimental Psychology, University of Oxford, Oxford OX1 3UD, UK

Robert I. Damper, Image, Speech and Intelligent Systems (ISIS) Group, Department of Electronics and Computer Science, University of Southampton, Southampton SO17 1BJ, UK

Steve Finch, Human Communication Research Centre, University of Edinburgh, 2 Buccleuch Place, Edinburgh, UK

David W. Glasspool, Department of Psychology, University College London, Gower Street, London WC1E 6BT, UK

Trevor A. Harley, Psychology Department, University of Warwick, Coventry CV4 7AL, UK

Charles Hulme, Department of Psychology, University of York, Heslington, York Y01 5DD, UK

Robert W. Kentridge, Department of Psychology, University of Durham, Durham DH1 3LE, UK

Joseph P. Levy, Department of Psychology, Birkbeck College Main Building, Malet Street, London WC1H 7HX, UK

Siobhan B. G. MacAndrew, School of Health and Social Sciences, University of Coventry, Coventry CV1 5FB, UK

James L. McClelland, Department of Psychology, Carnegie Mellon University, Pittsburgh, PA15213, USA

Jacob M. J. Murre, Medical Research Council – Applied Psychology Unit, 15 Chaucer Road, Cambridge CB2 2EF, UK

David C. Plaut, Department of Psychology, Carnegie Mellon University, Pittsburgh, PA15213, USA

Tim Preece, Department of Psychology, University of Wales, Bangor, Gwynedd LL57 2DG, UK

Martin Redington, Department of Experimental Psychology, University of Oxford, Oxford OX1 3UD, UK

Mark S. Seidenberg, Neuroscience Program, University of Southern California, Los Aneles, CA90089–2520, USA

Amanda J. C. Sharkey, Department of Computer Science, University of Sheffield, Sheffield S1 4DP, UK

Noel E. Sharkey, Department of Computer Science, University of Sheffield, Sheffield S1 4DP, UK

Richard Shillcock, Centre for Cognitive Science, 2 Buccleuch Place, Edinburgh EH8 9LW, UK

Leslie S. Smith, Centre for Cognitive and Computational Neuroscience, Department of Computing Science and Mathematics, University of Stirling, Stirling FK9 4LA, UK

MEMORY

Joseph P. Levy

Since the very beginning of interest in them, connectionist networks have been used at least as metaphors for human memory, if not as explanatory models. It is not difficult to pick out some of the historical highlights in the development of models of neural network memories. Hebb (1949) suggested that a short-term trace was held in the activation of a neural circuit while long-term information was held in the synaptic weights. Willshaw et al. (1969) were concerned to describe the formal properties of the information storage in a biologically plausible simple net. Hopfield (1982, 1984) demonstrated how to train a network that stored its memories as stable points in its dynamics – popularizing the theoretical jargon of *attractors*, *basins of attraction*, and optimizing training algorithms so as to store maximum information. Rumelhart & McClelland (1986) demonstrated the potential richness in the psychological explanatory power of simple attractor networks that could be trained using some form of learning algorithm. In particular, the way that a partial or noisy cue to an attractor network will, if within the appropriate basin of attraction, lead to the intuitively satisfying "recall" of the appropriate information. There has also been a great deal of work on connectionist models of various aspects of cognitive processing. These often need to appeal to notions of information storage, learning and memory (e.g. Elman & McClelland 1988, Seidenberg & McClelland 1989). The success of these processing models adds to the attraction of using connectionist systems to model memory since it seems likely that memory and processing are intimately connected.

It follows that neural networks and memory are natural partners because of the intuitively satisfying way that networks can learn and store multiple stimuli in the same set of weights; because of the properties of content addressability and noise resistance made possible by nets with attractor dynamics; and because of related successes in connectionist models of cognitive processes.

The chapters in this section describe some of the latest work on the successes and failures of this enterprise. Relationships between memory and processing are highlighted in Chapter 1 on short-term memory for verbal sequences and Chapter 2 on the manner in which groups of stimuli can be temporarily "chunked together". Chapter 3 covers important ideas on how memory abilities might *develop*, and the links between connectionist models and the mathematical memory model literature are discussed. A range of problems that back-propagation networks have in modelling human memory data – "catastrophic interference", "catastrophic discrimination" and unrealistic "hypertransfer" – are discussed in Chapters 4 and 5. This section ends with a discussion of models of the relationship between connectionist short- and long-term stores (Ch. 6).

In Chapter 1, Glasspool describes the Burgess & Hitch (1992) model of the articulatory loop from the working memory literature (e.g. Baddeley 1986). The model is successful at describing the kinds of order errors, word length effects and the detrimental effect of phonemic similarities between words that occur in the working memory literature. Glasspool describes an extension to the model that copes with lexicality effects – the fact that this component of short-term memory works better with words than non-words. The chapter also demonstrates the utility of the simple competitive queue mechanism of Houghton (1990).

In Chapter 2, Bairaktaris describes a low-level modular mechanism for "temporal chunking" serially ordered input sequences. It uses a recurrent back-propagation architecture described by Zipser (1991) to preprocess an input stream into chunks of temporally contiguous items. Bairaktaris stresses the ubiquitous need for this kind of mechanism throughout cognition.

In Chapter 3, Brown, Preece and Hulme describe the links between the current connectionist and mathematical techniques for modelling human memory. They argue that the different approaches have complementary strengths and weaknesses. They stress the need for a combination of the insights from psychological models of the *development* of memory with those from mathematical models of memory.

In Chapter 4, Sharkey & Sharkey describe the problem of "catastrophic interference" in back-propagation networks, where newly learned information completely overwrites previously learned information. They then describe the complementary problem of "catastrophic discrimination", the inability to distinguish items that have been learned from new ones. They stress the necessary trade-off between the two problems.

In Chapter 5, Murre demonstrates that back-propagation not only suffers from catastrophic interference but also from "hypertransfer", i.e. that in some circumstances performance on a set A actually *improves* when learning a second set B. The learning transfer effects are in disagreement with human learning data. Murre goes on to show that two-layer networks do not suffer from excessive transfer and are in fact in very close accordance with the human interference data as summarized in the classic paper by Osgood (1949).

In Chapter 6, Levy and Bairaktaris survey connectionist models of the interaction between short- and long-term stores. They describe systems with separate components and ones where each connection has both a short- and a long-term weight. They describe their own architecture of dual-weight connections where each weight acts independently. The chapter finishes by speculating on possible future modelling applications for this kind of architecture.

References

Baddeley, A. D. 1986. *Working memory*. Oxford: Oxford University Press.

Burgess, N. & G. L. Hitch 1992. Towards a network model of the articulatory loop. *Journal of Memory and Language* **31**, 429–60.

Elman, J. L. & J. L. McClelland 1988. Cognitive penetration of the mechanisms of perception: compensation for coarticulation of lexically restored phonemes. *Journal of Memory and Language* **27**, 143–65.

Hebb, D. O. 1949. *The organisation of behaviour*. New York: John Wiley.

Hopfield, J. J. 1982. Neural networks and physical systems with emergent computational abilities. *National Academy of Sciences of the USA, Proceedings* **79**, 2554–8.

Hopfield, J. J. 1984. Neurons with graded response have collective computational properties like those of two state neurons. *National Academy of Sciences of the USA, Proceedings* **79**, 2554–8.

Houghton, G. 1990. The problem of serial order: a neural network model of sequence learning and recall. In *Current research in natural language generation*, R. Dale, C. Mellish, C. M. Zock (eds), 287–319. London: Academic Press.

Osgood, C. E. 1949. The similarity paradox in human learning: a resolution. *Psychological Review* **56**, 132–43.

Rumelhart, D. E. & J. L. McClelland 1986. *Parallel distributed processing*. Cambridge, Mass: MIT Press.

Seidenberg, M. S. & J. L. McClelland 1989. A distributed, developmental model of word recognition and naming. *Psychological Review* **96**, 523–68.

Willshaw, D. J., O. P. Buneman, H. C. Longuet-Higgins 1969. Non-holographic associative memory. *Nature* **222**, 960–2.

Zipser, D. 1991. Recurrent network model of the neural mechanism of the short-term active memory. *Neural Computation* **3**, 179–93.

.

CHAPTER 1

Competitive queuing and the articulatory loop

David W. Glasspool

Introduction

This chapter reviews the psychological evidence for the articulatory loop. Two neural network implementations of the articulatory loop model are then discussed. The first is the model of Burgess & Hitch (1992). The second is an extended model which improves on that of Burgess & Hitch in its ability to operate with non-words as well as words, in its better simulation of the typical forms of order error made by human subjects, and in its ability to show both primacy and recency effects.

Both models use a competitive queuing architecture in order to achieve serial recall with single-shot learning. It is argued that competitive queuing is a particularly suitable paradigm for modelling the serial recall of verbal material due to its straightforward explanations for commonly observed error types and serial position effects.

The articulatory loop

There is evidence from studies on normal subjects and on neuropsychological patients that a specific dissociable system underlies the verbal component of short-term memory (e.g. see Baddeley 1990). This is the system presumed to be used in the immediate recall of word lists (span tasks), and which Baddeley & Hitch (1974) identified as the articulatory loop in their "working memory" model of short-term memory. The working memory framework proposes a modular system mediating short-term memory, in which a limited-capacity central executive is able to off-load information for short-term storage to at least two subsystems, the visuo-spatial scratchpad for spatial material, and the articulatory loop for verbally coded material.

The articulatory loop has been characterized by Baddeley (1986) as consisting of two components, a phonological store holding a rapidly decaying memory trace in the form of articulatory (motor) information, and a rehearsal mechanism, presumed to be formed from parts of the speech input and output mechanisms, which enables the system to maintain information for longer periods by continually refreshing the store's contents.

Auditorily presented material is presumed to gain immediate entry to the phonological store. Visually presented material gains access when it is articulated. In order to refresh the rapidly decaying trace it is necessary to re-articulate the store's contents. The rehearsal process places a limit on the amount of information which the store can hold – this is simply the amount of information which can be articulated by the speech mechanism in the time it takes for the trace in the store to decay to the extent that it is unusable. Notice that this limit is imposed by the fact that the process of articulation is inherently sequential, and not by a limited number of "chunks" or "slots" in the store. The store itself may be thought of as a constantly decaying trail left behind by all articulation, such that a speaker always has access to the last few seconds of speech.

Baddeley (1986) finds a limit of 1.5–2 seconds on the persistence of this trace, so the articulatory loop has a capacity set by the amount of speech which can be articulated in this time. This relationship between speech rate and span has been demonstrated several times (e.g. Baddeley et al. 1975, Ellis & Henneley 1980, Hulme et al. 1991).

The articulatory loop is specified at a very general level, but has nonetheless proved to be a very robust basis for the explanation of much of the data on human performance in this area. The word length effect (span is reduced for lists of longer words) and phonemic similarity effect (span is reduced for lists containing phonemically similar words) under different conditions of presentation modality and suppression are well accounted for (Baddeley 1986), the word length effect being due to articulatory rehearsal of the contents of a store which has a rapidly decaying trace, while the phonemic similarity effect is due to interference between similar items within the store. The model also predicts such phenomena as the effects of cross-linguistic differences in articulation rate on span. Ellis & Henneley (1980), for example, found that Welsh-speaking subjects had a shorter digit span than English speakers. The difference disappeared when the longer time required to articulate the Welsh digits was taken into account, as would be expected from a system with a limit on the articulatory duration of the material it can hold.

However, the model is specified at too general a level to account for other well established characteristics of verbal short-term memory. This is largely due to the fact that no architecture is specified for the "store" component. The representation of serially ordered information in a neural system is a far from trivial task, and it might well be expected that the need to store order information would place strong constraints on the architecture of the store, and hence on the detailed behaviour of the system. In particular, the basic articulatory loop model as outlined above fails to account for the following:

(a) The types and proportions of errors in immediate serial recall tasks. Between 70% and 80% of errors made by subjects in tasks involving the immediate recall of a list involve the order of items in the list rather than the identity of the items (Aaronson 1968). As Lashley (1951) pointed out, the form of typical errors argues against any form of "chaining" in the representation of stored lists. In fact, although the order of items may be incorrect they still tend to be produced only once, so a typical error is the exchange of two items in the list. This type of error is very difficult to explain on any intuitive symbolic account of the structure of the store.

(b) A long-term memory contribution to short-term memory. There is a good deal of evidence for the existence of a non-phonological component in short-term memory. For example, Besner & Davelaar (1982) find a lexicality effect – in tests on lists of non-words, recall is worse for non-homophones than homophones. In addition, Watkins (1977) reports that span for lists of high-frequency words is significantly greater than for lists of low-frequency words. Hulme et al. (1991) also report a lexicality effect, with consistently better recall for words than non-words. The articulatory loop model in itself cannot account for these effects, which are presumably due to interactions between long- and short-term memory.

(c) The precise origin of the phonemic similarity effect. The articulatory loop model is able to provide the seeds of an account for this effect, the suggestion being that phonemically similar items have similar articulatory representations in the store and are thus more likely to be confused. However, without constraining the form of the store, this does not constitute a full explanation.

(d) The form of the serial recall curve. It is well established that recall is better for the first few items in a list (the primacy effect) and the last few items (the recency effect) than for items in the middle of the list (e.g. see Murdock 1962). These effects are again generally seen as arising from the interaction between different memory systems, but there is a certain amount of evidence that the picture may not be so simple. For example, the recency effect, once thought to be entirely due to short-term recall, has been shown to extend to long-term memory (Tzeng 1973, Baddeley & Hitch 1977). Similar primacy and recency effects have been found in other serial recall situations, for example in certain pathologies of spelling (Caramazza & Miceli 1990, Jonsdottir et al. 1995), raising the possibility that it may be the serial recall mechanism itself which gives rise to these effects.

This chapter describes two related connectionist models which attempt to address these problems through implementation of the articulatory loop model. The first is a basic network model due to Burgess & Hitch (1992) which treats words as unitary items. The second is a considerably extended model which is able to handle non-words as well as words, and to temporally sequence phonemes within words. Both models remain at a fairly high conceptual level, but they nonetheless attempt to give a principled account for the areas of short-term memory performance outlined above. In particular, they define explicit structures for the "phonological store" and "articulatory control process".

Serial recall and competitive queuing

Connectionist models have had a great deal of success in areas such as the recognition and completion of patterns, where the entire stimulus is presented at once (in parallel), and the output is derived entirely by the spread of activation through the network over the weighted connections between nodes (e.g. see Rumelhart & McClelland 1986). However, as Lashley (1951) points out, much of human behaviour does not involve simultaneous patterns but is serially ordered in time.

Symbolic models based on serial digital computers are traditionally very good at modelling sequential behaviour, although this has been due in the main to a reliance on the underlying sequential control structures available in computer programming languages, which has resulted in a tendency to avoid addressing the fundamental origins of serial order in human and animal behaviour. Connectionist models are forced to address the problem of serial behaviour from first principles.

Although the majority of connectionist models have not been concerned with serial behaviour, a few have addressed this problem. One fairly simple way to introduce temporally ordered activity into a network is to include a time delay element in each inter-node connection. By adjusting the delay on the connections, the spread of activity across the network can be made to change with time. Networks of this type have been constructed for both the recognition and the generation of temporal patterns (e.g. Kleinfeld 1986, Tank & Hopfield 1987).

A different technique (developed by Jordan (1986) and Elman (1990) amongst others) uses a recurrent architecture in which the output from the network at time t is fed back as part of the input at time $t + 1$. This gives the network access to its past states, so that associations may be learnt between subsequent states. A system of this type is able to learn temporal sequences of states as a chain of associations.

Another approach (used by Rumelhart & Norman (1982), Dell (1988) and Houghton (1990) amongst others) uses the relative activation levels of items to order them sequentially. The most active of a set of items is selected by some selection mechanism, after which it is inhibited. The next most active item is now the most active and is selected. Repeatedly selecting and inhibiting the most active item results in a sequence of selections following the initial relative activation order of the items. This technique is attractive in its simplicity, and has the advantage that it suggests a natural mechanism for the high incidence of order errors found in psychological studies. An error occurs if a node has slightly too high or too low an activation, which results in the order of the sequence changing, rather than in the sequence breaking down completely as with typical errors in a "chained" system. This means that such models can show the type of chaining-independent behaviour which is typical of human recall (Lashley 1951) with errors such as the exchange of items, and can show a dominance of order over item errors (it is generally easier for the model to confuse two within-list items with similar activation levels than to confuse an active list item with an inactive out-of-list item).

This approach is central to Houghton's (1990) "competitive queuing" (CQ) model. Houghton extends the approach to give a complete serial recall model, outlined in Figure 1.1.

The CQ architecture consists of three main elements. First, a set of nodes representing items which may be incorporated into lists. A "local" representation scheme is used so each item is represented by a single node.

Secondly, the item nodes are activated by weighted connections to a "control signal" which varies with time during sequence production. This signal provides a cue to temporal position in the sequence to which recall of individual items can be tied. Allowing this control signal to vary with time is the central novelty of Houghton's approach, and it confers three advantages compared with simply allowing the suppression of already produced items to provide temporal ordering:

(a) Since activation levels of items can change during sequence production, the activation level of each selected item is not constrained to be lower than that of the previous selection. The items output by the system can thus be more evenly activated.

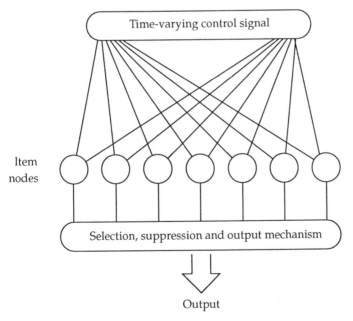

Figure 1.1   Houghton's (1990) CQ model.

9

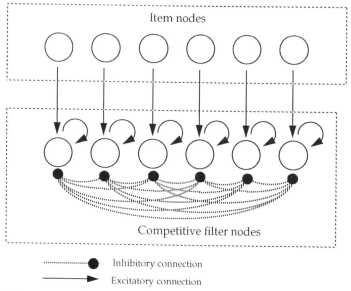

Figure 1.2 The competitive filter.

(b) Items which have been produced and suppressed do not receive strong top-down activation once their "correct" position in the sequence has passed, preventing spurious reselection of items.

(c) Sequences can contain repeats. This requires the reactivation of selected and suppressed items "to order".

The third element of the CQ approach is a "competitive filter" (Fig. 1.2), which selects and passes forwards the most active of the item nodes at each step during production.

This consists of a layer of nodes in parallel with the item nodes and with one-to-one excitatory connections from them. At the beginning of a recall time step the competitive filter nodes mirror the relative activations of the item nodes. However, this layer contains self-excitatory and strong mutually inhibitory connections between nodes, which are updated several times during each time step. The filter nodes quickly settle into a state where the node which started out with the highest activation level has inhibited all other nodes to a subzero level and is the only node left active. This constitutes the selection by the filter of the most active item node.

Houghton's CQ system was developed for word pronunciation from long-term memory. The time-varying control signal in this system consists of two nodes, a "start" node, which is activated at the start of recall and decays in activation during recall, and a "stop" node, which becomes more active during recall, to reach a peak at the end of recall. For sequences of more than two or three items (phonemes, in this case), correct recall requires a period of supervised learning. This is obviously a disadvantage in an immediate recall application. However, by increasing the complexity of the control signal (Houghton 1994) it is possible to obtain perfect recall for long sequences after a single presentation, using a Hebbian learning rule. This is the approach taken in the articulatory loop models described below, in which a more complex multinode control signal is used.

The Burgess & Hitch model

The Burgess & Hitch network model of the articulatory loop applies the CQ sequencing paradigm to the articulatory loop framework. The model is based around a four-layer feedforward network (Fig. 1.3). Phonemes and words are locally represented. All nodes take activations between –1 and +1 with a resting activation of zero. While their activation is positive, nodes' activations are dependent only on their current inputs, but they retain a proportion of their activation from one time step to the next when negative (so that normally nodes are immediately reactive to input, but inhibited nodes recover slowly).

A list of words is presented to the model one at a time. As each word arrives, the input phoneme nodes corresponding to the word's phonemes are activated simultaneously and feed activation forwards through a set of prelearned "recognition" weights to activate the corresponding word node. A set of temporary weights learns an association between the active word node and the current state of a set of "context" nodes, by one-shot Hebbian update. The context nodes provide a form of temporal context – their activation pattern changes slowly as each word is presented. In the initial form of their model, Burgess and Hitch use a sequence of patterns in which a proportion of the nodes randomly change state from highly active to inactive or vice versa at each time step during presentation. Each context state is thus likely to be more highly correlated with adjacent states than with distant states.

A second set of decaying temporary weights learns associations between the output phonemes of one word and the input phonemes of the next.

The sequence of context states used during presentation is repeated exactly during recall of the list, providing the model with a constant cue to its position in the list. As each context state appears during recall, the temporary weights tend to activate the word node which was associated with that state, and hence occupied the corresponding position in the list, during presentation. The output-to-input chaining weights also tend to activate each word node in the sequence as its predecessor appears at the model's output. The two sets of temporary weights thus both contribute to the sequential activation of word nodes at recall, the relative influence of the two systems being a free parameter of the model. In order to reproduce the correct sequence of words, it is thus simply necessary for a competitive filter to select the most active of the set of word nodes at each time step.

Activation feeds forwards from the filter via a set of prelearned "articulation" weights to a layer of output phoneme nodes, where the nodes corresponding to

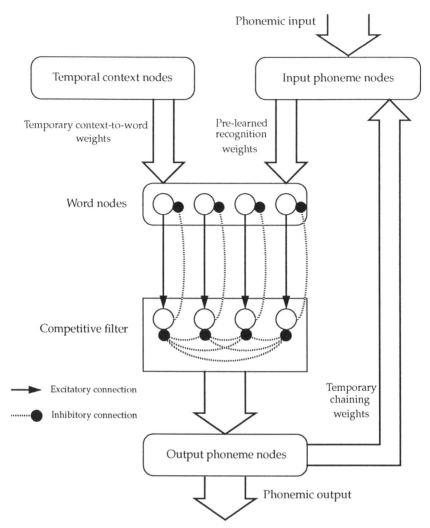

Figure 1.3 A general outline of the Burgess & Hitch articulatory loop model, adapted from Burgess & Hitch (1992).

the constituent phonemes of the winning word are activated in parallel. Simultaneously, strong inhibitory connections from filter nodes back to the corresponding word nodes serve to inhibit the winning word node following its production.

In order to induce the model to make errors, a small amount of random noise is added to the node activation levels, and the temporary weights from context to word nodes are allowed to decay with time during presentation and recall (modelled as one unit of decay for every phoneme presented to or recalled by the system). When the weights decay to the point where the random noise becomes significant in determining the winner of the "competition" the system will begin to make errors.

Performance of the Burgess & Hitch model

The Burgess & Hitch model performs well in the light of short-term memory data. In particular, the following aspects of human performance are well modelled:

(a) *Limited span.* The span of the model is limited due to the decay of the temporary weights and the addition of random noise to activation values.

(b) *The effect of word length.* Span is reduced for longer words, as in human subjects. This is due to the fact that the temporary weights decay once per phoneme rather than once per word, so that the capacity of the system is dependent on the phonemic length of the input list rather than the number of words it contains. The relationship between word length and span is close to that observed in human subjects.

(c) *The predominance of order errors.* Because the selection of words at recall is made on the basis of activation level, and close competitors usually come from elsewhere (close) in the list, order errors predominate to a similar extent to that observed in human data. The inhibition of words after their selection tends to prevent their spurious reselection, and thus each word in the list tends to be produced only once. This leads to correct modelling of exchange errors, which are otherwise difficult to explain.

(d) *The effect of phonemic similarity.* The model shows poorer recall for lists of phonemically similar words, in accordance with the human data. This effect is largely due to the operation of the second serial ordering system mentioned above – the chaining system – which operates on the phonemic representations of words. However, in the absence of this mechanism the model should still be expected to show some effect of phonemic similarity due to the partial activation of similar words by the recognition weights during learning. This should lead to the formation of weak associations between context states and partially active incorrect words, which would disrupt recall.

Interestingly, the output-to-input temporary weights were found to contribute nothing useful to the model's recall of word lists, and in fact prevent the correct modelling of certain aspects of the psychological data – notably, the occurrence of paired transposition errors, the zig-zag serial position curve for errors in lists of alternating similar and dissimilar words (Baddeley 1968), and the correct production of sequences containing repeats. The phonemic similarity effect is also partially obscured by the fact that chaining causes the similarity of two items to have as much effect on their successors as on the items themselves. These problems lead Burgess & Hitch to suggest that future models should rely completely on the temporal context rather than any form of item-to-item chaining. The form of the sequence of context patterns also caused problems, as their random nature led to occasional high correlations between temporally distant states. Burgess & Hitch suggest a modified context arrangement in which a "window" of activation is shifted across a field of inactive nodes, so that each

Figure 1.4 Form of temporal context suggested by Burgess & Hitch for use in future models.

context state correlates highly (0.66) with those immediately preceding or following it, less strongly with the next most adjacent states (0.33) and not at all with more distant states, as shown in Figure 1.4. With this form of context signal, the model shows a recency effect as well as a primacy effect, implying that this was previously obscured by the stronger effect of long-range context state correlations.

An extended model

The Burgess & Hitch model is a good first approximation to an articulatory loop implementation. However, the design is somewhat simplistic. Specifically, words lack internal structure (all the phonemes of a word are presented or recalled simultaneously), and non-words cannot be remembered – clearly an important aspect of human performance and language acquisition (e.g. see Gathercole et al. 1991). Burgess & Hitch make the reasonable assumption that such aspects of human performance are secondary to the main modelling problem and may easily be added to the basic model. Part of the rationale for developing an extended model was to test this assumption.

The intuition behind the extended model (Glasspool 1991) is that the phonemes constituting words and non-words should be sequenced by the same type of mechanism as that which sequences words within lists. This suggests a hierarchical arrangement based around a phoneme level memory system, which will remember non-words and words equally well as lists of phonemes, with a word-based system operating alongside it to provide support when recognized words are encountered. An outline of the model is shown in Figure 1.5.

The model uses a feedforward layered architecture with local representation for words and phonemes. To allow the word system to recognize words presented one phoneme at a time, and to articulate recalled words in the same manner, recognition and articulation is achieved using connections with time delays (Tank & Hopfield 1987, Elman 1990). This simplifies modelling, but the architecture is sufficiently modular that a different scheme could be substituted relatively easily. Both phoneme and word memory systems use a competitive queuing structure, each with a competitive filter and a time-varying context pattern similar to the "moving window" arrangement suggested by Burgess & Hitch.

In order to avoid buffering incoming items, the model does not attempt to distinguish between words and non-words as they are presented. Rather, both the phoneme and word level systems attempt to remember the presented list as best they can. The phoneme system will do equally well with words or non-words, but the word system will perform much better if words are recognized in the input. At recall, the word system will be able to support the phoneme system if the presented list contained words, but not if it contained non-words. This behaviour is consistent with the data of Hulme et al. (1991), which suggests that the advantage which the recall of words enjoys over that of non-words is due to additional support from lexical information in long-term memory over and above a basic phonological capability, rather than, for example, the use of wholly separate memory systems for words and non-words.

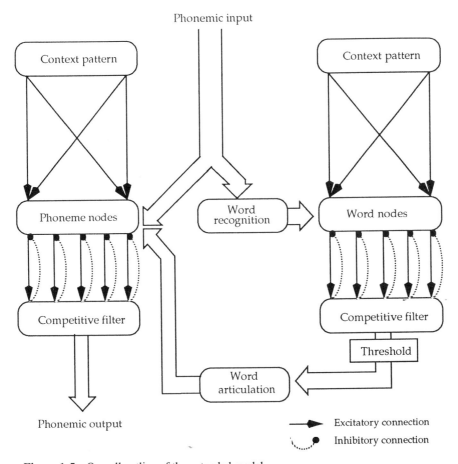

Figure 1.5 Overall outline of the extended model.

15

The input to the system is a stream of phonemes. As each is presented, the corresponding phoneme node is activated and a set of temporary weights learn an association between the phoneme nodes and the current phoneme context pattern using a one-shot Hebbian learning rule. The phoneme context pattern is then updated. The incoming stream of phonemes is segmented into words (or non-words) to simplify the task of the recognition network. Segmentation is achieved by inserting an interword marker into the phoneme stream. The marker is treated by the phoneme system as a phoneme like any other, and is stored for later recall. However, the arrival of the interword marker triggers the word system to learn the association between the current word context and the state of the word nodes, and to update its context pattern. If the immediately preceding phonemes formed a recognized word, one of the word nodes will be fully active at this point and a strong association will be learned, along with weak associations to any phonemically similar word nodes which will be partially activated by the recognition network. If, however, the preceding phonemes did not form a word, none of the word nodes will be fully active and only weak associations will be learned.

At recall, the phoneme system produces one phoneme (or interword marker) at each time-step, again following the sequence of competitive filter selection, subsequent inhibition and context shift. While the context pattern for the phoneme system is updated at each time step, the context pattern for the word system remains constant for the duration of each word or non-word, and is only updated when an interword marker appears in the output. The output from the word system is thresholded so that output will only be produced if a strong association has been learned to the current word context state. This output is produced one phoneme at a time by the articulation network, and strongly influences the outcome of the phoneme level competition.

In general at each time step during recall more than one word node will be active, for two reasons:

(a) Positions close together in the list have context states which share active nodes, so words close to the "correct" word in the presented list will be partially activated by their association with similar context patterns.

(b) At presentation, words which share phonemes with the word currently being presented are partially activated by the recognition weights, and form weak associations to the current context state. They are thus partially activated when this context state appears again at recall.

Any words which are both close to the "correct" word in the list and share phonemes with it will be particularly highly activated along with the correct word, although in the absence of noise the correct word will have the highest activation at the appropriate point during recall.

The first of these factors also holds for phoneme nodes, so that in general more than one phoneme node will also be active at each recall step. The addition of random noise to phoneme and word activation levels causes the model to make errors in selecting the correct word or phoneme for recall.

Both sets of temporary weights decay at the same rate, by a fixed amount every time a phoneme is presented to or produced by the system. The capacity of the system is thus related to the phonemic length of the stored material rather than the number of words/non-words it contains. However, since the phoneme system operates on smaller "chunks" than the word system, and therefore has to store more information, span for words should be greater than span for non-words.

During recall the system must utilize the chunked representations stored by the word weights whenever words were recognized in the input. As indicated above, the word system adds its articulated output to the phoneme nodes if the activation of the winning word node exceeds a threshold. Since the absolute activation level of a correctly recognized and recalled word varies according to the length of the list, this threshold must also vary with list length. However, the threshold may conveniently be set by the activation of the most active phoneme node, which must be scaled by an experimentally determined factor of around 0.75 for reliable operation. In practice, after the addition of noise, the threshold is still occasionally unreliable. To make selection completely reliable, a small bias is added to the threshold value to bias the model slightly towards word or non-word recall. This fits with the intuitive idea that human subjects are aware that the list they are trying to recall contains words or non-words. The model is still able to recall lists of mixed words and non-words when this bias is set to zero, although it is subject to occasional lexicalization errors. Hitch has found similar lexicalization errors in human subjects under this condition (personal communication).

Formal description of the model

Node activations
Nodes in the context fields take binary activation values of 0 or +1. Activation A_j at time t of any phoneme, word or competitive filter node j with connections to n other nodes is given by

$$A_j(t) = f\left(\delta A_j(t-1) + \sum_{i=1}^{n} W_{ij} A_i(t-1) \right) + \eta \qquad (1.1)$$

where δ is a decay constant ($\delta = 0$ for positive A_j, $0 \leq \delta \leq 1$ for negative A_j), η is a random noise value, W_{ij} is the weight on the connection to node j from node i, and

$$f(x) = \tanh(x) \qquad (1.2)$$

which acts as a "squashing" function to keep activations between ± 1 (noise is added after application of $f(x)$ so activations may occasionally slightly exceed the ± 1 limits).

Depending on the position of the node in the network, weights may impinge on a node conveying excitation from another layer, inhibition from another layer, inhibition from other nodes in the same layer, and excitation from the node itself. Magnitudes for all of these weights are given later.

Learning

During list learning, the input to the phoneme nodes is a vector in which all elements are zero except for that corresponding to the current incoming phoneme, which is 1. The input to the word system is a vector derived from the recognition network. The activation A_j of node j at time t is given by

$$A_j(t) = \delta A_j(t - 1) + V_j(t) \qquad (1.3)$$

where V_j is element j of the phonemic input or word recognition output vector. The recognition and articulation networks both use fixed weights from the phoneme nodes comprising a particular word to the corresponding word node. As in the Burgess & Hitch model, recognition weights have a magnitude of

$$\frac{1}{\sqrt{n_{\mathrm{ph}}}} \qquad (1.4)$$

where n_{ph} is the number of phonemes in the word. This allows the network to distinguish between long and short words ending with the same phonemes. Articulation weights have a magnitude of 1. The connections in these networks include delays arranged so that, in the recognition network, excitation from each of the serially presented phonemes of a word arrives at the appropriate word node simultaneously, and in the articulation network superthreshold excitation by a word node causes subsequent serial excitation of the appropriate phoneme nodes in the correct order.

Learning only occurs in the temporary context-to-phoneme and context-to-word weights, which are updated according to the modified Hebbian rule used by Burgess & Hitch:

$$W_{ij}(t+1) = \begin{cases} \varepsilon A_i(t) A_j(t) & \text{if} \quad W_{ij}(t) < \varepsilon A_i(t) A_j(t) \\ W_{ij}(t) & \text{otherwise} \end{cases} \qquad (1.5)$$

where ε is a learning rate parameter.

Weight decay

The weights from context-to-word and context-to-phoneme nodes are temporary and decay with time. Decay of temporary weights at each phoneme input or output is given by

$$W_{ij}(t+1) = \Delta W_{ij}(t) \qquad (1.6)$$

where Δ is a constant decay factor, $\Delta \leq 1$.

Parameter values

The following are the parameter values used for the simulations summarized below:

(a) Limits on random noise $\eta = \pm 0.0088$.

    (This is set by reference to human data to give the correct span.)

(b) Temporary weight decay rate $\Delta = 0.94$.

    (This must be close to 1.0 to give an appreciable advantage for words over non-words. If the weights decay too fast, the difference in activation between successive phonemes becomes so much larger than that between successive words that the greater distinctiveness of successive phonemes offsets the greater number of items in phoneme lists.)

(c) Item node to competitive filter excitation weight $= 1.0$.

(d) Competitive filter to item node inhibition weight $= 4.0$.

    (A large value is required to give strong inhibition of winning items after their production.)

(e) Competitive filter self-excitation weight $= 0.5$.

(f) Competitive filter mutual inhibition weight $= 1.3$.

    (These are experimentally determined to give efficient filter operation.)

(g) Learning rate $\varepsilon = 3.0$.

(h) Scaling factor for phoneme activations before comparison with word activations $F_p = 0.75$.

(i) Lexicalization bias $= 0.1$ (non-word lists) $= -0.1$ (word lists).

    (Added to phoneme activations after scaling by F_p.)

All parameter values are fully motivated in Glasspool (1991).

Performance of the extended model

The following results summarize the major aspects of the model's performance. Unless otherwise stated, all results were obtained from 500 trials.

Figure 1.6 shows the performance of the model for various list lengths and word lengths.

The model has a limited capacity due to the decay of temporary weights and noisy activations. The form of the "span" curve is similar to that for humans (Guildford & Dallenbach 1925), with good recall for short lists, poor recall for very long lists, and a smooth transition for intermediate lengths. The span was set to approximately seven two-phoneme words by adjusting the level of noise. All other parameter values were set essentially without reference to psychological data, however, values being selected simply to give qualitatively correct operation of the model.

The model shows a pronounced word length effect, with better performance on lists of shorter words or non-words. This effect is due to the fact that the temporary weights decay once for each phoneme, rather than each word or non-word, so that the capacity of the model is related to the phonemic length of the presented list rather than the number of words or non-words it contains. (Note

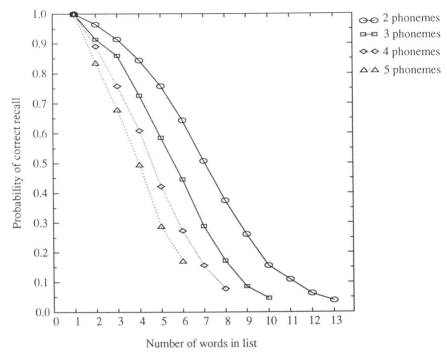

Figure 1.6 Probability of correct recall of entire lists against list length for lists of dissimilar words of two, three, four and five phonemes.

that phonemic length is used in the model to represent temporal length, purely for modelling convenience. Real phonemes are of course of different durations, and a more accurate model would have the weights decay once per unit of time, rather than once per phoneme.)

Figure 1.7 compares the performance of the model on lists of phonemically similar and distinct words. The model shows a clear effect of phonemic similarity, with shorter span for lists containing phonemically similar words. This is due to the particularly high activation at recall of words which are both close to the target word in the list and phonemically similar to it. These compete strongly with the target word, leading to a higher error rate than that for distinct words.

A problem for the model as it stands is that the mechanism by which the phonemic similarity effect arises for words (and that for the same effect in the Burgess & Hitch model) is incompatible with the architecture of the phonemic part of the model.

For words, the effect is due to the fact that similar words are strong competitors at recall, as they are partially activated during learning, and form weak associations with the context pattern. Non-words, however, are learned as sequences of phonemes, which are not partially activated, and hence are not subject to stronger competition, under conditions of phonemic similarity. However, the model does show a phonemic similarity effect for non-words, due to the

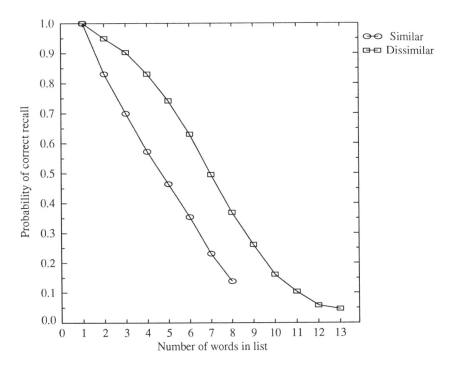

Figure 1.7 Probability of correct recall of entire lists against list length for phonemically similar and dissimilar two-phoneme words.

higher occurrence of repeated phonemes in lists of phonemically similar non-words. The competitive queuing mechanism is capable of correctly recalling lists with repeated items, but recall is not so robust, and the error rate is higher than for lists without repeats.

The fact that the same effect is produced for words and non-words by different mechanisms does not seem satisfactory. One way to resolve this problem would be to treat phonemes as collections of phonetic features rather than indivisible objects. An association network could then recognize phonemes at presentation, allowing the phonemic similarity effect for phoneme lists to arise in a similar way to that for word lists.

Figure 1.8 compares the performance of the model on lists of words and non-words. (The y axis shows the probability of error-free recall of an entire list. Note that both word and non-word lists include end-of-word markers.) The model stores lists of non-words as lists of phonemes, but is able to "chunk" lists of words. For a given phonemic length of list the word-based system stores fewer items than the phoneme-based system, and thus makes fewer errors. The result is a clear lexicality effect of the same order as that reported for experimental subjects (Hulme et al. 1991).

Figure 1.9 shows the probability of a correct response in each word position when recalling a word list.

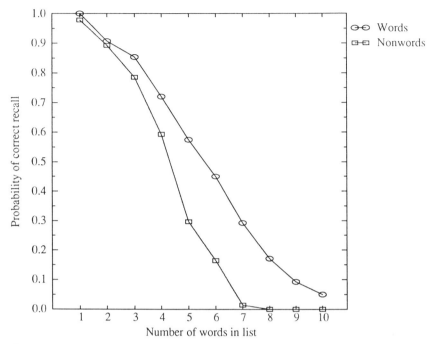

Figure 1.8 Probability of correct recall of entire lists against list length for three-phoneme words and non-words.

The model shows a clear primacy effect, which is largely due to its tendency to make order errors rather than item errors. This means that an error early in the list tends to lead to further errors later in the list.

The model also shows a small but robust recency effect. This is due to the fact that the last item in the list shares fewer active context nodes with other items than most other list items, and so is less susceptible to error. The last-but-one item is also more distinct than items in the middle of the list, but this is to a lesser extent, and is not enough to confer any appreciable recency effect. The same is of course true of the first and second items, but the effect at the beginning of the list is masked by the much stronger statistical effect already mentioned.

Primacy and recency effects very similar to these are seen in experimental studies when presentation is visual (e.g. Crowder 1972).

Finally, due to the use of a competitive queuing mechanism for serial recall, the model produces more order errors than item errors, and makes many exchange errors, most involving adjacent items. The number of item errors made by the model depends on the vocabulary of words given to the model before testing. Item errors occur usually when a word in the model's vocabulary but not in the learned list is phonemically similar to a word in the list. This causes the non-list word to form weak associations with the context during learning, and to be partially activated during recall. The rate at which the model

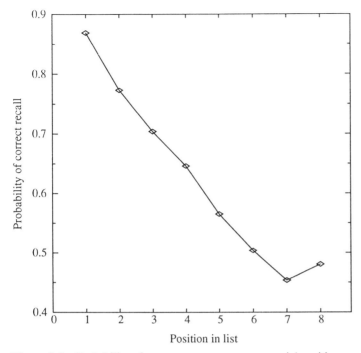

Figure 1.9 Probability of a correct response versus serial position in lists of eight two-phoneme distinct words. Data obtained from 5000 trials.

makes item errors thus depends on the number of words in its vocabulary which share phonemes with words in the list. Most item errors share phonemes with list items, which fits the psychological data (e.g. Conrad 1964).

Discussion

Relationship of these models to the articulatory loop framework

It was argued in the introduction that the articulatory loop model is specified at too general a level to account for several well known aspects of the immediate recall data, particularly the types of errors made, the effect of lexicality, the effect of phonemic similarity and the form of the serial error curves. The implementations discussed here improve on the basic articulatory loop framework by positing explicit mechanisms within a CQ architecture as analogues for its "phonological store" and "articulatory control process". The rapidly decaying phonological store corresponds to the decaying associative weights from a gradually changing temporal context, which produce a changing gradient of activation over a set of item nodes, and the articulatory control process corresponds to the competitive filter, which must repeatedly select and output the most active of a set of items. In the new model, both the phoneme and word filters contribute to the articulatory control process, with the final output selected by the phoneme filter, but under heavy influence from the word filter if words are recalled.

The main predictions of the articulatory loop model still hold within this new formalism. First, the models have a limited capacity set by the rate of decay of the temporary weights. Secondly, since these weights decay with time (modelled here by phonemic length for implementation convenience), the capacity of the system is set by the temporal length rather than syllabic length or word count of the stored material. The models thus both show the same effect of word length as is observed in the psychological data.

The articulatory loop framework holds that the phonemic similarity effect is due to confusion between similarly coded traces in the store. The models show a similar effect, the confusion being due to similar activation levels resulting from partial recognition of similar words at presentation (i.e. during encoding), although an interesting alternative is suggested by the treatment of non-words in the extended model, where the effect is due to the effect of repeats in the phonemic representation. By specifying an explicit structure for the store component the models show how such an effect can operate in practice.

The models both show effects of serial position within lists on the incidence of errors. The extended model shows both primacy and recency effects. These effects arise from the sequencing mechanism and do not require interaction with external memory systems. The primacy effect is largely due to the fact that errors early in a list tend to cause further errors later, while the recency effect arises from the greater confusability of context states near the middle of the list compared with those near the end of the list, which have fewer neighbours.

Finally, the extended model shows the expected effect of lexicality. It explains the superior recall of words compared with non-words in terms of the lower number of items in the word system's representation, which can be used if known words are recognized in the list, compared with the phoneme system's representation, which must be relied on in the case of non-word lists.

Sequencing

Sequencing by activation level

One of the central themes of the CQ approach is the idea that items in a list are sequenced on the basis of their relative activation levels. This approach has been used to good effect in other areas, for example typing (Rumelhart & Norman 1982), spelling (Houghton et al. 1994) and word production (Houghton 1990). The major advantage of the approach is the form of the errors which tend to result from the disruption of recall. These are generally order errors, in good agreement with the high incidence of order errors in psychological studies in the relevant areas. Models relying on chained associations between recall items cannot satisfactorily explain the tendency of subjects to confuse order rather than item identity, as Lashley (1951) pointed out.

The second basic feature of the CQ approach, the suppression of selected items following output, besides improving recall, introduces the possibility of paired transposition (exchange) errors. These are a common feature of human performance in many areas of serial behaviour (for example, speech (e.g. see Shattuck-Hufnagel 1979), impaired spelling (Caramazza & Miceli 1990) and typing (Rumelhart & Norman 1982)) as well as in short-term memory tasks, and cannot be adequately explained by chained association models.

The good results obtained using a completely chaining-free sequencing system in the extended model confirm Burgess & Hitch's conclusion that chaining is unnecessary for correct operation.

The context signal

The CQ mechanism in its original formulation by Houghton (1990) used the activation pattern across a pair of nodes to form the time-varying signal driving sequential recall. One of these (the I node) is maximally active at the start of recall, and its activation falls monotonically during recall. The other node (the E node) starts with a low activation and increases in activation to reach a maximum at the end of the sequence. Houghton (1994) argues that this is the simplest form of control signal which is able to give reliable operation. However, although it is able to learn sequences using a single-shot Hebbian update procedure, the CQ system using this simple two-dimensional control signal requires a short period of supervised learning for good performance on sequences of more than three or four items. This is because the control signal does not give a good level of discrimination between adjacent states, and the weights must be finely adjusted if many states are to be discriminated.

The main motivation for moving away from the simple control signal of Houghton's (1990) version of CQ in addressing the articulatory loop is the need for an extended period of supervised learning if sequences of a realistic length are to be stored. This is clearly problematic for a model of immediate recall. The approach taken by Burgess & Hitch in improving the discrimination between successive states in the control signal is to add more units, allowing successive states to be patterns with a lower correlation. This approach is also taken in the extended model. An advantage of this type of context signal is that it is very easy to control the correlation between states. A disadvantage is that the complexity of the control signal itself has been increased to the point where it is no longer easy to take its production for granted. With the form of context originally used by Burgess & Hitch a sequence of context states randomly generated during presentation must be replayed verbatim during recall, opening the model to the criticism that the problem of serial recall is simply being moved to a different part of the system rather than explained. The original two-node system of Houghton (1990) has the advantage that it is much easier to generate in a neurally plausible way. The activation of the start node merely has to decay exponentially with time from a high starting value, while the end node's activation pattern could be achieved by an inhibitory connection from the start node.

This simple signal can be explicitly generated within the model, avoiding an appeal to external sequencing mechanisms.

In some recent unpublished work, Houghton (1994) has shown that the discriminability of separate states of the control signal in this type of model is a function of two parameters: the correlation between successive context states, which can be improved by adding more units to the control pattern (in Houghton's terms, adding more dimensions to the control state-space), and the sensitivity of the net input rule employed by item nodes, which determines how responsive items are to changes in the control signal. There are thus two obvious approaches to the problem of the low discriminability of the two-node control signal. The first is to add more nodes, as Burgess & Hitch do with their complex context sequence, and the second is to increase the sensitivity of the net input rule to changes in the control signal. Houghton (1994) demonstrates that a simple two-node control signal, in conjunction with a more sensitive net input rule using a radial basis function, is capable of producing sequence recall after a single exposure with dynamics comparable to the Burgess & Hitch system. The some-what artificial context signals used in the models described here are thus not nec-essary for correct operation, although they may be viewed as an implementation convenience which allows the influence of the context signal on recall to be more easily analyzed.

Chunking

Both models chunk phonemes into words. The extended model carries this chunking over into the sequencing system itself, however, and is able to store both chunked and unchunked information.

The chunking scheme is rather simplistic, a fixed two-level hierarchy is employed where in reality the chunking of phonemes into words is no doubt a far more subtle process, perhaps involving various stages of pre-lexical chunking and quite rigorous constraints on the legal formation of words from phonemes (Treiman & Danis 1988, Hartley & Houghton 1995). The model thus, for exam-ple, completely ignores the fact that non-words are themselves chunks.

However, the simple chunking scheme has proven to work effectively, and offers an explanation for the advantage of words over non-words in immediate recall tasks in terms of the amount of information which must be stored by the system in order to remember a list of words compared with that required to store a list of non-words. The model thus directly addresses the interesting problem of interaction between long-term and short-term memory.

This raises the question of the validity of using two separate queuing systems to model the processing of words and non-words. While it is a practical proposi-tion to use a separate queue for each level in a two-layer hierarchy, this would clearly become impractical quite rapidly as complexity increased. There are several obvious alternative architectures for a model capable of handling words and non-words, including, for example:

(a) Storage could be limited to phonological information only, relying on a recognition system at recall to "clean up" the trace for words. This is the suggestion made by Hulme et al. (1991) to explain their data, and is certainly a proposition which deserves investigation, as it would lead to a simpler model. Since word identity information is available at presentation time it seems reasonable that it should be used to strengthen the memory trace as it is laid down, however, rather than attempting to re-extract this information at recall. It may be possible to separate these possibilities experimentally, as outlined below.

(b) Incoming phonemes could be chunked into words if possible as they arrive, before storage. This has the advantage of using only one queue, and also of being extensible to any number of hierarchical levels without increasing the number of queues. However, this proposal would require buffering at input until recognition could be completed, which introduces additional complications to the model.

The dual-queue model discussed here shows that the separate queue approach can be made to work well. A certain amount of complexity is introduced by the need to synchronize the two systems and to correctly switch between them, but this appears to be no greater than the complexity of the buffering, recognition and synchronization mechanisms required for the alternative strategies outlined above.

Further development may require experimental work to indicate the most promising directions. Experiments with mixed lists of words and non-words may be particularly interesting here. The incidence of subjects' lexicalization errors might, for example, provide evidence for storage of specific words in specific list positions and phonemes or syllables in others, as in the present model, for the marking of certain positions as word positions, with subjects relying on the phonological trace to provide word identity, or simply for memory that certain words occurred somewhere in the list, subjects relying on the phonological trace for both position and identity.

Conclusions

Although still specified at a fairly high level, these models constitute concrete implementations of Baddeley and Hitch's articulatory loop. In particular, by positing an explicit mechanism for the "phonological store" and "articulatory control process" the models provide a principled explanation for the observed pattern of errors and primacy and recency effects. The CQ paradigm is central to these explanations of otherwise problematic phenomena. It is interesting to note that essentially the same mechanism has now been used to explain the detailed structure of serial behaviour in domains as widespread as typing (Rumelhart & Norman 1982), speech production (Houghton 1990), spelling (Houghton et al. 1994), non-word repetition (Hartley & Houghton 1995) and immediate recall in the models discussed here. Such generality must indicate that an important aspect of the dynamics underlying these activities is captured by the idea of a competitive queue.

The extended model attempts a consistent account of the handling of non-words and structure within words by hierarchically extending the basic Burgess & Hitch model. Effectively, the model implements a crude "temporal chunking" system whereby phonemes can be chunked into words on the basis of a recognition system. The results so far suggest that it will be worthwhile continuing to refine the chunking operation in the light of psychological data on chunking processes.

Acknowledgements

I am grateful to Graham Hitch and Jon Shapiro for their supervision of the thesis on which this work is based, and to George Houghton, Tom Hartley, Joe Levy and an anonymous reviewer for their comments on the manuscript.

References

Aaronson, D. 1968. Temporal course of perception in an immediate recall task. *Journal of Experimental Psychology* **76**, 129–40.

Baddeley, A. D. 1968. How does accoustic similarity influence short-term memory? *Quarterly Journal of Experimental Psychology* **20**, 249–64.

Baddeley, A. D. 1986. *Working memory*. Oxford: Oxford University Press.

Baddeley, A. D. 1990. *Human memory: theory and practice*. Hove: Lawrence Erlbaum.

Baddeley, A. D. & G. L. Hitch 1974. Working memory. In *Recent advances in the psychology of learning and motivation*, vol. VIII, G. H. Bower (ed.), 47–90. New York: Academic Press.

Baddeley, A. D. & G. L. Hitch 1977. Recency re-examined. In *Attention and performance*, vol. VI, S. Dornic (ed.), 647–67. Hillsdale, New Jersey: Lawrence Erlbaum.

Baddeley, A. D., N. Thomson, M. Buchanan 1975. Word length and the structure of short-term memory. *Journal of Verbal Learning and Verbal Behavior* **14**, 575–89.

Besner, D. & E. Davelaar 1982. Basic processes in reading: two phonological codes. *Canadian Journal of Psychology*, **36**, 701–11.

Burgess, N. & G. L. Hitch 1992. Towards a network model of the articulatory loop. *Journal of Memory and Language* **31**, 429–60.

Caramazza, A. & G. Miceli 1990. The structure of graphemic representations. *Cognition* **37**, 243–97.

Conrad, R. 1964. Acoustic confusions in immediate memory. *British Journal of Psychology* **55**, 75–84.

Crowder, R. G. 1972. Visual and auditory memory. In *Language by ear and by eye*, J. F. Kavanagh & J. G. Mattingley (eds), 251–76. Cambridge, Mass.: MIT Press.

Dell, G. S. 1988. The retrieval of phonological forms in production: tests of predictions from a connectionist model. *Journal of Memory and Language* **27**, 124–42.

Ellis, N. C. & R. A. Henneley 1980. A bilingual word-length effect: implications for intelligence testing and the relative ease of mental calculation in Welsh and English. *British Journal of Psychology* **71**, 43–51.

Elman, J. L. 1990. Finding structure in time. *Cognitive Science* **14**, 179–211.

Gathercole, S. E., C. Willis, H. Emslie, A. D. Baddeley 1991. The influences of number of syllables and wordlikeness on children's repetition of non-words. *Applied Psycholinguistics* **12**, 349–67.

Glasspool, D. W. 1991. *Competitive queuing and the articulatory loop: an extended network model*. MSc thesis, Department of Psychology, University of Manchester.

Guildford, J. P. & K. M. Dallenbach 1925. The determination of memory span by the method of constant stimuli. *American Journal of Psychology* **36**, 621–28.

Hartley, T. & G. Houghton 1995. A linguistically constrained model of short-term memory for nonwords. *Journal of Memory and Language*, in press.

Houghton, G. 1990. The problem of serial order: a neural network model of sequence learning and recall. In *Current research in natural language generation*, R. Dale, C. Mellish, C. M. Zock (eds), 287–319. London: Academic Press.

Houghton, G. 1994. *Some formal variations on the theme of competitive queuing*. University College London, Internal Technical Report UCL-PSY-CQ1.

Houghton, G., D. W. Glasspool, T. Shallice 1994. Spelling and serial recall: insights from a competitive queuing model. In *Handbook of normal and disturbed spelling*, G. D. A. Brown & N. C. Ellis (eds), 365–404. Chichester, England: John Wiley.

Hulme, C., S. Maughan, G. D. A. Brown 1991. Memory for familiar and unfamiliar words: evidence for a long-term memory contribution to short-term span. *Journal of Memory and Language* **30**, 685–701.

Jonsdottir, M., T. Shallice, B. W. Wise 1995. Language specific differences in graphemic buffer disorder. Submitted.

Jordan, M. I. 1986. Attractor dynamics and parallelism in a connectionist sequential machine. *8th Annual Meeting of the Cognitive Science Society, Proceedings*, 531–46. Hillsdale, New Jersey: Lawrence Erlbaum.

Kleinfeld, D. 1986. Sequential state generation by model neural networks. *National Academy of Sciences of the USA, Proceedings* **83**, 9469–73.

Lashley, K. S. 1951. The problem of serial order in behaviour. In *Cerebral mechanisms in behaviour*, L. A. Jefress (ed.), 112–36. New York: John Wiley.

Murdock, B. B. 1962. The serial position effect of free recall. *Journal of Experimental Psychology* **64**, 482–88.

Rumelhart, D. E. & J. L. McClelland (eds) 1986. *Parallel distributed processing: explorations in the microstructure of cognition*, vol. 1. *Foundations*. Cambridge, Mass.: MIT Press.

Rumelhart, D. E. & D. A. Norman 1982. Simulating a skilled typist: a study of skilled cognitive-motor performance. *Cognitive Science* **6**, 1–36.

Shattuck-Hufnagel, S. 1979. Speech errors as evidence for a serial-ordering mechanism in sentence production. In *Sentence processing: psycholinguistic studies presented to Merrill Garrett*, W. E. Cooper & E. C. T. Walker (eds), 295–342. Hillsdale, New Jersey: Lawrence Erlbaum.

Tank, D. W. & J. J. Hopfield 1987. Neural computation by concentrating information in time. *National Academy of Sciences of the USA, Proceedings* **84**, 1896–1900.

Treiman, R. & C. Danis 1988. Short-term memory errors for spoken syllables are affected by the linguistic structure of the syllables. *Journal of Experimental Psychology: Learning, Memory and Cognition* **14**, 145–52.

Tzeng, O. J. L. 1973. Positive recency effect in a delayed free recall. *Journal of Verbal Learning and Verbal Behavior* **12**, 436–9.

Watkins, M. J. 1977. The intricacy of memory span. *Memory and Cognition* **5**, 529–34.

CHAPTER 2

Temporal chunking and synchronization using a modular recurrent network architecture

Dimitrios Bairaktaris

Introduction

This chapter presents a neurally inspired mechanism for the temporal chunking of serially ordered input sequences. The proposed mechanism is a modular recurrent network architecture based on the back-propagation through time (BPTT) algorithm and previous research work on short-term active memory by Zipser (1991). The mechanism is aimed at providing a low-level information-processing account of the processes involved in the activation maintenance and temporal synchronization of temporally distributed input. It can be applied as a general preprocessing temporal chunking device to a number of existing models of dynamic binding, such as the system proposed by Shastri & Ajjanagadde (1993).

Unlike visual object recognition, where all the features of the target object are simultaneously present in the input, there are a number of tasks, most notably in natural language processing and working memory, where the target object consists of a sequence of individual features arranged over a limited period of time.

Consider, for example, immediate serial recall list experiments. A subject is presented with a list of words, one word at a time. At the end of the list the subject has to recall the words in the original order of presentation. Experimental evidence (Baddeley 1966, Baddeley et al. 1975) clearly indicates that temporal length and phonological similarity of the words in the list affect recall performance. These findings indicate that each list item is represented by a time-varying signal which is initially decomposed into primitive phonemic features and then reassembled into a single representation. Under these conditions, a temporal chunking mechanism becomes an essential part of the encoding and recall processes in serial list recall.

Another example where dynamic chunking is a critical part of a cognitive task comes in the form of a reflexive reasoning task. Consider the following two texts:

(1) John gave Mary a red ball. Mary gave the ball to George.

To answer the question who is the owner of the red ball, a sequence of simple inferences has to be made:

Give (John, Mary, red ball) → Owner (Mary, red ball)
Give (Mary, George, (red) ball) → Owner (George, red ball)

Now consider the following text:

(2) Mary gave John a red ball. John gave the ball to George.

Although the actors (John, Mary, George) and the subject (red ball) involved in (2) remain identical to those of (1), the actual sequence of presentation affects the answer to the above question. Shastri & Ajjanagadde (1993) in their account of dynamic binding and reflexive reasoning proposed a connectionist architecture which requires that the actors and the subject of the sentence above are simultaneously present. They claim that a strong categorization mechanism is required to provide appropriate input to their system. An essential part of this categorization preprocessor must be a mechanism of temporal chunking which supports the formation of the required synchronized input. Such a mechanism requires two separate processes in order to perform the task.

Activation maintenance is required to sustain the trace of the individual features until completion of the sequence, and output synchronization is required to enable the individual activation maintenance mechanisms to emit an output signal simultaneously, thus creating a trace which represents the sequence as a whole.

Short-term active memory

In a previous study, Zipser (1991) has shown that a simple recurrent network architecture trained using the BPTT algorithm as described by Williams & Zipser (1989), was capable of performing a range of delayed response tasks (Gottlieb et al. 1989, Shintaro et al. 1990). In the network model shown in Figure 2.1, stimulus presentation causes a single node of the model to enter a period of an elevated firing rate and to sustain this rate until the time when a second stimulus appears.

The network architecture shown in Figure 2.1 comprises two hidden units and one output unit which receive two input signals and emit a single output signal. The Analogue In input line conveys information about the stimulus, while the Gate In line is set to On only at the beginning of the stimulus presentation and set to Off immediately afterwards. During training, the network starts from a set of random weights and learns to maintain as output the value present at the Analogue In line, whenever the gate goes from active to inactive (Zipser 1991). Close observation of the model's behaviour revealed that the patterns of activity observed on the hidden nodes have a close resemblance to the firing patterns of real single neurons in monkeys engaged in a delayed-response task. From an

OUT

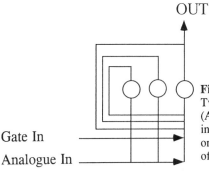

Gate In

Analogue In

Figure 2.1   A recurrent network architecture. Two input signals are employed. The (Analogue In) signal conveys quantitative information about the stimulus, the (Gate In) on/off binary signal is used to indicate the start of the experiment.

anatomical point of view, the proposed network architecture resembles a number of neuronal assemblies with localized recurrent synapses in the brain, most notably area CA3 in the hippocampal formation.

While the model in Zipser (1991) maintains the output during stimulus presentation, in our task output must be suppressed until the end of sequence, at which point all outputs from the individual stimuli must simultaneously emit an output signal. In the model presented in the following sections, a sequence of stimuli appears, one at a time, over a very short period of time. A separate subnetwork similar to the one shown in Figure 2.1 is allocated for each of the stimuli. At the end of the sequence all the stimuli are bound together to form a single representation. To enable all the subnetworks to emit a synchronized output signal, the activation caused by each stimulus must be sustained and delayed until the sequence is complete.

Activation maintenance

To accommodate the special requirements of arbitrary length input sequences the original network architecture shown in Figure 2.1 was modified as shown in Figure 2.2.

The network shown in Figure 2.2 comprises two hidden units and an output unit. It receives input from three different sources. The Analogue In channel conveys information about the presence of a particular feature stimulus in the input sequence. The Gate In channel serves to convey a global synchronization signal. The Gate In line is set to On at the beginning of the input sequence and Off when the last stimulus of the input sequence is presented. The Noise In channel conveys a random signal which represents information irrelevant to the task but nevertheless present while the sequence of stimuli is presented. The Noise In channel can be very useful when used to model an experimental situation where the subject is asked to remember a list of visually presented words at the presence of unattended speech. It is important to emphasize at this point that the network architecture shown in Figure 2.2 is only responsible for represent-

OUT

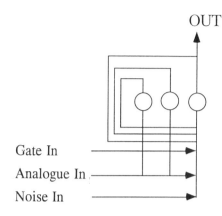

Figure 2.2 A recurrent network architecture. Three input signals are employed. The Analogue In signal conveys quantitative information about the stimulus, the Gate In on/off binary signal is set to On at the beginning of the input sequence and to Off at the end. The Noise In signal represents information about other stimuli which although concurrently present are irrelevant to the task.

Gate In

Analogue In

Noise In

ing one of the features of the target object. As shown in the following section, a number of identical copies of the basic network block shown in Figure 2.2 will be employed in a modular design architecture to represent all the features of the input sequence.

The value ranges for each of the three input signals was set as follows:

(a) All three input channels take continuous input values which range between 0 and 1.

(b) The Analogue In (I_A) line can be in either of the following two states. When the input feature to which the particular network is specifically tuned to is present it conveys a randomly selected input I_A such that $I_A > 0.7$. When this input feature is not present, I_A is randomly set such that $I_A < 0.4$. The presence of a positive signal over the Analogue In line even when the corresponding stimulus is not present is there to reflect the fact that even at the absence of the corresponding feature, similar stimuli are capable of exciting the network but to a lesser degree.

(c) The Gate In (I_G) line is set to the value $I_G = 0.1$ at the beginning of the sequence, and maintains this value until the end of the sequence, while $I_G = 0$ at all other times. The present model does not address the question of the origin of the Gate In signal. However, it is reasonable to assume that onset and offset of the input sequence are detected by supplementary attention-sensitive mechanisms which generate this signal every time a change in the environment is detected. The relatively low value of the Gate In signal was chosen to reflect the relative weakness of such a global synchronization signal.

(d) The Noise In (I_N) line takes a random value such that $0 < I_N < 1$ at all times. The introduction of unstructured noise into the network is motivated by the fact that biological systems are inherently noisy and also to account in part for the situation where the original input sequence is embedded within other sequences or blended with general background noise in the environment.

The task at hand is to test whether the proposed network architecture can perform the required function. For this purpose, a set of training input/output signal patterns was developed. Two of these signals are shown in Figure 2.3. Each input signal was generated in such a way that over a random period of time it exceeded the 0.7 threshold exactly once. The training pattern shown in Figure 2.3a has a sequence onset at time step 4 ($I_G = 0.1$), the stimulus arrives at time step 5 ($I_A = 0.84$), the sequence terminates at time step 7 (sequence duration = 3 time steps). The network was trained to learn to anticipate the end of the sequence and emit an output signal (Target) $O = 1$ at time step 8. Noise levels vary randomly for the total length of the training pattern. Similarly, the training pattern shown in Figure 2.3b has an sequence onset at time step 2 ($I_G = 0.1$), the stimulus arrives at time step 7 ($I_A = 0.86$), the sequence terminates at time step 8 (sequence duration = 6 time steps) and the network learns to emit a Target signal of 1 at time step 9. The total length of the training pattern is always greater but otherwise independent of the total length of the sequence. The time step at which the stimulus appears is random but always within the time window of the sequence.

A total number of 300 training patterns were generated and were shown to be sufficient for the network to learn the task. Recurrent neural network architectures allow the user to choose the depth of recurrence (the number of times steps the current hidden unit outputs can be used to influence future network behaviour), and the number of hidden units in the network. Both these parameters were investigated using the same training set. The nature of the task does require that some depth of recurrence is available as information about past input signals must affect the behaviour of the network in the future.

Our simulations have shown that four levels of recurrence are sufficient for the network to learn the task. The effect of the number of hidden units on the network's learning performance was significant only with respect to the speed of convergence; given enough iterations the network achieved the same level of learning performance independently of the number of hidden units. These results confirm the temporal nature of the task. Modified versions of the task where more than one analogue input signal is involved will increase the spatial complexity of the task and probably require more than one hidden unit. When the network is given the opportunity to propagate its hidden unit output signals in time for ten time steps, it performs significantly better than when no time delay on its hidden-to-hidden connections is allowed.

Approximately 1000 test patterns were generated to test the network's performance. All the test patterns were generated according to the signal specification described above. In all cases the tests were absolutely successful. The network has learned:

(a) to detect the onset and offset of random length input sequences provided that the analogue input signal has exceeded threshold at some point between onset and offset;

(b) to emit an output signal at the offset only when a stimulus was present at some point during the sequence presentation.

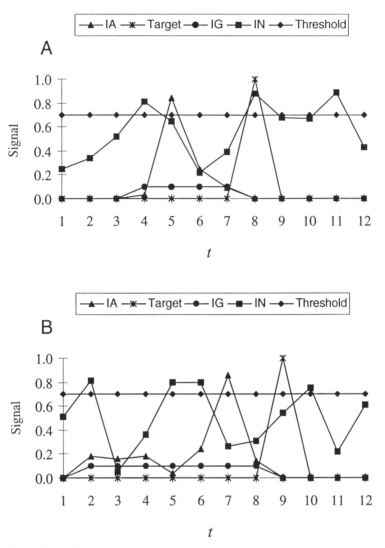

Figure 2.3 Two examples from the training set used to train the network. Each graph shows the Analogue, Gate, Noise and Target signals for each pattern. A total of 300 training patterns were generated for the training phase.

When the network was given just the onset and offset signals via the Gate In line but no stimulus was present, it failed, as expected, to generate an output signal at the offset. The basic building block for our model of temporal chunking is now in place. A simple recurrent network architecture can detect the beginning and the end of a time-varying signal, sustain the activity caused by the input stimulus and emit an output signal at offset.

Modular network architecture

The final architecture of the proposed model comprises n identical copies of the trained network described in the previous section, where n is the number of all the possible input features. No further network training will take place. Each subnetwork is specifically tuned to respond to the strongest only when a specific feature is present in the input sequence.

It is important to emphasize that given a specific input sequence not all the subnetworks in the system receive input through their analogue input channel. It is only those subnetworks which correspond to the feature currently present at the input sites which receive analogue input. This modular network organization is in line with evidence from various cortical areas (i.e. visual cortex, somatosensory cortex) where a topographic organization of the input space is observed. In these cortical areas, relatively small areas of neurons are highly tuned to respond to a very narrow range of stimuli.

The Noise In line is local to each subnetwork, to reflect levels of local background noise. The Gate In line of every subnetwork, however, receives the same input as all the other subnetworks, regardless of the presence or not of the stimulus the subnetwork is tuned to respond, via a global synchronization line.

The proposed architecture achieves temporal synchronization as follows. A sequence of features is generated. The sequence can be of arbitrary length. Each feature occurs at some random time step within the duration of the sequence, passes a signal to the corresponding subnetwork, and it is immediately removed from the input. The model allows two or more features to occur at the same time in order to accommodate cases where multimodal information processing is involved. This situation may arise in experimental design where the subject is shown a light and simultaneously hears a sound. The synchronization line carries information about the onset and offset of the input sequence. All the signals involved take continuous values in the ranges described in the previous section.

To test the model, 100 different input sequences of varying length were constructed. In every test sequence the model achieved perfect synchronization of the output of each subnetwork; at the offset of each sequence every subnetwork emitted an output signal. At this point, all the features of the sequence are now present simultaneously and may allow a higher-order representation to be formed.

Figure 2.4 shows the subnetwork input signals and output responses for a sequence which extends over six time steps, comprising two individual features. The Gate In signal is set to On at time step 2. At time step 4 the first feature appears (Fig. 2.4a) and at time step 8 the second feature appears (Fig. 2.4b) while the Gate In signal is set to Off.

During the next time step both subnetworks emit an output signal to indicate sequence completion and enable output synchronization to take place. It should be emphasized that all the subnetworks in the model receive the same onset/offset Gate In signal but only those three which received the elevated (>0.7) Analogue In signal emit an output at the termination of the sequence.

37

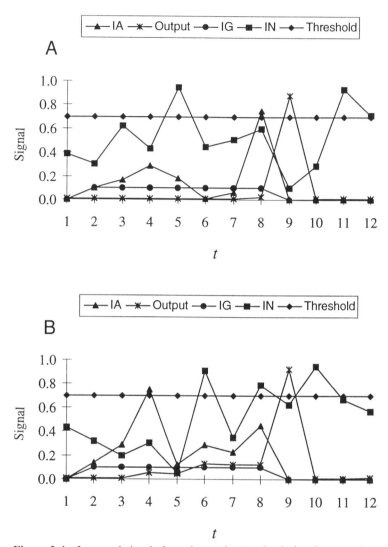

Figure 2.4 Input and signals from three subnetworks during the presentation of a three-feature list. The individual features occur at different times during the sequence but no output is emitted until the sequence is completed.

Figure 2.5 shows the response of a subnetwork which received the onset signal at time step 2 and the corresponding offset signal at time step 8. Because the feature corresponding to this subnetwork actually occurs at time step 11, where $I_A > 0.7$, the network does not emit an output signal at the offset of the Gate In signal. If there was another offset signal occurring at any point after time step 11, the subnetwork would emit an output signal immediately after this second offset.

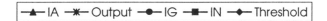

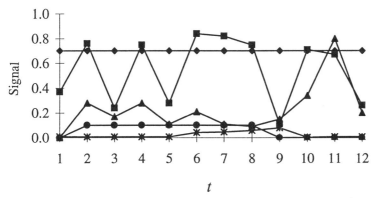

Figure 2.5 When the Analogue In signal in a given subnetwork never rises above the threshold during the onset/offset period the subnetwork does not emit an output immediately after the offset.

The proposed model provides a general temporal chunking mechanism where a sequence of individual features from an input sequence are actively sustained until the end of sequence. This is followed by synchronized output activity. The majority of information in the proposed model is encoded locally, in the nodes of each subnetwork. Activation maintenance is also sustained locally by means of fixed-weight synapses. The only computationally, and to some extent biologically, expensive element of the model is the existence of the global synchronization line which has to run across all the subnetworks and provides general information about the start and the end of the input sequence. From a theoretical point of view this single global line is the absolute minimum requirement in order to perform the synchronization task.

Conclusions

The proposed model successfully tackles the problem of temporal chunking and synchronization by providing a general-purpose biologically relevant modular network design. All the modules of the system have been pretrained using the BPTT algorithm to perform an idealized delayed-response task. The proposed model does not provide a developmental account as to how these networks may develop, although their localized recurrent synapse organization is often met in a number of cortical areas. The particular training regime which supports the training phase must be viewed only as a convenient tool which enables us to obtain, relatively quickly, a network configuration suitable for the final task. The proposed model emphasizes synchronized spike emission as the means for temporal synchronization. There is growing evidence (Abeles 1988) that a single spike emitted from a single neuron is not sufficient to sustain a higher-order representation of the input. Current work focuses on a model where some form of sustained elevated firing rate enables the output nodes to enter a phase of synchronized oscillatory behaviour. In this respect the model may function as a first approximation to a more complex mechanism which will incorporate further detail about information passing between neighbouring neurons in the brain. Future discoveries in the area of single-neuron information processing are unlikely to affect the functional and theoretical contribution of the current model. More detailed and biologically plausible models of neurons will most definitely emerge in the future; however, the solution to the temporal chunking problem in working and short-term memory will always have to rely on a mechanism which combines activation maintenance and temporal synchronization.

References

Abeles, M. 1988. *Corticonics: neural circuits of the cerebral cortex*. Cambridge: Cambridge University Press.

Baddeley, A. D. 1966. Short-term memory for word sequences as a function of acoustic, semantic and formal similarity. *Quarterly Journal of Experimental Psychology* **18**, 302–9.

Baddeley, A. D., N. Thomson, M. Buchanan 1975. Word length and the structure of short-term memory. *Journal of Verbal Learning and Verbal Behavior* **14**, 575–89.

Gottlieb, Y., E. Vaadia, M. Abeles 1989. Single unit activity in the auditory cortex of a monkey performing a short-term memory task. *Experimental Brain Research* **74**, 139–48.

Shastri, L. & V. Ajjanagadde 1993. From simple associations to systematic reasoning: a connectionist representation of rules, variables, and dynamic bindings using temporal synchrony. *Behavioural and Brain Sciences* **16**(4), 417–51.

Shintaro, F., C. J. Bruce, P. S. Goldman-Rakic 1990. Visual spatial coding in primate prefrontal neurons revealed by oculomotor paradigms. *Journal of Neurophysiology* **63**, 814–31.

Williams, R. J. & D. Zipser 1989. A learning algorithm for continually running fully recurrent neural networks. *Neural Computation* **1**, 270–80.

Zipser, D. 1991. Recurrent network model of the neural mechanism of the short-term active memory. *Neural Computation* **3**, 179–93

Learning to learn in a connectionist network: the development of associative learning

Gordon D. A. Brown, Tim Preece, Charles Hulme

Introduction

In this chapter we describe an approach to modelling human memory that attempts to link current mathematical and connectionist approaches. We argue that such a bridge is necessary if progress is to be made towards a comprehensive formal model of human memory, for the connectionist models of human memory that exist at present are best seen as complementary to current formal mathematical models, in that they have been used to account for different types of empirical data and have a complementary pattern of strengths and weaknesses when viewed as psychological accounts.

More specifically, we argue that the ability of connectionist approaches to underpin psychological models of memory that are intrinsically *developmental* must be combined with the insights incorporated in current mathematical models of memory, for these mathematical approaches have not generally been used to address data concerning the development of human memory. We illustrate this with new simulation work that takes a developmental approach to paired associate learning, and shows that the proportion of different error types in paired associate learning can be predicted to vary over developmental time. In particular, the model is shown to make one very specific prediction: that the proportion of cue intrusion errors in paired associate learning should decrease during development.

The plan of the chapter is as follows. In the first section, we review some of the existing mathematical models of memory, focusing particularly on those that use convolution as the basic associative mechanism, and correlation as the basic retrieval operation. Current mathematical models provide a good account of a wide range of empirical data concerning adult human memory performance. However, we also refer to some of the developmental data that any complete model of human memory must address, and show that neither mathematical nor connectionist models can easily be extended to account for even the basic developmental phenomena.

The second part of the chapter introduces DARNET (developmental associative recall network). DARNET is a connectionist-like architecture that "learns to learn". Using a gradient descent learning algorithm, DARNET gradually develops the ability to form associations between items (represented as vectors) with a single exposure to the items. It therefore combines the ability to perform one-shot learning (like convolution or Hebbian learning) with a developmental aspect – the one-shot learning ability is not present from the start, but gradually develops. Thus the new architecture allows examination of the development of associative learning in a way that has not previously been possible.

The third part of the chapter reviews some of our earlier work (Brown et al. 1993, Brown et al. 1994, Brown et al. 1995) which shows that when DARNET has "learned to learn" it can store and retrieve single associations better than the convolution/correlation operations that are widely used in mathematical models of memory, and that it is also flexible (in that it can make use of different memory trace sizes) and can be used, in a similar way to mathematical convolution-based models, to account for psychological data (e.g. from serial list learning and the study of paired associate learning for similar and dissimilar items).

The next part of the chapter makes use of the developmental aspect of the model to show that certain phenomena exhibited by other models (e.g. the nature of the errors in paired associate learning) may change over developmental time. These simulations illustrate the need for a developmental approach. It is concluded that DARNET does indeed provide a bridge between connectionist and mathematical models of human memory, in that it shows how a connectionist-like architecture can learn, using gradient descent learning, to combine two item vectors together into a third, memory trace vector in just the way that is required by some current mathematical models. This may facilitate the eventual production of genuinely developmental accounts of human memory that are as wide-ranging (in terms of the empirical data they account for) as are current mathematical models of adult memory.

Mathematical models of human memory

Any formal model of memory must incorporate some mechanism for forming associations between different representations. These representations are often vectors of features, where individual features may or may not be seen as psychologically meaningful objects (e.g. Metcalfe Eich 1982, Gillund & Shiffrin 1984, Metcalfe Eich 1985, Hintzman 1986, Lewandowsky & Murdock 1989).

Candidate associative mechanisms include, for example, Hebbian learning and back-propagation (from within a connectionist framework) and convolution/correlation from mathematical models of memory. Both TODAM (theory of distributed associative memory; e.g. Murdock 1982, Lewandowsky & Murdock

1989, Murdock 1993) and CHARM (composite holographic associative recall model; Metcalfe Eich 1982, Metcalfe Eich 1985) use convolution (described below) as the fundamental method of forming associations. Other mathematical schemes have also been used: for example, the matrix model (e.g. Humphreys et al. 1989) represents associations as the matrix products of the vectors to be associated. Other mathematical models are less specific about exactly how associations are implemented, but are more concerned with, for example, whether associations are formed between items and list or chunk code representations (e.g. Estes 1972) or between items and some representation of an item's position in an ordered list (e.g. Johnson 1991). In SAM (Gillund & Shiffrin 1984), for example, a strength is simply assigned to each inter-item association, and hence there need be no commitment regarding the low-level mechanism of association. MINERVA 2 (e.g. Hintzman 1986) assumes that vectors representing events are stored separately. Probes are assumed to lead to parallel activation of pre-existing memory traces, and two associated items can be represented as different sets of elements within one longer vector. Thus here again there is no need for a more complex mechanism for combining two vectors into a third.

It is important to note that the choice of basic associative mechanism is to some extent independent of the architectural assumptions of the memory model within which the associative mechanism is embedded. For example, either convolution or back-propagation could be used to form item-to-item associations in the context of a chaining model (e.g. Lewandowsky & Murdock 1989) or item-to-context associations (e.g. Houghton 1990, Burgess & Hitch 1992) or item-to-list-position associations.

However this does not of course mean that the choice of associative mechanism is arbitrary, because different associative mechanisms have different patterns of strengths and weaknesses (Brown et al. 1993, Brown et al. 1994, Brown et al. 1995). More specifically, gradient descent learning algorithms such as back-propagation (Rumelhart et al. 1986) are able to extract the underlying regularities from an input/output mapping system with some underlying structure, and hence have been widely used for modelling the development of such processes as verb past tense learning (Rumelhart & McClelland 1986) or reading (Seidenberg & McClelland 1989). However the very aspect of gradient descent learning algorithms that is partially responsible for allowing them to extract out these underlying regularities, namely the fact that only small amounts of learning take place in response to each exposure to the data set, means that (at least with normal learning rates) they are fundamentally unsuited for characterizing the single-exposure learning that is characteristic of human memory performance in many situations.

The wide range of mechanisms that can form associations within a single trial, such as Hebbian learning or convolution, can in contrast be used to model single-trial learning situations, but there has been no account forthcoming of just how a *developmental* account could emerge within such a framework (see Brown et al. (1993) for extensive discussion of this point).

We now discuss this in the context of mathematical models that use convolution as the basis for associative storage. It is assumed in such models that items (such as words in a to be remembered list) can be represented as vectors of features, and it is these vectors that can be associated via convolution. More specifically, convolution is used to associate two item vectors together into a memory trace vector in such a way that either of the two item vectors can be used as a probe to retrieve an approximation to the other item vector from the trace vector using correlation (mathematically the approximate inverse of convolution). Thus, convolution can be used as a basic associative mechanism that associates two item vectors, e.g. a and b, into a memory trace vector T in such a way that T can be probed with either a or b leading to the retrieval of an approximation to the other item. We denote a retrieved item as a' or b'. Thus, for example, T can be probed with a, whereupon b' is retrieved. Metcalfe Eich (1982) describes the convolution and correlation operations in detail.

Several different associations (e.g. a with b, c with d, e with f) can be formed in the same way, and the resulting convolutions superimposed into the composite memory trace T. T can then be probed with any of the original associates, leading to the retrieval of an approximation to the other member of the associated pair. So, in the example above, T could be probed with e and the result would be an approximation to f, i.e. f'. Convolution therefore provides a mechanism of one-shot association that can be used as the basis for a memory architecture. Note that in this type of system an association between items is represented in the form of a vector, although this could alternatively be conceived of as a vector of connection strengths.

Convolution-based models have been described and evaluated in a series of papers by Murdock and his co-workers (e.g. Murdock 1982, Murdock 1983, Lewandowsky & Murdock 1989, Murdock 1993) and by Metcalfe Eich (e.g. Metcalfe Eich 1982, Metcalfe Eich 1985, Metcalfe 1991). Several models with different architectures have been used to model a wide variety of data from paired associate learning (Metcalfe Eich 1982), levels-of-processing (Metcalfe Eich 1985), recognition failure (Metcalfe 1991) and serial recall (Lewandowsky & Murdock 1989) as well as other paradigms. These models have different architectures, but all use convolution as the basic mechanism for storage, and correlation as the basic mechanism for retrieval.

The need for a developmental account

Convolution-based models have the advantage of being applicable to both one-trial learning and incremental learning (the latter via either "closed-loop" learning, equivalent to delta-rule learning; or probabilistic encoding; Murdock & Lamon 1988). However, one problem arises from the need to model the *development* of memory capacity in children. There is much data on the development of short-term memory in children (e.g. Dempster 1981, Nicolson 1981, Case et al.

1982, Hulme et al. 1984, Hitch et al. 1989, Hulme & Tordoff 1989, Henry 1991a,b, Henry & Millar 1991, Gathercole et al. 1993, Gathercole & Adams 1992, Hitch et al. 1993). Much of this work has examined the development of memory span for words or non-words, but there is rather less work on the development of memory using other experimental paradigms. It is clearly desirable that formal models of memory should eventually be able to address the developmental data that exist, and that they should be able to make clear predictions about the developmental course of, for example, serial position effects and the development of different types of associative learning. However, it is far from clear that current approaches, from within either the mathematical or connectionist modelling traditions, yet have the potential to provide natural accounts of single-exposure learning.

Within connectionism, gradient descent learning models have provided useful developmental accounts of a number of cognitive phenomena in virtue of their learning capacity. However, such models are not simultaneously able to model one-trial learning, and such an ability is necessary if phenomena such as the development of memory span are to be evaluated. Connectionist learning architectures that are able to perform one-shot association (such as Hebbian learning) have not hitherto been used to show how such an ability could develop.

Similar limitations apply to current convolution-based models, which must have the mechanisms for performing convolution and correlation hardwired into the system. It is therefore difficult to envisage an account of the development of memory emerging naturally from such a framework (Brown et al. 1993), although memory span could be made to increase in such a model by increasing the probability with which to be remembered items were encoded (Murdock & Lamon 1988). An alternative approach to modelling the development of memory capacity could involve increasing the dimensionality of the memory trace. However, the nature of the convolution operation places constraints on the maximum size of memory trace that can usefully be used to store input vectors of a given size.

Yet another possibility is that the nature of the to be associated representations changes over developmental time, and that this explains developmental increases in memory performance. However, capacity increases are apparent even for simple and highly familiar material, where it seems unlikely that such representational changes would be of practical significance.

A connectionist network that learns to learn

What is needed, therefore, is an approach that combines the developmental aspects of connectionist architectures with the one-shot associative learning capacity that is characteristic of Hebbian learning or convolution. In this section, we review some of our own work that has been directed towards this goal (Brown et al. 1993, Brown et al. 1994, Brown et al. 1995).

The particular architecture that we have developed is referred to as DARNET (developmental associative recall network). DARNET is a multilayer connectionist-like architecture that learns, using a gradient descent learning procedure similar to back-propagation, to perform single-trial learning. Having developed this ability, its ability to learn single or multiple associations can then be separately examined. We begin with a brief description of these two separate phases of learning that are involved in modelling with DARNET, and then describe the storage and retrieval operations.

Items to be remembered are represented as vectors of features. (Feature values are chosen from a normal distribution with a mean of zero and a variance of one.) DARNET learns to combine pairs of these vectors into a memory trace vector in such a way that probing the trace vector with either of the originally associated pair of vectors produces an approximation to the other. In this respect the operation is similar to convolution, but the mechanism of association is different in our approach. Thus the model can learn in two distinct phases. In phase 1 learning, DARNET uses gradient descent learning to learn connection strengths (marked "learned weights" in Fig. 3.1) such that at the end of phase 1 learning it

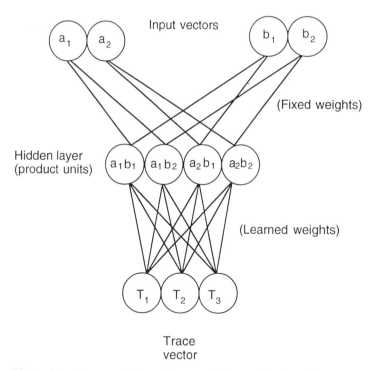

Figure 3.1 The association, using convolution, of two two-element vectors (*a* and *b*) into a three-element composite memory trace vector (*T*). Each element of *T* is the sum of products, where each product is made up of an element from *a* and an element from *b*.

can form an association between two input vectors in a single trial. In phase 2 learning, the weights are all held constant and the model's ability to learn single or multiple associations in a single trial learning paradigm is examined. Thus it is worth noting that what is stored in the connection strengths of the network is not associations between items, but, rather, the ability to form single-trial associations.

The details of the model's operation are described in detail in Brown et al. (1993), but the basic principles are as follows. Figure 3.1 shows how two to be associated input vectors are connected to the trace vector and how storage is carried out. For the purposes of illustration, two-element input vectors and a three-element trace vector are shown. (Larger vectors were used in all the simulations reported below, however.)

The input vectors are connected to the trace vector through an intermediate layer of hidden units. Each unit in this intermediate layer is connected, with a fixed weight, to an element of one input vector and an element in the other input vector (as illustrated in Fig. 3.1). Each unit in the intermediate layer computes the product of the values of the two elements it is connected to, and hence we term them "product units".

Every product unit is connected to every memory trace vector unit with a learned weight that has been fixed at the end of a "learning to learn" phase (phase 1 learning). To form an association between two input item vectors during phase 2 learning, they are simultaneously presented to the network. Then the pairwise products described above are calculated, forming a pattern of activity over the intermediate layer of product units. Finally, each trace vector unit activation is calculated on the basis of the inputs from each product unit modulated by the connection strength from that product unit to the trace vector unit.

The memory trace vector produced in the manner described above can be used as the basis for recall. Either one of two item vectors that have been associated together in the composite memory trace vector can be used as a probe to retrieve an approximation of the other one.

In recall, the trace vector and the probe vector are combined using a different set of weights. This process is illustrated in Figure 3.2, which shows input vector a being used to probe the trace vector T that is assumed to contain the association a–b. Elements of a and T are connected with fixed connection strengths to an intermediate hidden layer. The value of each element of this intermediate layer is the product of an element of a and an element of T. Every product element is connected to every retrieval vector element with a weight previously learned by the network in the learning to learn phase.

The learning to learn algorithm is described in detail in Brown et al. (1993); it adjusts the strengths of the connections between product units and memory trace vector units, using a gradient descent learning procedure. Phase 1 learning thus involves training the model with pairs of novel randomly generated input vectors, examining the error produced when the model attempts to recall one of the input pair of vectors when probed with the other, and adjusting the weights to

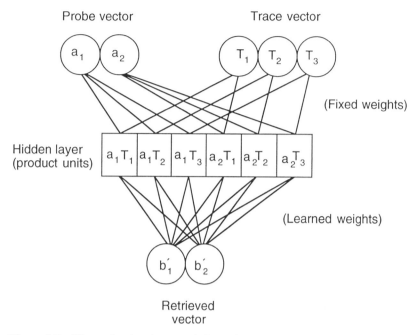

Figure 3.2 The retrieval, using correlation, of an approximation (b') to one of the input vectors (b) by "probing" the composite memory trace vector (T, as in Figure 3.1) with the other vector (a) that entered into the association as illustrated in Figure 3.1.

minimize the resulting error. Each learning trial consists of the presentation of a new pair of previously unseen input vectors. At the end of this learning to learn stage (phase 1), DARNET can form and retrieve associations between item vectors within a single trial.

DARNET's single-trial learning ability can then be separately examined. In this phase 2 learning, multiple associations are superimposed into a composite memory trace. Thus, phase 2 learning is similar to what takes place in many of the convolution-based models described above. However, although the basic approach of the model to association and retrieval is similar to that of the convolution models discussed earlier, DARNET is able to perform successful storage and recall using a set of weights that it has learned itself. It does not need to have this ability hardwired into it.

The development of associative learning

We have previously examined DARNET's performance in a wide range of circumstances. Brown et al. (1993) describe the results of training DARNET to perform single-trial association.

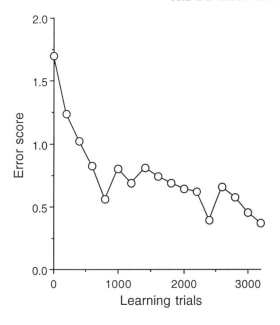

Figure 3.3 Reducing error score during learning to learn. The model is using input vectors of dimensionality 10 and a composite trace vector of dimensionality 19.

Figure 3.3 shows the results of training DARNET to associate item vectors with ten elements into a memory trace vector of dimensionality 19 elements. The network was initialized with small random weights, and for each learning trial a new pair of vectors was generated. These vectors were input to the network and a forward pass made through the storage weights to produce a trace vector. One of the two input vectors was then randomly chosen to be the probe (and the other, by default, became the target). The probe vector was then passed through the retrieval weights together with the trace. This gave a retrieved approximation to the target vector, which was compared to the true target vector to give an error score. The learning rules were then applied to calculate a small change for each weight in both the storage and retrieval subnetwork.

Figure 3.3 (after Brown et al. 1993) shows the error score for DARNET as a function of learning. (The error score is calculated as the square root of the sum of the squared differences between retrieval and target element values.) It can be seen from Figure 3.3 that the error score of the network reduced over time, and that after 3500 learning trials the error score reached about 0.05. (Note that DARNET was continuously being presented with new pairs of item vectors rather than a specified set of vectors.) The reducing error indicates that DARNET can indeed learn to associate pairs of novel item vectors into a memory trace vector in such a way that the memory trace can be probed with either of the original input vectors and the other can be retrieved. This is an ability similar to that of the convolution/correlation models discussed earlier, although DARNET has learned to form the associations for itself.

Brown et al. (1993) also demonstrated that DARNET can do better than convolution at learning single associations, in that the retrieved vectors are less variable (i.e. the variance in error score between target and actual retrieved vector was lower). They also showed that DARNET can learn to perform the operations of convolution and correlation, although this solution to the one-trial learning problem is not the one the model chooses if it is not constrained to do so. Thus, they argued that DARNET combined some of the advantages of both convolution and connectionist approaches as described above. DARNET can learn to learn (thus potentially providing a basis for developmental accounts).

We note at this point that DARNET is not finding the same solution as used by the convolution/correlation algorithms. The fact that there is less variability in the outputs produced during retrieval suggests this, as does the fact that when DARNET is forced to learn convolution it takes much longer to do so than it does to learn its own solution (Brown et al. 1993). Also, inspection of the weights (possible on small versions of the network only) that are learned reveals that a solution is learned which is different from convolution, for the learned weights are not those that would be required to implement convolution.

DARNET can also handle memory trace vectors of arbitrary dimensionality (allowing capacity to be varied in a systematic way; Brown et al. 1994). Thus, for example, if DARNET is required to learn associations between pairs of ten-element input vectors, there is no reason why DARNET could not use a five-element memory trace vector or a 50-element memory trace vector. Brown et al. (1994) demonstrated that DARNET can indeed make efficient use of increased numbers of trace vector elements, by assessing DARNET's ability to make use of memory trace vectors with eight through 64 elements when learning to associate pairs of random eight-element input vectors. Thus, DARNET can make use of additional trace vector elements if it is provided with them. However, it is of course the case that DARNET must be retrained if the dimensionality of the vectors to be associated changes, and this limits its flexibility.

We also investigated DARNET's ability to account for psychological data from a variety of experimental paradigms (Brown et al. 1995). These simulations involved the use of the single-trial associative mechanism embodied in DARNET as a substitute for convolution/correlation in various models developed by Metcalfe Eich (1982), Lewandowsky & Murdock (1989) and Lewandowsky (1991) to account for paired associate learning, serial position effects and gradual unlearning, respectively. In other words, we adopted the same model architectures as had previously been used by others, but using DARNET's associative mechanism instead of convolution. Brown et al. (1995) show that, under these circumstances, the DARNET-based models did as well as the convolution-based models at accounting for the relevant psychological data.

A developmental approach to paired associate learning

The simulations described above have shown that DARNET can use a gradient descent learning algorithm to gradually learn a one-shot associative learning capability. However, it would be particularly attractive to show that DARNET gives rise to particular predictions about the developmental time-course of specific psychological phenomena.

In the present section, therefore, we address one such issue, concerning the proportion of cue intrusion errors as a function of item similarity in a simple paired associate learning paradigm. To anticipate our conclusion: the adoption of a developmental process demonstrates that the high proportion of cue intrusion errors observed both experimentally and in a composite, convolution-based memory model (Metcalfe Eich 1982) may be seen as characteristic of the particular level of associative ability embodied in the model, or alternatively of a particular developmental stage in a process of learning to learn.

Metcalfe Eich (1982) examined the ability of her composite convolution-based model (CHARM) to model data from the paired associate learning of similar and dissimilar items. CHARM was required to learn (by convolution) three paired associates. The six item vectors that these three pairs were drawn from were either similar (half the vector elements had identical values) or dissimilar (element values randomly chosen from a normal distribution).

In order to assess recall, one member of each pair (the cue) was correlated with the composite memory trace which contained the three convolved item vector pairs. The resulting vector was then compared with a "vocabulary" of 12 potentially recallable items, made up of the six similar or dissimilar items that made up the three paired associates, with a further six independent items. Whichever one of the 12 vocabulary items that most closely resembled the item vector that resulted from convolving the cue item with the composite memory trace was assumed to be the "recalled" item.

Metcalfe Eich found that CHARM did well on this task, and that performance was worse for similar than for dissimilar item vectors. However the finding of primary interest for current purposes was that there was a high proportion of "cue intrusion" errors when the to be associated items were similar to one another. More specifically, when similar item vectors were associated she found that the target item was (correctly) recalled 42% of the time, with the cue item being incorrectly recalled on 14.6% of occasions, and incorrect recalls of non-target stimuli and non-target responses each accounting for about 22% of responses. As she points out, the overall levels of recall are arbitrary, but the model nevertheless makes the interesting (and counterintuitive) prediction that the proportion of cue intrusion errors should be particularly high when the items for the paired associate learning task are drawn from a similar set.

This prediction was not confirmed in an initial paired associate learning experiment conducted by Metcalfe Eich (1982), probably, as she suggests, because subjects could use their knowledge that cue and target were never the same to suppress recall of the cue item itself in response to the cue as probe. However, a second experiment, in which some of the paired associates were identical, did find that a higher proportion of the errors were cue intrusion errors in a synonym condition compared to an unrelated condition (specifically: 55% of the intrusion errors were cue intrusions in the synonym condition; 39% in the unrelated condition). This is an attractive finding, because it is in line with the counterintuitive prediction made by CHARM.

We investigated the ability of the basic associative mechanism embedded in DARNET to account for the same data when DARNET's associative ability was used in exactly the same way as convolution was used in Metcalfe Eich's examination of CHARM. The initial results of this are reported in Brown et al. (1995), and show that a version of DARNET that has "learned to learn" performs in a very similar manner to an equivalent convolution-based model, and also gives rise to the high proportion of cue intrusion errors when the items to be associated are similar to one another.

However it is also possible to examine the performance of DARNET on the same task when it has only partially completed its learning to learn phase. Thus, it will be less good at storing and retrieving single associations between item vectors after a single exposure, and could be thought of as being at an earlier stage in development.

We therefore examined the performance of both a partially learned and a completely learned version of DARNET on paired associate learning of items drawn from similar sets. Both versions had been given training, using DARNET's gradient descent learning to learn algorithm, in associating pairs of ten-element item vectors. The learning of the partially learned version was stopped when it had reached a level of about 50% error, and the other version continued learning to learn until it was performing single-trial association with about 10% error. The two versions were then compared on their performance on the learning of three pairs of item vectors drawn from a set of six similar vectors. The procedure adopted was identical to that used by Metcalfe Eich, and described above, with one significant difference in the way the similar vectors were created. Metcalfe Eich created similar item vectors by first generating a set of normalized vectors and then changing alternate items of the members of the similar set so that they were identical for all members of the set. Thus the item vectors in the similar set had 50% of their element values in common. However, one consequence of this is that the similar item vectors lose their perfect normalization. However, the convolution/correlation procedure for storage and retrieval, as well as that learned by DARNET, relies partly on the normalization of the item vectors, and so it is theoretically possible that part of the reduced level of performance that Metcalfe Eich observed with similar item vectors was due to the less complete normalization of those vectors (i.e. in addition to their similarity). We therefore adopted a slightly more complex method, which involved a further stage of changing the non-identical elements of each similar item vector so that the vectors each regained perfect normalization.

A further difference between our procedure and Metcalfe Eich's was that we had to use low-dimensionality item vectors: each of our item vectors contained only ten elements. Both versions used a memory trace vector with 19 elements. (The need to use item vectors with fewer elements arises from the computational

Table 3.1 Performance of two versions of DARNET on paired associate learning of similar items (results are averaged over 100 runs).

DARNET version	Target (correct)	Cue intrusion error	Non-target response error	Non-target stimulus error	Extra-list intrusion error	Errors that are cue intrusions
	%	%	%	%	%	%
Well learned	39	3	30	27	2	4.9
Partially learned	27	16	25	25	8	22

constraints of learning the requisite versions of DARNET within a reasonable timescale.)

The results of using the two versions of DARNET (partially learned and inaccurately learned) to model paired associate learning of similar item vectors is illustrated in Table 3.1.

It can be seen that, not surprisingly, the overall level of performance of the well learned version of DARNET was higher than for the partially learned version (39% correct versus 27% correct). We do not regard the overall level of performance as particularly significant; higher levels of accuracy would result if item vectors (and/or a memory trace vector) of higher dimensionality were used (Brown et al. 1994), and also, as Metcalfe Eich herself points out, it is difficult to know how to interpret the similarity of item vectors in this kind of model in relation to the similarity of items used in relevant experiments with human subjects.

Of more interest is the relative proportion of different error types in the two versions of the model. In particular, it can be seen that the majority of additional errors that occur in the partially learned model are cue intrusion errors (with errors increasing from 3% of responses to 16%). In other words, if the three associations to be learned were A with B, C with D, and E with F, with the first member of each pair being used as the probe item, probing with A (or C or E) would lead to the erroneous recall of A (or C or E) for a much higher proportion of the time in the partially learned model. Indeed, the number of non-target response errors (e.g. probe with A, recall D or F) and the number of non-target stimulus errors (e.g. probe with A, recall C or E) actually *decreased* in the partially learned version of the model. Thus, the proportion of the errors that were cue intrusion errors was only 4.9% for the well learned model, but 22% for the partially learned model. We draw three main conclusions from these data.

First, it is advantageous (and perhaps even necessary) to adopt a developmental perspective in deriving predictions from this type of model. Predictions about the proportions of particular error types may reflect the particular level of associative ability exhibited by a model, and this is something that might be expected to change across developmental time in human subjects.

Secondly, we can derive developmental predictions; specifically, we predict that the proportion of cue intrusion errors in paired associate learning will decrease developmentally. We know of no extant data that bear on this prediction; we are currently planning an experiment to test it.

Thirdly, the increase in the proportion of a particular type of error may be seen as a general reflection of the performance of a model in a more difficult learning task. More specifically, a higher proportion of cue intrusion errors in paired associate learning may result either from the use of more similar item vectors, or from a model with less well developed associative ability. Thus the behaviour of a model such as DARNET or CHARM may be seen as a reflection of the precise level of associative ability of the model, and in that sense is not intrinsic to the architecture within which the mechanism of association is embedded.

Summary and conclusions

The approach that we have described is intended to develop a link between connectionist learning algorithms and convolution-based models of memory, by showing that a connectionist-style learning algorithm, DARNET, can learn not only to combine two vectors (representing items to be associated) into a third, memory trace vector using the convolution operation, but can also find its own solution to the problem of single-trial learning that is actually better (in the sense of leading to more consistently accurate recall) than convolution itself.

The results we have obtained also show that DARNET can do as well, or better than, convolution-based models at accounting for data from ordered serial recall (primacy and recency effects), paired associate learning and the effects of similarity on overall recall (including some counterintuitive data also explained by Metcalfe), and can overcome catastrophic interference effects in the same ways as can convolution-based models.

Most importantly for the purposes of the present chapter, they show that qualitatively different results may emerge in networks at different stages of "learning to learn" – the high proportion of cue intrusion errors observed by Metcalfe Eich in her convolution-based model are reproduced in DARNET, but only at certain stages of the simulated developmental process of learning to form associations. Furthermore, specific developmental predictions can be derived from DARNET.

However, it is important to note that DARNET (seen as a basic mechanism of association) inherits many of the limitations of convolution-based models when embedded in a similar architecture. The importance of DARNET lies, rather, in its provision of a *general* mechanism that allows the examination of the development of single trial learning. This is clearly necessary if the development of short-term memory capacity is ever to be understood.

Acknowledgements

This research was supported by grant no. R000232576 from the Economic and Social Research Council to the first and third authors.

References

Brown, G. D. A., C. Hulme, P. Dalloz in press. Modelling human memory: connectionism and convolution. *British Journal of Mathematical and Statistical Psychology*.

Brown, G. D. A., P. Hyland, C. Hulme 1994. The effects of varying memory vector size in a network that learns to learn. *Proceedings of the World Congress on Computational Intelligence*, IV, 2291–96. Piscataway, New Jersey: IEEE.

Brown, G. D. A., P. Dalloz, C. Hulme 1995. Mathematical and connectionist models of memory: a comparison. *Memory* **3**(2), 113–45.

Burgess, N., & G. J. Hitch 1992. Towards a network model of the articulatory loop. *Journal of Memory and Language* **31**, 429–60.

Case, R., D. M. Kurland, G. Goldberg 1982. Operational efficiency and the growth of short-term memory span. *Journal of Experimental Child Psychology,* **33**, 386–404.

Dempster, F. N. 1981. Memory span: sources of individual and developmental differences. *Psychological Bulletin* **89**, 63–100.

Estes, W. K. 1972. An associative basis for coding and organization in memory. In *Coding processes in human memory*, A. W. Melton & E. Martin (eds), 161–90. Washington, DC: Winston.

Gathercole, S. E. & A-M. Adams 1992. Phonological working memory in very young children. *Developmental Psychology* **29**, 770–8.

Gathercole, S. E., C. Willis, H. Emslie, A. Baddeley 1993. Phonological memory and vocabulary development during the early school years: a longitudinal study. *Developmental Psychology* **28**, 887–98.

Gillund, G. & R. M. Shiffrin 1984. A retrieval model for both recognition and recall. *Psychological Review* **91**, 1–67.

Henry, L. A. 1991a. The effects of word length and phonemic similarity in young children's short-term memory. *Quarterly Journal of Experimental Psychology* **43A**, 35–52.

Henry, L. A. 1991b. Development of auditory memory span: the role of rehearsal. *British Journal of Developmental Psychology* **9**, 493–511.

Henry, L. A. & S. Millar 1991. Memory span increase with age – a test of two hypotheses. *Journal of Experimental Child Psychology* **51**, 459–84.

Hintzman, D. L. 1986. "Schema abstraction" in a multiple-trace memory model. *Psychological Review* **93**, 411–28.

Hitch, G. J., M. S. Halliday, J. E. Littler 1989. Item identification time and rehearsal rate as predictors of memory span in children. *Quarterly Journal of Experimental Psychology* **41A**, 321–38.

Hitch, G. J., M. S. Halliday, J. E. Littler 1993. Development of memory span for spoken words: the role of rehearsal and item identification processes. *British Journal of Developmental Psychology* **11**, 159–69.

Houghton, G. 1990. The problem of serial order: a neural network model of sequence learning and recall. In *Current research in natural language generation*, R. Dale, C. Mellish, M. Zock (eds), 287–319. London: Academic Press.

Hulme, C. & V. Tordoff 1989. Working memory development: the effects of speech rate, word length and acoustic similarity on serial recall. *Journal of Experimental Child Psychology* **48**, 1–19.

Hulme, C., N. Thomson, C. Muir, A. Lawrence 1984. Speech rate and the development of short-term memory span. *Journal of Experimental Child Psychology* **38**, 241–53.

Humphreys, M. S., J. D. Bain, R. Pike 1989. Different ways to cue a coherent memory system: a theory for episodic, semantic and procedural tasks. *Psychological Review* **96**, 208–33.

Johnson, G. J. 1991. A distinctiveness model of serial learning. *Psychological Review* **98**, 204–17.

Lewandowsky, S. 1991. Gradual unlearning and catastrophic interference: a comparison of distributed architectures. In *Relating theory and data: essays on human memory in honor of Bennet B. Murdock*, W. E. Hockley & S. Lewandowsky (eds), 445–76. Hillsdale, New Jersey: Lawrence Erlbaum.

Lewandowsky, S. & B. B. Murdock 1989. Memory for serial order. *Psychological Review* **96**, 25–57.

Metcalfe, J. M. 1991. Recognition failure and the composite memory trace in CHARM. *Psychological Review* **98**, 529–53.

Metcalfe Eich, J. M. 1982. A composite holographic associative recall model. *Psychological Review* **89**, 627–61.

Metcalfe Eich, J. M. 1985. Levels of processing, encoding specificity, elaboration, and CHARM. *Psychological Review* **92**, 1–38.

Murdock, B. B. 1982. A theory for the storage and retrieval of item and associative information. *Psychological Review* **89**, 609–26

Murdock, B. B. 1983. A distributed memory model for serial-order information. *Psychological Review* **90**, 316–38.

Murdock, B. B. 1993. TODAM2: a model for the storage and retrieval of item, associative, and serial-order information. *Psychological Review* **100**, 183–203.

Murdock, B. B. & M. Lamon 1988. The replacement effect: repeating some items while replacing others. *Memory and Cognition* **16**, 91–101.

Nicolson, R. 1981. The relationship between memory span and processing speed. In *Intelligence and learning*, M. Friedman, J. P. Das, N. O'Connor (eds), 179–84. New York: Plenum Press.

Rumelhart, D. E. & J. L. McClelland 1986. On learning the past tenses of English verbs. In *Parallel distributed processing: explorations in the microstructure of cognition*, vol. 2. *Psychological and biological models*, J. L. McClelland & D. E. Rumelhart (eds), 216–71. Cambridge, Mass.: MIT Press.

Rumelhart, D. E., G. E. Hinton, R. J. Williams 1986. Learning internal representations by error propagation. In *Parallel distributed processing: explorations in the microstructure of cognition*, vol.1. *Foundations*, D. E. Rumelhart & J. L. McClelland (eds), 318–62. Cambridge, Mass.: MIT Press.

Seidenberg, M. S. & J. L. McClelland 1989. A distributed, developmental model of word recognition and naming. *Psychological Review* **96**, 523–68.

CHAPTER 4

Interference and discrimination in neural net memory

Noel E. Sharkey & Amanda J. C. Sharkey

Introduction

Back-propagation training of feedforward neural nets (Rumelhart et al. 1986) has been applied widely in modelling a number of cognitive phenomena such as category formation, phoneme recognition, word recognition, speech perception, language acquisition and so on. However, a number of recent simulation studies have shown that there are aspects of human learning and memory that are difficult to train realistically using back-propagation (e.g. when memory sets are presented to a net for training in sequential blocks or when nets are given no negative exemplars during training). This has most recently been posed as the sequential learning problem (McCloskey & Cohen 1989, Ratcliff 1990) and previously as the stability–plasticity problem (e.g. Grossberg 1987). The most serious shortcomings that have arisen concern the ability of a net to retain earlier memories when it is trained on new memory sets, and the ability of a net to discriminate between items on which it has been trained and items on which it has not been trained.

The first of these shortcomings occurs when feedforward neural nets are trained, using back-propagation, to memorize sets of items in sequential blocks. Under these circumstances, newly learned information *catastrophically* interferes with, and overwrites, previously learned information entirely (e.g. Hetherington & Seidenberg 1989, McCloskey & Cohen 1989, Ratcliff 1990, McRae & Hetherington 1993). For example, McCloskey & Cohen (1989) used the back-propagation learning rule to train a net on the arithmetic problem of +1 addition. Subsequently they found that after training the same net to add +2, it had forgotten how to add +1. In an extension of this work to the detailed modelling of human recognition memory, Ratcliff (1990) found similar interference problems in a large number of simulation studies. For instance, when 16 items were trained sequentially, only the final item was retained. We refer to this problem as *catastrophic interference*. The second shortcoming of nets trained with back-propagation is *catastrophic remembering*, i.e. the inability to tell items that appeared in the memorized material from items that did not appear. It is clear that to model human memory, nets must be able to learn by training only on positive examples, i.e. without negative feedback from erroneous patterns. They should then be able to determine which patterns they have and have not seen before. However, a discrimination problem has been shown for back-propagation training under two different sets of circumstances: (a) Ratcliff (1990) found that with sequential block training (and buffer rehearsal training) that old–new discrimination broke down as a function of the amount of training on the items in the memory sets; (b) Ratcliff found that when a net is trained on a classification task without negative exemplars it exhibits catastrophic remembering problems. Ratcliff used back-propagation to train nets, with one output unit, on a yes/no task. During training a net was required to output a +1 ("yes") on its one output unit for all patterns in the memorization set. Without further training, the net was tested on new items and required to output a zero ("no"). The discrimination performance was poor and the net eventually produced +1 output for all patterns.

These interference and discrimination problems have serious implications for many aspects of psychological modelling. Apart from list learning and accurate recognition memory, it is clear that much of human and animal learning is sequential, or incremental, in nature. For example, in adult learning, it is clearly possible to continually refine and update knowledge of some domain without forgetting all that was previously known about it. Similarly, during development many abilities, such as grammatical inference, emerge gradually and are not acquired in an all or none fashion as a result of being taught the entire training set each time a new item is learnt. (See also Lewandosky (1991) and Goebel & Lewandowsky (1991) for a review of other psychological areas in which interference and discrimination are important.)

A number of variants of back-propagation have been proposed as solutions for catastrophic interference with varying degrees of success (e.g. French 1991, French 1992, Kruschke 1992, Sloman & Rumelhart 1992, Kruschke 1993). These methods are most successful when they involve the localization of input patterns onto the hidden units (see Sharkey & Sharkey (1994) for a formally guaranteed solution). Such localization has already been demonstrated in research on other training techniques that do not suffer catastrophic interference (e.g. Aleksander et al. 1984, Kanerva 1988, Murre 1992). However, our goal here is to find out if the problems can be overcome with standard back-propagation.

In this chapter we examine the relationship between catastrophic interference and discrimination. Previous investigators, using standard back-propagation, have tended to concentrate *either* on methods for reducing interference *or* on detailing the loss of discrimination. As a result, a coherent picture of the whole problem has failed to emerge. For example, a number of recent studies have indicated that severe interference can be partially alleviated by ensuring that a net trained with back-propagation is first prestructured rather than being trained with random weights (Hetherington & Seidenberg 1989, Pratt 1992, McRae & Hetherington 1993, Pratt 1993, Sharkey & Sharkey 1993a). However, these papers do not discuss the potential loss of old–new discrimination. Others (e.g. Ratcliff 1990) have focused on the issue of old–new discrimination, whilst failing to give sufficient experience of the tasks to the nets for them to perform well. Unlike the human experimental subject, each net enters the experiments as a *tabula rasa* (or at least with random initial conditions). We begin with a formal analysis of the twin problems and then examine the trade-off between interference and discrimination in nets that have either been pretrained or been allowed to extract knowledge from a large number of sequential blocks.

An analysis of the problems

The problems of interference and discrimination can be described with reference to an empirical study presented by Ratcliff (1990). Ratcliff conducted a series of simulation experiments to test the notion that a feedforward net with two layers of weights, trained using the standard back-propagation learning rule, could be employed to model human recognition memory. He took the view that recognition performance could be simulated by an auto-encoder net that reproduced the inputs as outputs. Thus, an attempt was made to train each net to be an accurate encoder memory in which the binary input vectors, v_m, in the memorization set, M_i, are reproduced on the outputs, $f(v_m) = v_m$. Recognition accuracy is then measured by the match between the input and the output vectors. This is a relatively easy task for a feedforward net trained with back-propagation. However, in order to be a minimal model of human recognition memory, Ratcliff set two further conditions on the method of training for which we give a more formal treatment. First, the net must not receive any negative feedback, i.e. $v \in V^n \notin M_i$ (i.e. the union of all the memory sets) as input, where V^n is the set of all binary vectors in n space. This has a strong psychological basis, since negative examples are not required by humans in memory experiments. Second, the net must be able to learn memory sets in sequence. That is, to reproduce subsets of V^n in blocks so that the subsets do not interfere with one another. More formally, let $M_1 \subset V^n$ and $M_2 \subset V^n$ be two memorization sets such that $M_1 \cap M_2 = \varnothing$, $M_1 \cup M_2 \neq V^n$. The net is first required to learn $f(v) = v$, $\forall v \in M_1$ and $f(v) \neq v$, $\forall v \notin M_1$. Then the same net is required to learn $f(v) = v$, $\forall v \in M_1 \cup M_2$ and $f(v) \neq v$, $\forall v \notin M_1 \cup M_2$. The eventual function should then be

$$f(v) = \begin{cases} v, v \in \cup M_i \\ -v, v \notin \cup M_i \end{cases} \qquad (4.1)$$

Building on the results from several simulations, Ratcliff suggested that two of the problems that arise from using back-propagation learning with a standard feedforward net to model recognition memory are: (a) retroactive interference in which, after training a net to reproduce its inputs as outputs, that net is retrained on a second memorization set, $f(v) = v$, $\forall v \in M_2$, and consequently forgets some, or all, of the results in the first set, $f(v) \neq v$, for some $v \in M_1$; and (ii) poor old–new discrimination by which a net's performance indicates that some inputs that it has never seen before were part of the original training set, i.e. $f(v) = v$ for some $v \notin M_1 \cup M_2$.

We suggest that the discrimination problem arises because a memory set can be a subset of a set of the input/output pairs of some more general function. For example, the successive memory sets for the encoder task used by Ratcliff are all consistent with the more general autoassociation or identity function. Since a back-propagation net is a universal approximator (see White 1992), it will approximate any function given a sufficiently representative sample of the input/output pairs of the function. If an encoder net has been trained on $M \subset V^n$ with no negative feedback, it may begin to approximate the identity function. This is because the only patterns that the net has been exposed to are a subset of V^n as shown in Figure 4.1. The net may then exhibit good generalization performance on the identity function. However, according to the task that the experimenter desires (accurate recognition memory), the net will suffer from *catastrophic remembering*; the inability to discriminate old items from new. In the worst case it will recognize all $v \in V^n$.

In this analysis, the interference problem arises because the sample sets are too small to extract the more general function underlying all of the memory sets. If M_1 is a very small sample, as in the list-learning studies, the net may provide a very poor approximation of the general function, and so when trained on M_2 will suffer catastrophic forgetting for M_1. This is because, given a very small sample, M_i, of the general function, the net may learn to represent any one of a number

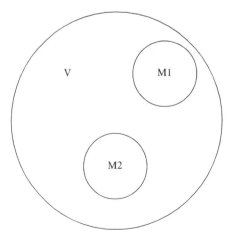

Figure 4.1 A Venn diagram showing two memorization sets, M, as subsets of V^n.

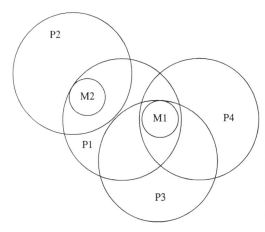

Figure 4.2 A Venn diagram of some of the sets of ordered pairs possible for a given sample memorization set. The set of ordered pairs for the identity function is shown as P1.

of different possible functions, with ordered pairs P_i, all of which have M_i as a subset. The number of possible functions varies with the number of patterns, p, in the training sample and the number of outputs. For singular binary functions (i.e. with one output unit), if there are n input units, the number of possible functions is $2^{(2^n - p)}$, e.g. with $n = 10$, and $p = 4$, the number of possible functions is 2^{1020}, and with multiple outputs the number is much larger. Figure 4.2 shows a Venn diagram of the ordered pairs (P1–P4) for a number of functions, where P1 is the set for the general function. A net trained on M_1 may have extracted the function with ordered pairs P4 instead of P1. When it is subsequently trained on M_2 it may learn the function with ordered pairs P2. Since P2 $\cap$ P4 = $\varnothing$, i.e. they are mutually exclusive, a net representing P2 could not represent any of the ordered pairs in P4. Such a net would be said to be suffering from catastrophic interference. Thus the likelihood of interference is inversely proportional to the number of pairs in P1 $\cap$ P4.

Training on task structure

The foregoing analysis makes it clear that the degree of forgetting and the degree of discrimination ability are intimately related. This is inevitable because generalization to the general (identity) function is a measure of old–new discrimination; each percentage increase in generalization means a percentage decrease in old–new discrimination. In order to illustrate this trade-off, we can use a net consisting of a single matrix (layer) of weights, W, between the inputs and the outputs. With such a net the whole Identity function, for a given n, can be represented. This is obvious, since any identity matrix I, a matrix of 0s with 1s on the diagonal, multiplied by any vector v (of appropriate dimension) will be $Iv = v$, and therefore, if $W = I$, then $Wv = v$, $\forall v \in V^n$. Since an identity matrix of weights will, by definition, reproduce all inputs accurately on the outputs, there

61

will be no interference from successive memory sets, but there will be a total collapse of discrimination, i.e. zero discrimination ability. Of course, if what is being considered is training a net on memorization sets $M \subset V^n$, for accurate recognition and not on the total identity function, so $W \neq I$, it is unlikely that a single-layer net will suffice in all cases.

The identity net can be trained in one step using perceptron convergence. All it requires is that the training sample be the basis vectors for the space (see Sharkey & Sharkey (1994) for a fuller account and geometric visualizations). Thus there will be a training sample of n input = output pairs, where n is the input dimension of the net. Moreover, the ideal geometry of the identity net can be approximated very well by a net with two weight layers when trained on the basis vectors (Sharkey & Sharkey 1994). In this section we investigate whether the conclusions of our analysis generalize to simulation research on pretraining nets to alleviate interference effects.

A number of authors have attempted to train nets in a more realistic manner by departing from strictly sequential training (Hetherington & Seidenberg 1989), or by combinatorially structuring the learning domain (Brousse & Smolensky 1989). Moreover, McRae & Hetherington (1993) have suggested that one of the main causes of catastrophic interference is that, prior to training on a task, nets are not given knowledge of the domain. Unlike human subjects who enter a memory experiment with rich prior knowledge, a net typically begins a simulation as a *tabula rasa* with random initial weights. Indeed, the value of pretraining nets on one task has been shown by the facilitation of the convergence of the nets on other tasks (e.g. Pratt 1992, Pratt 1993, Sharkey & Sharkey 1993a,b). Unfortunately, in previously reported studies in the literature, it was not possible to get satisfactory estimates of the discrimination performance of the pretrained nets, and so two new simulation studies were conducted to demonstrate that the interference/discrimination trade-off still occurs when prior training patterns are selected randomly.

Ratcliff (1990) investigated the effects of domain knowledge and suggested that list learning may really be a matter of retraining existing knowledge. Three nets with differing numbers of hidden units (16–48) were pretrained on 100 associations to a learning asymptote. The memorization sets each consisted of 16 patterns randomly drawn from the pool of pretrained patterns. The result was that there was little interference on performance of the old patterns as new patterns were trained. However, it was difficult to estimate the discrimination properties of the nets since all "new" test items were drawn from the pool of items on which the net had been pretrained. So the old–new discrimination tests are really telling us about forgetting of the pretraining set. The match on new items was always close to the pretrained learning asymptote, and so the pretraining had never been unlearned. Ratcliff's investigation of the use of pre-experimental knowledge in training was concerned only with very specific pre-existing memory information, i.e. replication of items between pre- and post-training. Another, less specific, way to provide a net with pre-experimental knowledge is to give it prior training on the general nature of the task. In this case, a net would be pretrained on one pool of associations and the memorization sets would be drawn from a different pool. Thus any reduction in interference could only result from knowledge of the task. However, if the goal is to build an accurate recognition memory, pretraining a net on the task may produce an undesirable side-effect. As the net extracts more information about the task its generalization performance (on the identity function) is likely to increase to the point were there is little, if any, discrimination between old and new items.

In one investigation of task knowledge on the reduction of interference, McRae & Hetherington (1993) pretrained a 400–150–400 encoder net on 2897 associations and then ran simulations of free recall and serial list learning using new associations. All of the distributed binary input patterns were generated by using the Seidenberg & McClelland (1989) naming model as a preprocessor. This constrained the number of available input patterns from the 2^{400} possible. In the free recall study (this is similar to the recognition task described earlier), two memorization sets each consisting of eight new associations were chosen for sequential training. Following training on the second memorization set, there was no evidence of interference for the first memorization set. By contrast, a net that had not been pretrained on the original associations did suffer from severe interference. One of the most interesting aspects of McRae & Hetherington's results was naturally occurring node sharpening in the pretrained nets. From our analysis, this could be evidence that the nets were separating the representations to enable extraction of the identity function. Unfortunately, since the authors were not concerned with accurate recognition memory, no error results were provided for the second memorization set prior to training and so we cannot ascertain the old–new discrimination characteristics of the net. If the net had extracted sufficient structure from the pretraining set, it may have had a high degree of generalization on the autoassociation task. One clue that this may have been happening comes from the authors' two statements that: ". . . as well as reducing overlap, pretraining preserved similarity information . . ." [on the hidden units] ". . . Despite the absence of interference, similar items still overlapped at the hidden unit layer, so generalization should still occur" (Seidenberg & McClelland 1989). Of course, this is not really a problem for these authors as they are not concerned with accurate recognition memory.

There was a similar problem for the serial-list learning study. A naive and a pretrained net was sequentially trained on 16 associations by repeating each association until it had learned to criterion before proceeding to the next. The result was that the naive net suffered from severe interference in all but the final pattern, whereas the pretrained net exhibited good retention over all 16 associations. Again, since no error results were given for the memorization sets prior to training, we cannot determine the extent to which the retention results were due to generalization in the pretrained net. It is unlikely that the preprocessing constraints on the possible input patterns limit generalization while eliminating interference.

In two new simulation studies reported here, the relationship between interference and generalization effects was investigated systematically in pretrained and block-trained nets. In the first study, the amount of exposure to a fixed-size pretraining set was used as the independent variable to examine the effect of prior knowledge on retroactive interference and the relationship between interference and discrimination. In the second study, knowledge was introduced to the net through sequential training. In previous studies, nets were typically trained up to a maximum of 16 memory sets (e.g. Ratcliff 1990, French 1991, Murre 1992, French 1992). This was insufficient to eliminate interference, and so in our second simulation study the net was trained with over 1000 memory sets.

Simulation 1: the effects of pretraining

In the first simulation study, nets were given varying amounts of knowledge about the general task of autoassociation. This enabled an evaluation of the effects of the amount of task knowledge on both retroactive interference and old–new discrimination. In the McRae & Hetherington (1993) study, the nets were pretrained to a learning criterion on a small proportion of the total set. In the simulation study reported here, all nets were pretrained on 25% of the total number of possible input patterns. The amount of task knowledge was varied by exposing the nets to differing numbers of training cycles: from 0 to 1750 training cycles in increments of 10.

The nets were trained as autoencoders. All of the feedforward nets employed were 10-10-10 ($\eta = 0.1$). Prior to training on a memorization task, nets were pretrained on a randomly selected 25% of the total possible number of binary input patterns ($2^{10} = 1024$), for increasing numbers of cycles from 0 to 1750 in increments of 10. After pretraining was completed, each of the nets was used in a standard interference experiment to test for interference and discrimination; they were trained first on a four-pattern memorization set and then on a two-pattern interference set. None of these sets had appeared in the pretraining set. The root mean square (rms) error was used as a measure of the match between the actual and required output of the net. This is a standard method for measuring the match between outputs and targets in connectionist research:

$$\text{rms} = \sqrt{\left(\sum_{i=1}^{p}\sum_{j=1}^{n}(t_{ij} - o_{ij})^2\right)(np)^{-1}} \qquad (4.2)$$

where t is the target value for an output unit, o is the actual value, n is the number of elements in an output vector (10 in this simulation) and p is the number of patterns in the memorization set. Discrimination was tested by measuring the rms error for the four-pattern memorization set on each of the pretrained nets before memorization had started; the higher the rms error, the better the discrimination. Interference was tested by measuring the rms error for

Table 4.1 Distributed memorization and interference sets.

	Distributed memory set	Orthogonal memory set
Memorization set	1 0 1 0 1 1 1 0 1 1	0 0 0 0 0 0 0 0 1 0
	0 0 1 0 0 1 1 1 1 0	0 0 1 0 0 0 0 0 0 0
	0 0 1 0 1 0 0 0 0 1	0 0 0 0 0 0 0 1 0 0
	0 0 0 1 1 0 0 1 1 0	0 0 0 0 1 0 0 0 0 0
Interference set	1 0 1 0 0 0 0 0 0 1	0 0 0 1 0 0 0 0 0 0
	0 1 0 0 0 0 1 1 0 0	0 0 0 0 0 1 0 0 0 0

each of the nets on the memorization set after training had completed on the interference set; the lower the rms error, the better the retention.

Two types of memorization and interference memory sets were employed in two separate simulations: (a) randomly chosen distributed pattern sets, and (b) selected orthogonal pattern sets, shown in Table 4.1.

Results
(a) *Distributed input patterns*. The results for the distributed memory and interference sets are plotted in Figure 4.3. The dashed line at the bottom of the graph is the error for the memorization set at the completion of training.

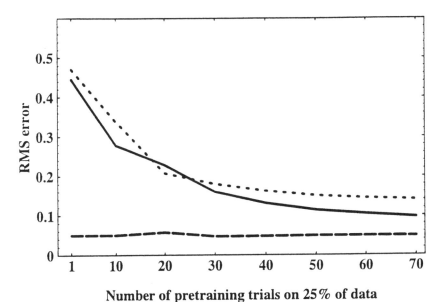

Number of pretraining trials on 25% of data

Figure 4.3 A plot of the effect of pretrained nets on interference and discrimination for distributed input patterns. The dashed curve shows the RMS error after training on the memorization set, the solid curve shows the error for the memorization set after training on the interference set, and the dotted curve shows the error on the memorization set prior to training.

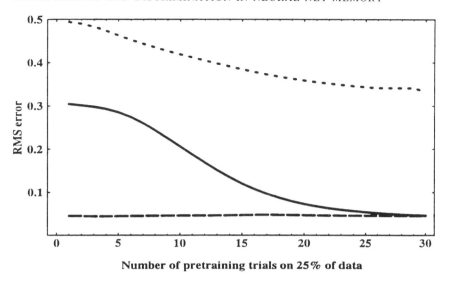

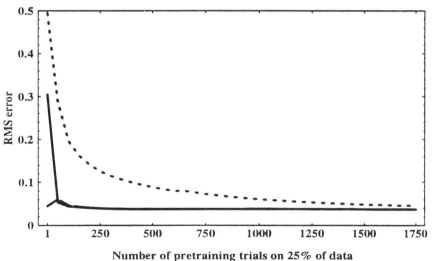

Figure 4.4 Plots of the effect of pretrained nets on interference and discrimination for orthogonal input patterns. The dashed curve shows the RMS error after training on the memorization set; the solid curve shows the error for the memorization set after training on the interference set, and the dotted curve shows on the memorization set prior to training. (a) Pretraining from 0 to 30 cycles and (b) Pretraining from 1 to 1750 cycles.

This shows the training asymptote of rms < 0.1. The solid plot shows the retention error for the memorization set after training on the interference set. With small amounts of pretraining, severe interference is observed, but as the amount of pretraining increases from 0 to 40 cycles, interference is almost eliminated. This replicates and extends the pretraining results of McRae & Hetherington (1993). However, as anticipated, reductions in interference were accompanied by reductions in old–new discrimination. The dotted plot shows performance on the memorization set before it had received any training. This closely follows the interference plot. Thus, interference reduction and the reduction of discrimination go hand in hand.

(b) *Orthogonal input patterns.* The results for the orthogonal memory set appear to be more promising at first. In Figure 4.4a the retention error (solid plot) decreases more rapidly than discrimination (dotted plot). After only 15 cycles of pretraining the net is already showing good retention while maintaining its old–new discrimination ability. However, this effect is short-lived. In Figure 4.4b the plot is extended to 1750 cycles by sampling the errors every 50 pretraining cycles. This plot shows a sudden and dramatic reduction in discrimination so that after 250 pretraining cycles the net is unable to tell the difference between old and new items. With orthogonal input patterns, there is a window of pretraining from around 25 to 150 cycles in which the nets can perform the encoder task accurately. It would require a lot of fine tuning to get the window in the right place for particular tasks. It is not very robust and would be affected both by the random initial conditions of the net and the nature of the pretraining sample.

Simulation 2: sequential structure extraction

In the previous simulation study, the nets were pretrained using a single block of 25% of the possible binary patterns. They were successful in extracting the structure of the autoencoder task – it generalized well to the identity function. However, it is unlikely that pre-experimental knowledge would be acquired by humans in such large blocks. Thus, in the second simulation study, a net was trained sequentially on successive memory sets (randomly selected) until all possible associations were exhausted. This was unlike previous studies in the literature which used up to a maximum of 16 memory sets. The question here is, could back-propagation extract the nature of the autoencoder task from sequentially presented training sets? In other words, could interference be eliminated by sequentially pretraining a net? Both interference and discrimination were measured in the same way as in the previous simulation study. It was expected that any reduction in interference would be accompanied by a similar reduction in discriminative ability.

The memorization sets each consisted of four ten-dimensional binary patterns. These were drawn at random without replacement from the total (2^{10}) set of possible unique ten-dimensional binary vectors, yielding 256 memorization sets in all. Each of these was trained, in turn, to completion (error tolerance = 0.1) on a 10–10–10 encoder net using the back-propagation algorithm ($\eta = 0.3$). The memorization set numbers 1, 50, 100, 150, 200, 246 were selected for detailed testing of interference and discrimination using the RMS error as a measure.

Results

The results showed clearly that even with sequential presentation of randomly selected memory sets the net was able to extract the structure of the identity function. One way to test this was by examining the learning time for the net as it progressed through the training sets as shown in Figure 4.5. After presentation of the 20th training set, there was a very marked reduction in learning time, and this continued to decrease throughout. These results conform to the idea that the net is learning the nature of the task and gradually extracting the identity function through successive approximations.

To examine the interference effects, each of the selected memorization sets was tested eight times. First, in order to measure discrimination/generalization each memory set was tested on the net before training. Next the net was tested just after it was trained, and subsequently after learning each of the next six memorization sets, i.e. after delays of 0 to 6 intervening memory sets. The results are shown in Figure 4.6. The RMS errors for the delay conditions are labelled $d0$ to $d6$ according to their sequential distance from the selected memory sets. The first memorization set in the plot in Figure 4.6 shows the classic signs of catastrophic interference: the error for $d0 < 0.1$, and the error for $d1 > 0.4$. However, the interference begins to diminish across the other five delay conditions. For the next selected memory set, number 50, there is no longer any evidence of retroactive interference across the delay conditions; the net has learned the structure of the task.

The discrimination ability of the net was examined by presenting each memory set to the net prior to training. The rms error is shown in the plot labelled as GEN. Again there is an observed reduction in old–new discrimination over time, and it closely follows the reduction in interference. Thus the elimination of interference comes at the cost of losing the power of discrimination.

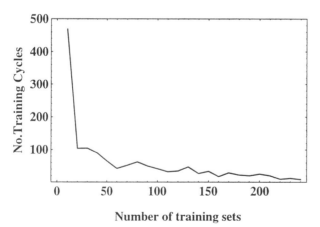

Figure 4.5 Incremental training and reduction in the number of cycles to train sets of four binary patterns as a function of the number of sets previously trained.

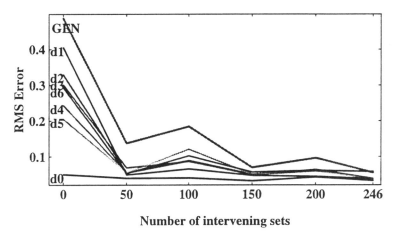

Number of intervening sets

Figure 4.6 Reduction in RMS error as a function of the number of sets of four binary patterns previously trained. *d0* shows the RMS error for a memory set immediately after it was trained, *d1* after one intervening pattern set, *d2* after two intervening pattern sets, and so on. GEN shows the RMS error for the pattern set prior to training on it.

Back-propagation is not good at learning to be an accurate recognition memory under these conditions. But the results show that, even with sequential training, it is quick to extract task structure.

Discussion and conclusions

We have presented an analysis of the causes of both the interference and discrimination problems that have been reported recently in the literature on back-propagation learning of sequentially presented memory sets. In the first section we discussed how the problems arose from an account of the function approximation properties of neural net training and showed an inevitable trade-off between interference and discrimination. These analyses were supported and extended by the results from two simulation studies.

In the first simulation study it was shown that, for both distributed and orthogonal pattern sets, pretraining on the encoder task for as little as 30 or 40 cycles on 25% of the possible training data was enough to eliminate retroactive interference. It appears that by pretraining with a large enough sample the net forms an approximation to the more general identity function. This conclusion was also borne out by the discrimination results. For distributed patterns the reduction in interference was closely followed by a concomitant reduction in discrimination. For orthogonal input patterns the reduction in discrimination lagged behind the interference reduction and thus allowed a small window of accuracy. In the second simulation study, back-propagation quickly picked up the structure of the task even though it was trained sequentially (50 memory

69

sets). Under these conditions there was an absence of retroactive interference. This extends previous research that only considered up to a maximum of 16 small memory sets, which is, in the main, insufficient to eliminate interference. But, as in the first study, the elimination of interference found in the second simulation study was accompanied by a disappearance of old–new discrimination. These findings show that a net can induce the total identity function through a sequence of successive approximations created by successive training sets.

In conclusion, having examined the causes of the interference and discrimination problems for back-propagation that have been reported in the literature, it is clear that one or other of these is unavoidable if Ratcliff's restrictions are applied to training, i.e. that it be sequential, and without negative exemplars. Without these restrictions neither problem will arise. The interference problem will not occur when old memory items are refreshed during training of new ones and the discrimination problem will not occur when the negative exemplars are mixed with the positive. It was shown here that interference does not cause difficulties in realistic circumstances where the successive memory sets are consistent with some more general function; or when the memory set is *representative* of the general function as, for example, the basis vectors for the total identity function. This is good news if the goal is to build a generalizable memory without interference rather than a good discriminator; then the discrimination error becomes percentage correct generalization, and so it can be shown that as interference diminishes, generalization increases (see also Hetherington 1990). However, discrimination is still a problem for those tasks in which negative exemplars are not available, as is the case in laboratory experiments on memory. If a representative training sample exists for the general function and it appears in any of the memory sets, there will be immediate catastrophic remembering of patterns that have never been seen before.

Acknowledgements

We would like to thank the ESRC R-100-23-3441 for funding this research.

References

Aleksander, I., W. Thomas, P. Bowden 1984. WISARD, a radical new step forward in image recognition. *Sensory Review*, 120–4.

Brousse, O. & P. Smolensky 1989. Virtual memories and massive generalization in connectionist combinatorial learning. *11th Annual Conference of the Cognitive Science Society, Proceedings*, 380–7. Hillsdale, New Jersey: Lawrence Erlbaum.

French, R. M. 1991. Using semi-distributed representations to overcome catastrophic forgetting in connectionist networks. In *13th Annual Conference of the Cognitive Science Society, Proceedings*, 173–8. Hillsdale, New Jersey: Lawrence Erlbaum.

French, R. M. 1992. Semi-distributed representations and catastrophic forgetting in connectionist networks. *Connection Science* **4**, 365–77.

Goebel, R. P. & S. Lewandowsky 1991. Retrieval measures in distributed memory models. In *Relating theory and data: Essays on human memory in honor of Bennet B. Murdock*, W. E. Hockley & S. Lewandowsky (eds), 509–27. Hillsdale, New Jersey: Lawrence Erlbaum.

Grossberg, S. 1987. Competitive learning: from interactive activation to adaptive resonance. *Cognitive Science* **11**, 23–64.

Hetherington, P. A. 1990. Interference and generalization in connectionist networks: within-domain structure or between-domain correlation? *Neural Network Review* **4**, 27–8.

Hetherington, P. A. & M. S. Seidenberg 1989. Is there catastrophic interference in connectionist nets? In *11th Annual Conference of the Cognitive Science Society, Proceedings*, 26–33. Hillsdale, New Jersey: Lawrence Erlbaum.

Kanerva, P. 1988. *Sparse distributed memory*. Cambridge, Mass.: MIT Press.

Kruschke, J. K. 1992. ALCOVE: an exemplar-based model of category learning. *Psychological Review* **99**, 22–44.

Kruschke, J. K. 1993. Human category learning: implications for backpropagation models. *Connection Science* **5**, 3–36.

Lewandowsky, S. 1991. Gradual unlearning and catastrophic interference: a comparison of distributed architectures. In *Relating theory and data: essays on human memory in honor of Bennet B. Murdock*, W. E. Hockley & S. Lewandowsky (eds), 445–76. Hillsdale, New Jersey: Lawrence Erlbaum.

McCloskey, M. & N. J. Cohen 1989. Catastrophic interference in connectionist networks: The sequential learning problem. In *The psychology of learning and motivation*, vol. 24, G. H. Bower (ed.), 109–65. New York: Academic Press.

McRae, K. & P. A. Hetherington 1993. *Catastrophic interference is eliminated in pretrained networks*. University of Ohio Electronic Archives, Technical Report.

Murre, J. M. J. 1992. *Learning and categorization in modular neural networks*. Hemel Hempstead, England: Harvester Wheatsheaf.

Pratt, L. Y. 1992. *Non-literal transfer of information among inductive learners*. Computer Science Department, Rutgers University, Working Paper, May 1992.

Pratt, L. Y. 1993. Discriminability-based transfer between neural networks. In *Advances in neural information processing systems*, C. L. Giles, S. J. Hanson, J. D. Cowan (eds), vol. 5. San Mateo, Calif.: Morgan Kaufmann.

Ratcliff, R. 1990. Connectionist models of recognition memory: constraints imposed by learning and forgetting functions. *Psychological Review* **96**, 523–68.

Rumelhart, D. E., G. E. Hinton, R. J. Williams 1986. Learning internal representations by error propagation. In *Parallel distributed processing: explorations in the microstructure of cognition*, vol. 2. *Physiological and biological models*, D. E. Rumelhart & J. L. McClelland (eds), 318–62. Cambridge, Mass.: MIT Press.

Seidenberg, M. S. & J. L. McClelland 1989. A distributed, developmental model of word recognition and naming. *Psychological Review* **96**, 523–68.

Sharkey, N. E. & A. J. C. Sharkey 1993a. Adaptive generalization. *AI Review* **7**, 313–28.

Sharkey, N. E. & A. J. C. Sharkey 1993b. Prestructured neural nets and the transfer of knowledge. In *Neural computing research and applications*, G. Orchard (ed.), 157–65. Bristol: IOP Publishing.

Sharkey, N. E. & Sharkey, A. J. C. 1994. *Understanding catastrophic interference in neural nets*. Department of Computer Science, University of Sheffield, Research Report CS-94-4.

Sloman, S. A. & Rumelhart, D. E. 1992. Reducing interference in distributed memories through episodic gating. In *From Learning theory to cognitive processes: essays in honor of William K. Estes*, A. S. Healy, S. M. Kosslyn, R. N. Shiffrin (eds), 227–48. Hillsdale, New Jersey: Lawrence Erlbaum.

White, H. 1992. *Artificial neural networks: approximation and learning theory*. Oxford: Blackwell.

CHAPTER 5

Transfer of learning in back-propagation and in related neural network models

Jacob M. J. Murre

Limits of back-propagation

Given the range of neural network paradigms available at the moment, we might ask why anyone would still want to use back-propagation. An important argument for using this learning algorithm seems to be its popularity. Back-propagation has become one of the standard technologies in connectionist modelling. Although it was invented by Werbos in 1974, it has only been with the publication of the so-called PDP volumes in 1986 that it has been applied to a variety of cognitive modelling tasks (Rumelhart & McClelland 1986). Since then, back-propagation has served an important purpose as a "generic" and easy to understand error-correction learning mechanism. There are, however, strong limits to its suitability to modelling human learning and memory as we shall argue and demonstrate in this chapter. After evaluating some of these limits, as well as some of the advantages, we present a number of simulations investigating interference and transfer of human learning. It will be shown that back-propagation shows hypertransfer: learning an interfering list B may actually improve performance on the original list A in some circumstances, including conditions where human subjects do not show such behaviour.

We will also investigate transfer with delta rule learning and compare the results with the human data. These networks are shown to exhibit transfer that is fully compatible with classic findings in transfer theory (Osgood 1949).

A major drawback, in general, is the very slow learning speed of the back-propagation algorithm. Several thousands of iterations (of the entire learning set) are often needed to obtain the desired behaviour. This prompted the original developers of back-propagation to include a "momentum" term in their learning rule (Rumelhart et al. 1986) in order to increase the learning speed. Momentum itself is not part of the actual gradient descent method that underlies the back-propagation algorithm. It is simply a heuristic that works well to speed up the convergence in most cases. Since 1986, many other authors have proposed ways to improve the algorithm's speed (e.g. Jacobs 1988, Mirchandani & Cao 1989, Kruschke & Rodriquez-Movellan 1991, Wang et al. 1993). Introduction of these methods, however, complicates the algorithm and thus diminishes one of its major attractions: its simplicity and ease of use.

The slow learning rate limits the application of back-propagation to relatively small learning tasks. The learning speed is not, however, psychologically implausible. Below, we will show that even with as few as five to ten learning trials psychologically plausible learning behaviour can be obtained, if we apply a winner-take-all mechanism to the output layer.

The main focus in this chapter will be on the plausibility of variants of back-propagation networks as models for human long-term memory. It has been demonstrated repeatedly that three-layer back-propagation networks (i.e. with a hidden layer) are very sensitive to interference caused by additional training without rehearsing patterns learned earlier. A simulation study by McCloskey & Cohen (1989) showed that back-propagation networks may forget most learned memories, if they are trained on a new list of items without rehearsing the old ones. They coined the term "catastrophic interference" for this learning behaviour. A study by Ratcliff (1990) showed essentially the same phenomenon. The implication of this behaviour is that back-propagation is unsuitable for modelling real-time learning, which had already been argued by Grossberg in 1987. In the following sections we will further examine the extent of interference in back-propagation networks. We will argue that the hidden-layer representations in back-propagation are responsible for implausibly large interference effects, but that two-layer networks are, in fact, well suited for modelling interference in human subjects. In particular, we will demonstrate that such networks can model the classic Osgood surface (Osgood 1949), which summarizes some of the major findings in human interference.

There are still other aspects that make back-propagation less useful for modelling human cognition. The paradigm has been criticized by Kruschke (1990) as a model for concept learning in humans. He has introduced a model, called ALCOVE, that is similar to back-propagation but with a hidden layer consisting of nodes that have a radial-basis activation function rather than the logistic function used normally. The model's structure is based on Nosofsky's generalized context model (Nosofsky 1985, Nosofsky 1986, Nosofsky 1992). Each node in ALCOVE has a limited receptive field. Only a few hidden nodes become activated in response to any given stimulus. ALCOVE has shown to be a good model for a range of experiments in concept formation (see also Nosofsky et al. 1992). Models based on two-layer back-propagation networks (i.e. without a hidden layer; these are essentially delta rule networks, see below) have also proven to be successful in modelling data in concept formation. Gluck & Bower have developed a model for binary inputs (Gluck & Bower 1988, Gluck & Bower 1990, Gluck 1991), based on Shepard's (1987) theory of stimulus generalization and the Rescorla–Wagner (1972) learning rule (this is essentially the Widrow–Hoff (1960) learning rule, of which back-propagation is a generalization). Values of the inputs are recoded in terms of all possible combinations. If there are three dimensions, 1, 2 and 3, there will be seven combinations: 1, 2, 3, 12, 13, 23 and 123. With many input dimensions the model is not very plausible because this input coding scheme gives an explosion of combinations. But for low dimensions it fits the experimental data well. A related model, for continuous values, has recently been developed by Shanks & Gluck (1994). This model also has three layers. The "hidden layer" receives coded inputs over non-adaptable connections, so that all learning takes place in the second layer of weights (i.e. weights to the output layer) using a delta rule.

We can conclude that whereas straightforward back-propagation may not be a good model of concept formation, models based on variants of back-propagation (including two-layer networks with a simple delta rule) are more promising. A more fundamental critique can be directed, however, to any model of human learning that is exclusively reliant on error correction. It can be argued that in real life we often build up categories in the absence of explicit feedback about their "correctness" and that this regularity learning is crucial for our cognitive functioning (Grossberg 1976, Grossberg 1982, Murre 1992a). Error-correcting learning is a powerful learning method, but it cannot develop categories autonomously. For a complete model of human learning and memory this is a great limitation, suggesting that error correction learning will form a necessary but not sufficient part of such a model.

In addition to the above considerations, we can add that from a neurobiological point of view, back-propagation networks are not plausible either. The back-propagation aspect of the learning rule, whereby an error measure is transmitted in reverse through the synapse has not been observed in the brain. Also, back-propagation is usually employed in a simple feedforward structure with one hidden layer. In the brain, however, we can observe a great deal more initial, modular structure. By ignoring these modular architectures in the brain we may do ourselves a great disservice by making the learning tasks we are pursuing much more difficult that they need to be (Murre 1992a). Having said this, however, it should be pointed out that nothing in the back-propagation algorithm forbids the use of modular architectures. Indeed, the studies in which such architectures were used have been successful in accounting for certain overall aspects of brain architecture (e.g. Rueckl et al. 1989). In addition, there have been several studies showing that back-propagation can be used as an inference mechanism to arrive at biologically plausible structures, even though the method by which they are obtained is not biologically plausible (e.g. Zipser & Anderson 1988, Lockery et al. 1989).

Merits of back-propagation

Notwithstanding these disadvantages, there might still be good reasons to use back-propagation. First, the method is easy to understand, which is important for conveying a model to an audience of specialists in a certain field (e.g. memory psychology) who are not connectionist modellers. This trivial fact will continue to inhibit the success of more powerful and more plausible – but also more complicated – models.

Secondly, especially for the novice modeller it may be important that a large choice of neurosimulators is available for back-propagation (Murre 1995a). Even for an experienced modeller it may take several weeks or even months to implement a complicated neural network model from scratch.

Thirdly, back-propagation is a powerful method that has been shown to be able to approximate all "well behaved" functions (Hornik et al. 1989). This generic aspect leads many to expect that the method may work well as a first approximation to a given learning problem. This approach to modelling, however, harbours the danger of describing something that is not well understood (i.e. the data) by something that is even less well understood (i.e. a working back-propagation "model" with difficult to interpret hidden-layer representations). Elsewhere (Murre 1992a) we have advocated a more informed approach to modelling whereby the initial architecture represents the constraints of prior knowledge about the modelling domain. Notwithstanding back-propagation's shortsighted approach to modelling, the generic aspect seems to continue to have a great appeal to the research community.

Fourthly, in contrast to other learning models, back-propagation does not force the modeller to specify many parameters. The main parameter is the learning speed, which for small values gives satisfactory learning behaviour. The Boltzmann machine (Ackley et al. 1985), for example, necessitates specification of an annealing schedule. The same is true for Kohonen's self-organizing maps (Kohonen 1982, Kohonen 1990).

Fifthly, back-propagation possesses most of the basic elements of a "prototypical" neural network: distributed representations, graceful degradation, pattern completion and adequate generalization of learned behaviour. In its simplest form (i.e. without recurrent connections) it does not, however, have an interesting dynamic behaviour that can be utilized in modelling, say, reaction time. Such dynamic measures must be derived indirectly.

As a final merit we might mention the fact that originally motivated the developers of back-propagation: it can learn nonlinearly separable pattern sets, which the simple delta rule, for example, cannot. We shall return to this issue in the discussion.

In summary, despite its drawbacks, back-propagation appears to be widely accepted as a good first approximation for many modelling tasks. Therefore, it is a useful enterprise to investigate its limitations and to try to remedy them without unduly complicating the algorithm. Also, as was pointed out above, two-layer models based on the original delta rule (Widrow & Hoff 1960, Rescorla & Wagner 1972) have successfully modelled a large range of experiments in human and animal concept formation. As we shall see below, this also applies to modelling interference in human memory.

The interference problem in neural networks

Due to an inherent learning capacity neural network models seem eminently suitable for modelling processes of memory and learning. It is, therefore, somewhat surprising that so few studies have appeared that explore their potential for modelling processes of memory and forgetting. One of the reasons for this limited interest may be found in the negative results produced with one neural network paradigm: back-propagation. McCloskey & Cohen (1989) coined the term "catastrophic interference" for the results obtained with a simple model for learning arithmetic. Ratcliff (1990) describes similar results in a series of systematic attempts to improve interference behaviour in a very small back-propagation network. In both these studies the problems arise when the network has been trained on a first set of stimuli, and is then trained on a second set without simultaneous retraining on the first set. After having learned the second set, recall of the first set is near zero: the network seems to have lost all earlier memories. This behaviour is not psychologically plausible and is only approached, perhaps, in patients with severe forms of anterograde amnesia. In this chapter, we will not focus on this problem of catastrophic sequential interference in back-propagation, but rather examine more broadly what the human data on this subject are and how these can be modelled by neural networks.

Interference in normal human subjects has been thoroughly investigated in a large number of studies, particularly in the 1930s and 1940s. Though this literature is now largely forgotten, it is well suited to modelling with neural networks or with mathematical models of associative memory (e.g. Mensink & Raaijmakers 1988). In this paper we will concentrate only on some of the classic results, summarized by Osgood (1949). These represent some of the most reliable results in associative interference theory. Most of his conclusions are still accepted today and can be found in many textbooks on memory since the 1950s (e.g. Baddeley 1976). Osgood's summary generalizes to experiments with both meaningful and nonsense materials. Osgood's review focuses on paired associated learning. At training, a subject is presented repeatedly with word pairs (e.g. "church"–"table", or a pair containing nonsense syllables). At testing, only the first word is presented and the subject is required to recall the second word. The first word of a pair is usually called *stimulus*, the second word *response*. In the standard interference study, the subject is first trained on a target list A. After a pre-set criterion has been reached, training on an interfering list B follows. Finally, the subject is retested on the target list A. The difference in performance before and after the interfering list is an indication of the level of interference caused by learning the second list. This type of interference is called retroactive interference, because something new acts on something old. (In a complete experimental design, one or several control conditions would be included. We will not discuss these here.)

Interestingly enough, the effect of the interfering list is by no means always negative: the performance on the first list can actually increase due to learning a

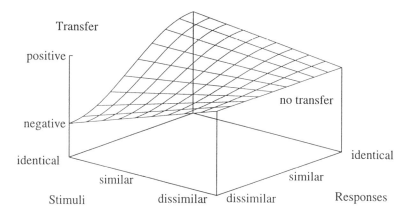

Figure 5.1 The Osgood transfer surface.

second list. Such "positive interference" is usually called "transfer". In this paper, we will use the term transfer to refer both to positive transfer (i.e. increased recall) and to negative transfer (i.e. decreased recall or interference). The conditions that describe positive and negative changes in performance are governed by the between-list similarity of stimuli and responses, as follows:

(a) If stimuli in lists A and B are very dissimilar there is zero transfer.

(b) If stimuli in A and B are similar, with identical or very similar responses in A and B there is strong positive transfer, with dissimilar responses in A and B there is strong negative transfer (i.e. interference).

There is, thus, a clear asymmetric relation between the effects of stimuli and responses. If the second list has very dissimilar stimuli from the first, it will not cause any positive or negative transfer, no matter what the responses are. If the stimuli are similar, however, strong transfer may be expected. The response similarities determine what the direction of this transfer will be: negative or positive.

Figure 5.1 gives a graphical impression of the effect of similarity on transfer. This is the classic Osgood transfer surface (Osgood 1949). In the trivial case where the "interfering" list B is exactly equal to the target list A we simply have a case of additional learning of list A and we would expect improved performance on the target list. That is, we expect strong positive transfer. But even when the items are not identical but merely highly similar we may still expect some positive transfer. With highly similar stimuli but dissimilar responses we would expect strong interference, which might perhaps be called "catastrophic". It should be pointed out that the Osgood transfer surface does not give a quantitative description of the empirical data. It is purely a qualitative summary of experiments on transfer behaviour with a large range of data, including both nonsense stimuli and meaningful English words. Our aim in this study is thus not a quantitative fit, but rather a robust qualitative agreement.

The implausible behaviour of back-propagation lies not just in the fact that it shows such strong interference (McCloskey & Cohen 1989), but more so in the fact that it exhibits this behaviour even with fully dissimilar stimuli (e.g. Ratcliff 1990). McCloskey & Cohen (1989) showed that learning the interfering list causes very rapid forgetting of the learned list, as was evidenced by intermittent retesting on the first list (with learning temporarily switched off). This situation is typically compared to the data on AB–AC learning by Barnes & Underwood (1959), which show (a) that there is a gentle rather than a steep forgetting curve for the target list A with prolonged learning on list B, and (b) that forgetting on the first list seems to reach a non-zero asymptote. The comparison with Barnes & Underwood (1959) must be made with some caution, however, as they did not use a straightforward cued recall technique. On the contrary, this paper intro-duces a new technique, MMFR or "modified modified free recall", aimed at eliminating certain difficulties in previous interference experiments. With this technique the subject is asked to give both the B and D response (i.e. in an AB–CD paradigm). Though it is routinely assumed that response probabilities in MMFR follow the same pattern as normal cued recall, mathematical analysis indi-cates that this assumption is most probably wrong (Murre 1995a, b). The two main characteristics of the Barnes & Underwood data remain, nevertheless, largely valid in this particular case.

In the next section we will first investigate the interference behaviour of net-works without a hidden layer. With these networks we have less of a reason to expect interference in the dissimilar-stimuli condition, as earlier analyses have shown that most of the "catastrophic" interference behaviour in back-propaga-tion can be attributed to overlap of hidden-layer representations (French 1991, French 1992, Murre 1992a,b). We will show that these simpler networks are indeed able to model the Osgood surface, and are therefore able to account for some major characteristics of human memory interference. Following that we will proceed to investigate transfer of learning in back-propagation.

Simulating the Osgood surface

This section contains a number of rather extensive simulations that enabled us to plot the transfer behaviour of the networks studied in some detail. All results, both observed and simulated, concern paired associated learning. In this chapter we concentrate on the "left part" of the Osgood transfer surface. This part cov-ers all continuous variations in stimulus and response similarity. Other cases, such as using antonyms of the responses as an interfering condition, were not pursued. Nor did we look at the effects of recombining A–B pairs to create A–Br pairs in the second list (i.e. pairing each A with a different B, also selected from the first list).

Procedure

In the model, stimulus/response pairs were simulated with fully random four-letter sequences, such as BGTE or IPHR. Every simulation followed the following three-phase format: (a) presentation of list A; (b) presentation of the interpolated list B; (c) retesting with cued recall on list A, by presenting the first letter sequence (stimulus) to the input layer which was to be completed to the correct second letter sequence (response) at the output layer.

Model

A letter was coded by activating a single node (activation of 1.0) in a module of 26 such nodes. Node 1, coded an "A", node 2 a "B", etc. The input layer consisted of four such modules, to accommodate four-letter sequences. Before each simulation experiment, all weights were initialized uniform randomly in [–0.3, 0.3]. This meant that at first presentation of a stimulus, the spontaneous output produced by the network before any learning had taken place was random. Targets were presented in the same format as the input, so that the input layer and output layer both consisted of 104 nodes (four modules of 26 nodes each). The total number of weights in the two-layer network was 10 816 (i.e. full interlayer connectivity). At each trial all items in the batch were presented and the weights were updated for each item in the batch immediately after presentation (i.e. stimulus learning). After a predetermined number of learning trials, a response was derived from the output by imposing a winner-take-all scheme on the output modules. In each module, the node with the highest activation was set to 1.0, the other activations were set to 0.0. The response consisted of the letters corresponding to the (position of the) nodes with the highest activations.

Materials

All lists consisted of ten pairs of letter sequences. The interlist similarity of both stimuli and responses was systematically increased in steps of 0.1 to obtain a full transfer surface. Interlist similarity was manipulated as follows. A letter sequence OPQR in list A was copied to a sequence in list B, whereby the copying error was varied from 0.0 to 1.0 in order to generate sequences with approximate similarities of 0.0 to 1.0, respectively. With a copying error of 0.0 the correct letter was inserted; with an error setting of 0.5 this occurred only in half the cases. A copying error only inserted a letter that did not already occur in the list (at that position). If the error is zero, copying is perfect, and the resulting sequence in list B is identical. If the error is low, the B sequence is similar (e.g. OWQR or OPUZ compared with OPQR). If the error is close to 1.0 the sequence is dissimilar. (The latter condition is called neutral by Osgood (1949).) In this manner, lists were generated whereby the approximate stimulus similarity and response similarity were varied from 0.0 to 1.0. A list was presented for a fixed number of learning trials with the word pairs kept in a fixed order.

Scoring

Recall was scored by comparing the total number of correct letter-at-position responses before and after learning the interpolated list B. Transfer was derived by calculating the difference between recall after and before learning list B. If, for example, recall after learning list B was 15 letters correct, and before 19, transfer was 15 – 19 = –4 (i.e. negative transfer, retroactive inhibition or interference). For all interlist similarities, full replications of the simulation experiment were carried out, and the average over the replications was taken in order to reduce the variance in the results.

Simulation 1: Osgood's transfer surface with a two-layer network using the delta rule

A two-layer network (no hidden layer) with a simple delta rule was used with a logistic activation function:

$$\Delta W_{ij} = \mu a_j (a_i - t_i) \qquad (5.1)$$

where

$$a_i = \frac{1}{1 + \exp\left(-\sum_h w_{ih} a_h\right)} \qquad (5.2)$$

where W_{ij} is the weight from node j to node i, a_i is the activation of node i and t_i is its target stimulus, μ is the learning rate, and h is a subscript that addresses all nodes connected to node i. No bias weights were used. This learning method is similar to two-layer back-propagation, but the rule of Rumelhart et al. (1986) also includes multiplication with an additional term $a_i(1 - a_i)$. The rule used by us is more similar to the Widrow–Hoff (1960) delta rule. Other simulations by us (not reported here) showed that using the slightly more complex back-propagation activation rule of Rumelhart et al. (1986) gave essentially similar results.

Stimulus and response lists were each presented for five trials. The learning parameter μ was set at 0.5 (no momentum was used). This gives error scores on list A in the range 15–22 (out of 40). Every data point is the average of 100 replications, each starting with new random weights (uniform in [–0.3, 0.3]) and with new lists. Scores are based on the number of letters-on-position correct, after having applied a winner-take-all schema in each letter module.

Discussion of simulation 1

The results are shown in Table 5.1. All important features of the empirical transfer surface of Osgood (1949) are reproduced in the results of the simulation.

Table 5.1 Simulation of the Osgood transfer surface, using a two-layer delta rule network.

Transfer

											Response similarity
19.82	18.90	17.10	15.38	13.66	12.05	9.47	7.68	5.27	2.50	0.00	**1.0**
17.30	16.49	14.47	13.74	12.49	10.28	8.96	6.36	4.74	2.29	0.00	**0.9**
14.18	13.43	13.06	11.49	11.08	8.91	7.64	5.70	3.85	2.03	0.00	**0.8**
11.78	10.95	10.87	9.29	8.55	7.66	6.26	4.77	3.67	1.46	0.00	**0.7**
8.65	8.63	8.15	7.72	7.02	6.63	5.45	4.03	2.86	1.51	0.00	**0.6**
6.35	6.59	6.17	5.94	5.55	4.52	4.60	3.16	2.49	1.33	0.00	**0.5**
3.57	4.28	4.31	4.48	3.84	3.38	3.24	2.44	1.97	0.82	0.00	**0.4**
1.19	1.77	2.13	2.48	2.35	2.23	1.88	1.43	1.35	0.72	0.00	**0.3**
−1.68	−1.31	−0.23	0.81	0.57	1.00	0.96	0.80	0.77	0.43	0.00	**0.2**
−4.42	−3.54	−2.62	−1.13	−1.49	−0.34	−0.08	0.25	0.15	0.02	0.00	**0.1**
−6.69	−5.36	−4.48	−3.57	−2.81	−1.87	−1.30	−0.66	−0.57	−0.23	0.00	**0.0**
1.0	**0.9**	**0.8**	**0.7**	**0.6**	**0.5**	**0.4**	**0.3**	**0.2**	**0.1**	**0.0**	
			Stimulus similarity								

With fully dissimilar (neutral) stimuli, response similarity has no effect and there is neither positive nor negative transfer, but only zero transfer. The effect of response similarity is maximal when the interlist similarity of the stimuli is maximal, that is, with identical stimuli. Negative transfer occurs only with dissimilar responses. There exists a certain response strength that yields zero transfer for all stimulus similarities. A slight difference from the original Osgood surface is that Osgood postulated a straight isotransfer line for some intermediate value of response similarity. In Figure 5.1 it can be seen that estimated isotransfer contours around zero are not straight, but slightly curved. It should be pointed out that the amount of positive and negative transfer can be manipulated by increasing the number of learning trials. Clearly, if the error on list A is very low, due to prolonged learning, maximum positive transfer will be very low, whereas in the strongly interfering condition, strong negative transfer can be expected (also see below).

Simulation 2: Osgood's transfer surface with a three-layer network using back-propagation learning and activation rules

To compare the behaviour of "standard" back-propagation with the two-layer network, we repeated the above simulation with a three-layer network, using the algorithm described in Rumelhart et al. (1986). A difference was that no bias weights were used, nor did we use any momentum. Earlier simulations had suggested that momentum might cause undesirable side-effects in this type of simulation by increasing hidden-layer overlap of consecutive stimuli. The back-propagation algorithm was used in the stimulus learning mode, whereby the weights are updated after each stimulus presentation. Mimicking the typical

Table 5.2 Simulation of the Osgood transfer surface, using a three-layer network with back-propagation learning.

Transfer

											Response similarity
10.60	9.60	9.06	7.92	7.26	5.98	4.74	3.15	2.16	1.84	0.35	**1.0**
8.71	7.23	7.06	5.47	4.74	3.77	3.38	2.17	1.41	0.67	−0.32	**0.9**
6.76	5.15	4.81	4.42	2.77	2.43	1.96	0.86	−0.16	−0.77	−1.80	**0.8**
3.39	2.94	2.70	1.75	1.15	0.37	0.55	−0.61	−1.52	−2.41	−3.16	**0.7**
1.31	1.02	0.37	−0.56	−0.36	−0.90	−2.40	−2.88	−2.86	−3.83	−4.08	**0.6**
−1.62	−0.93	−1.98	−2.76	−3.53	−3.38	−2.85	−3.73	−4.60	−5.58	−5.94	**0.5**
−3.96	−4.95	−3.79	−4.26	−5.06	−5.48	−5.63	−6.00	−6.68	−6.18	−7.82	**0.4**
−6.38	−6.76	−7.31	−7.14	−7.72	−8.18	−8.09	−8.13	−8.20	−8.83	−8.76	**0.3**
−8.87	−9.73	−10.27	−9.96	−10.56	−9.94	−10.60	−11.09	−10.66	−10.58	−11.94	**0.2**
−12.65	−12.68	−12.96	−12.81	−12.73	−13.23	−12.95	−13.58	−13.63	−13.73	−13.92	**0.1**
−16.65	−16.52	−16.51	−16.62	−16.00	−16.24	−17.03	−16.50	−16.51	−15.92	−16.74	**0.0**
1.0	**0.9**	**0.8**	**0.7**	**0.6**	**0.5**	**0.4**	**0.3**	**0.2**	**0.1**	**0.0**	
	Stimulus similarity										

verbal learning experiment, in all simulations reported in this paper stimuli were presented in list order (i.e. not randomized). The hidden layer contained 30 hidden units.

Stimulus and response lists were each presented for ten trials. The learning parameter h was set at 0.5. Every data point is the average of 100 replications, each starting with new random weights (uniform in [−0.3, 0.3]) and with new lists. Scoring was identical to simulation 1.

Discussion of simulation 2

The results of this simulation are shown in Table 5.2. For identical stimuli and dissimilar responses there is relatively more negative transfer than in simulation 1 (−16.65 compared to −6.69), but this may be due to the fact that initial learning might have been more effective in simulation 2. This lowers the transfer scores for all entries in the table. For example, in the extreme case where list A is learned until perfection, we could not have any positive transfer because the error scores could not be improved by further learning. Thus, both for the AB–AB and for the AB–AC condition, there is no important qualitative difference between simulation 1 and simulation 2. Nor is there such a difference between the simulations and the empirical Osgood transfer surface.

The most important discrepancy between the two simulations is that with back-propagation we do not obtain zero transfer for dissimilar (neutral) stimuli. On the contrary, for dissimilar responses we find that there is equally strong negative interference no matter what the stimulus similarity is. This behaviour is strongly at odds with the empirical data summarized in the Osgood surface. Moreover, we see that for identical responses there is even some positive transfer with dissimilar stimuli. This was a surprising result, because on the basis of published studies on catastrophic interference we expected a disruptive effect of intermediate learning in these conditions. Here, we seem to find that interpolated, sequential learning may actually improve scores. This hypertransfer is not observed in human or animal learning and represents another major difficulty of back-propagation as a model for human learning and memory.

Because the transfer for identical responses and dissimilar stimuli in Table 5.2 is only marginally above zero, we decided to run a number of very small-scale simulations to investigate the characteristics of this effect. In these simulations, lists of only three item pairs were used. Experimentation with different learning times showed that convincing transfer in this condition could be obtained with training list A for three iterations followed by intermediate training on list B for five iterations. An illustrative example of results typically obtained is shown in Table 5.3. The network used is the same as in simulation 2. A modest positive transfer is obtained. Output 3 in list A, for example, changes from the incorrect "twup" to the correct target "twub" due to learning the intermediate list.

Simulation 3: exploring hypertransfer with back-propagation

We carried out a large number of small simulations of the type reported in Table 5.3. These showed that hypertransfer can be reproduced with a great variety of parameters: high or low learning speed, with or without momentum, many or few iterations, short lists or long lists. Based on the small-scale simulations, we

Table 5.3 Illustrative simulation with identical responses and dissimilar stimuli. List A was presented for three iterations, list B for five iterations. Transfer due to learning the intermediate list was 2. "Output" means the spontaneous output of the network after learning. The network's output responses are printed in italics.

	Stimuli	Responses	Output
List A training			
1:	rist	munk	*twup*
2:	gork	gomp	*toup*
3:	wemp	twub	*twup*
			7 errors
List B training			
1:	yupe	munk	*muup*
2:	maws	gomp	*twup*
3:	drin	twub	*twub*
			5 errors
List A retesting			
1:	rist	munk	*goub*
2:	gork	gomp	*tomp*
3:	wemp	twub	*twub*
			5 errors

Table 5.4   Hypertransfer in back-propagation. See text for details.

Transfer

											Response similarity
24.08	23.54	23.50	23.96	23.56	23.75	23.30	21.58	21.26	19.71	16.33	**1.0**
19.47	19.47	19.55	19.74	18.76	19.03	19.31	18.01	17.27	14.42	11.29	**0.9**
16.37	14.90	14.64	16.18	15.29	15.39	14.68	13.66	12.90	0.20	7.01	**0.8**
11.46	11.34	11.64	11.27	11.43	11.17	10.97	8.36	8.15	5.60	2.29	**0.7**
7.92	7.50	7.81	7.77	7.48	7.14	6.55	4.64	3.70	1.84	-1.86	**0.6**
3.67	3.00	3.96	4.32	2.61	3.45	1.64	1.32	0.27	-2.13	-5.16	**0.5**
-0.48	-0.47	0.06	-0.56	-0.58	-0.94	-1.61	-1.64	-2.92	-6.11	-9.02	**0.4**
-4.26	-5.16	-4.82	-4.64	-5.22	-4.82	-5.43	-6.61	-7.51	-8.76	-10.94	**0.3**
-8.40	-8.34	-8.02	-9.03	-9.28	-8.78	-9.42	-10.06	-10.63	-12.03	-13.90	**0.2**
-12.38	-12.61	-11.92	-12.47	-12.67	-13.15	-13.45	-12.66	-13.26	-14.60	-15.39	**0.1**
-15.70	-16.81	-16.17	-17.14	-16.38	-16.88	-16.52	-16.19	-16.46	-15.98	-16.58	**0.0**
1.0	**0.9**	**0.8**	**0.7**	**0.6**	**0.5**	**0.4**	**0.3**	**0.2**	**0.1**	**0.0**	
Stimulus similarity											

ran simulation 2 once more, but now with training list B for 50 iterations instead of 10. We chose these parameters because the small-scale simulations indicated that hypertransfer is particularly strong when learning on the second list is extensive. All network parameters were the same as in simulation 2 (list A was still presented for ten iterations). The results are shown in Table 5.4.

Discussion of simulation 3

When training on the second list is sufficiently long, a very strong hypertransfer of 16.33 is obtained. This hypertransfer is still relatively strong when compared to the maximal transfer of 24.08 obtained in the condition with both identical stimuli and identical responses. Thus, catastrophic interference in back-propagation is only one side of the coin; hypertransfer is the other. And as we shall argue below, they both have the same underlying cause: hidden-layer overlap.

General discussion

To the main implausibilies of back-propagation as a model for human and animal learning and memory we must add another: hypertransfer or positive transfer of learning with fully dissimilar stimuli and identical responses. More generally, we have shown that for back-propagation most transfer behaviour with dissimilar stimuli is implausible, including the case with dissimilar responses, in which case we observe strong catastrophic interference where we expect to fine none.

Another finding of interest this study is that two-layer networks with a delta rule show transfer of learning that fits the classic data summarized by Osgood (1949) very well. Recently, Lewandowsky (1991) has observed that the time-course of unlearning of AB associations in the AB–AC paradigm can be modelled with a two-layer network with a delta rule. Together with our findings, this makes two-layer networks with a delta rule much better candidates as models for human memory than back-propagation networks.

The results with the delta rule can be extended to a class of Hebbian rules with unlearning. In this study we combined the delta rule with sparse binary representations. A letter was represented by 25 activations of 0 and one activation of 1, the position of the 1 determining the letter. This combination of delta rule and representation causes the following effective learning rule:

(1) When the inputs are zero, the weight change is zero.

(2a) When input and target are both 1, then:
– when the output is higher than 1, the weight change is negative;
– when the output is lower than 1, the weight change is positive.

(2b) When input is 1 and the target is 0, then:
– when the output is higher than 0, the weight change is negative;
– when the output is lower than 0, the weight change is positive.

Starting with low random weights, (2) equals (2´) in practice:

(2a´) When the target and input are both 1, the weight approaches some high value.

(2b´) When the input is 1 and the target is 0, the weight approaches some low value (zero or negative).

An important implication of this is that the combined rules (1) and (2´) are identical to a Hebbian learning rule whereby the weights increase with both positive presynaptic and positive postsynaptic activation, and whereby they decrease with positive presynaptic activation and zero postsynaptic activation, the weights remaining unchanged if the presynaptic activation is zero. Based on biological concerns, such a rule has been proposed by Singer (1990).

What causes the implausible transfer with dissimilar stimuli in back-propagation? A detailed analysis of the first series of simulations in Ratcliff (1990) provides an answer to this question. On the basis of such analyses reported in Murre (1992a), it can be concluded that learning in the input-to-hidden weights is remarkably insensitive to the structure of the input and output patterns in that the similarity structure of, say, the input patterns is almost completely lost in the corresponding representations in the hidden layers (see also Sharkey & Sharkey 1995). On the contrary, in the hidden layer one tends to find representations in which around half of the nodes are activated, even if the input and output patterns are sparse. This causes strong overlap between hidden-layer representations, even if the input patterns are sparse and fully orthogonal.

With the type of sparse input patterns used here, only those input-to-hidden weights connected from an activated input can change. This means that during learning of the interfering list B the input-to-hidden weight representation of list A does not change in the case of fully dissimilar patterns, because those input nodes all have zero activations in list B. In other words, most of the learning takes place in the upper layer of weights (hidden-to-output weights).

Because hidden-layer representations are very similar in that they overlap strongly, we are in fact dealing with a condition on the Osgood surface where similar stimuli (hidden-layer representations) are associated with responses (output patterns). We would thus expect to obtain a level of transfer equal to the middle columns of Table 5.1 (stimulus similarity of around 0.5). If the hypothesis of relative insensitivity of the hidden-layer representations to input representations holds, we would furthermore expect to find only a weak effect of varying the similarity of the (true) input patterns. This can indeed be observed in Table 5.3. Hypertransfer in back-propagation can thus be explained as the result of associating similar hidden-layer representations (despite dissimilar input representations) with identical responses.

Some authors have pointed out that back-propagation is unrealistic for modelling human learning, because it assumes that the subject comes into the experiment without any prior knowledge (e.g. Hetherington 1991, McRae & Hetherington 1993). All internal (i.e. hidden-layer) representations must be formed from scratch during the experimental session. When the stimuli are drawn from a structured or combinatorial domain, such as all English words (i.e. forming a subset of all possible strings), pretraining a network can eliminate the interference (see also Brousse & Smolensky 1989). This can lead to sharpening of the hidden-layer activations in that fewer nodes are activated, but is not likely to completely eliminate overlap of between-pattern representations in the hidden layer. In particular, with random letter strings the effectiveness of pretraining is less than when a more limited combinatorial domain is used, such as English words. A more direct approach to reducing interference is to force orthogonal hidden-layer representations by imposing a k-winner-take-all mechanism on the hidden layer (French 1991, French 1992), where k is small (e.g. only two nodes are allowed to remain activated ultimately). Murre (1992a) has shown that this may indeed eliminate interference in the AB–CD condition (i.e. dissimilar stimuli and dissimilar responses). The main problem, however, is not simply to eliminate hidden-layer overlap but to have it reflect the similarity structure of the pattern set. Only in that case can we expect to be able to model the Osgood surface. And only then can we expect psychologically plausible behaviour with back-propagation. Hypertransfer was also found in a recent study by Sharkey & Sharkey (1993), although they do not focus on its psychological implausibility. They also present an analysis of the problem, based on a new visualization technique. Their conclusions about the causes of interference in back-propagation largely agree with ours.

Back-propagation has as a major advantage over the simple delta rule that it can learn nonlinearly separable patterns. If such patterns were frequently encountered in human learning, delta rule models (i.e. with only a single layer of weights) are clearly insufficient. We want to argue, however, that such patterns are in fact uncommon and that the need for a very powerful learning algorithm for modelling human learning is limited. A good example is the experiments on human associative learning, where we can distinguish two important situations:

(a) association of completely new materials (e.g. random letter strings) and (b) association of materials that are known (e.g. English words). In neither case is it clear that we need a learning algorithm that is as powerful as back-propagation. Of course, if pairs with the following non-separable (XOR-type) structure need to be associated, a two-layer network will fail:

Stimulus	Response (not linearly separable)
AA	CC
BB	CC
AB	DD
BA	DD

But in the case of sufficiently random and sufficiently long nonsense strings, such as used in this experiment, the probability of such dependencies is very low. In most cases the rest of the string will disambiguate the XOR-type structure:

Stimulus	Response (made separable by string context)
AAE	CCI
BBF	CCJ
ABG	DDK
BAH	DDL

In case of meaningful words, it is even more difficult to imagine how we could possibly encounter associations that are not linearly separable.

Associative learning experiments provide very interesting clues about possible underlying cortical representations of words. A surprising aspect of the Osgood surface, for example, is that it describes well experiments with both nonsense strings and meaningful words. In the former case similarities are governed purely by surface characteristics without much regard for meaning or any pre-existing representations whatsoever (except of course for the phonological level), whereas in the latter case similarity is defined with relation to deep semantic aspects of the word.

With nonsense strings, it might be that we form direct input/output associations between the phonemes, as in simulation 1. It is not at all obvious that subjects form elaborate internal representations, as in back-propagation, when associating totally nonsense materials, although with repeated exposure such internal representations may eventually be formed. But in contrast to back-propagation the internal representations will most likely be related to the structure of the input set. An example of more plausible internal representations than in the case of back-propagation is provided by the models of Shanks & Gluck (1994) and Gluck & Bower (1988, 1990). Their studies still do not explain how such representations may emerge; they are assumed (albeit on very solid grounds, e.g. Shepard (1987)), rather than explained. Further investigation into the emergence of internal representations during prolonged learning will be necessary to elucidate this process. A further step would be to use the stimuli used in the model of Shanks & Gluck (1994) in an associative interference paradigm. Such work could lead to a possible integration of the domains of concept formation and of association theory. Whatever the final process is that might be uncovered by such an undertaking, on the basis of the above results we can be certain that the process will not resemble straightforward back-propagation learning.

In the case where meaningful stimuli such as "church" and "table" are associated, we can infer that the internal or semantic representations of the stimuli and responses follow a similarity pattern that "conforms to our intuitions". When "church" is associated with "table" in list A ("church"–"table"), and "car" is also associated with "table" in list B ("car"–"table"), we typically find that there will be little or no interference, because the stimuli in this cases are dissimilar. If we assume that these representations have a cortical basis and that some associative learning rule is used in establishing connections, we can conclude that these cortical representations have a similarity structure (with respect to the learning rule) which mirrors that of dissimilar nonsense patterns: the semantic cortical representations cannot possibly overlap much. If they did overlap, we would expect transfer behaviour as displayed by back-propagation. This is a somewhat surprising conclusion because one expects a semantic representation to comprise many different aspects of a stimulus, a rich representation that is very likely to overlap with other representations. Yet, this does not seem to be the case. With respect to establishing associative bonds, any such overlap is eliminated by the cortical learning process and only the essence of the cortical representation of the word meaning is retained. Thus, our intuition about the dissimilarity of the meaning of "church" and "car" is supported by their transfer behaviour, and this behaviour in turn allows us to infer that their cortical representation must be distinct (non-overlapping) with respect to the formation of associative bonds.

This view of encoding and learning is strongly suggestive of a process whereby the word is somehow interpreted or reduced, so that only a certain aspect of its meaning is effective in inducing a cortical representation. Such a memory process has been proposed by Tulving (e.g. 1983). He postulates that the memory representation is combined with relevant aspects of the learning episode (e.g. retrieval cues) to produce a specific trace that includes only certain aspects of the encoded item. In the concept formation literature a process has been proposed that bears at least a superficial resemblance to such a process. In well controlled experiments whereby stimuli are presented that can vary on a number of continuous dimensions (e.g. angle, colour, etc.), subjects pay more attention to dimensions that are relevant to the specific categories that must be formed. The extent of this effect has been demonstrated convincingly in simulations based on a multidimensional scaling paradigm by Nosofsky (e.g. 1985, 1986). The general category structure seems to guide the representations of the (novel) stimuli formed during the experiment. The studies by Tulving and Nosofsky both show that semantic representations are much more active during learning than is usually assumed in connectionist modelling of human learning and memory. Such accounts fit in well with the above conclusions that call for distinct cortical, underlying representations, in that they provide evidence for the fact that interpretive or reduction processes are indeed operative during learning.

A reinterpretation of studies in associative interference combined with new experimental studies that are informed by connectionist or mathematical models may result a much more detailed knowledge of both the nature and the dynamics of semantic and cortical representations. Work by Mensink & Raaijmakers (1988) has already shown that even an associative model that does not include explicitly inter-item similarity can model a large number of studies in transfer theory, including the Osgood surface. Our general ignorance about the underlying representations of words and other items has been a major stumbling block for developing effective neural network models. A detailed analysis of the enormous but largely neglected database of studies in associative interference built up the 1930s and 1940s, combined with persistent modelling efforts, will lead to deeper insights into the problem of representation.

Acknowledgements

I want to thank Alan Baddeley and Jeroen Raaijmakers for helpful suggestions when preparing this article, and Roger Ratcliff for his help in the initial stages of this project.

References

Ackley, D. H., G. E. Hinton, T. J. Sejnowski 1985. A learning algorithm for Boltzmann machines. *Cognitive Science* **9**, 147–69.

Baddeley, A. D. 1976. *The psychology of memory*. New York: Harper International Edition.

Barnes, J. M. & B. J. Underwood 1959. "Fate" of first-list associations in transfer theory. *Journal of Experimental Psychology* **58**, 97–105.

Brousse, O. & P. Smolensky 1989. Virtual memories and massive generalization in connectionist combinatorial learning. *11th Annual Conference of the Cognitive Science Society, Proceedings*, 380–87. Hillsdale, New Jersey: Lawrence Erlbaum.

French, R. M. 1991. Using semi-distributed representations to overcome catastrophic forgetting in connectionist networks. *13th Annual Conference of the Cognitive Science Society, Proceedings*, 173–8. Hillsdale, New Jersey: Lawrence Erlbaum.

French, R. M. 1992. Semi-distributed representations and catastrophic forgetting in connectionist networks. *Connection Science* **4**, 365–77.

Gluck, M. A. 1991. Stimulus generalization and representation in adaptive network models of category learning. *Psychological Science* **2**, 50–55.

Gluck, M. A. & G. H. Bower 1988. From conditioning to category learning: an adaptive network model. *Journal of Experimental Psychology: General* **117**, 227–47.

Gluck, M. A. & G. H. Bower 1990. Component and pattern information in adaptive networks. *Journal of Experimental Psychology: General* **119**, 105–9.

Grossberg, S. 1976. Adaptive pattern classification and universal recoding, II: feedback, expectation, olfaction, and illusions. *Biological Cybernetics* **23**, 187–202.

Grossberg, S. 1982. *Studies of mind and brain: neural principles of learning, perception, development, cognition, and motor control.* Boston: Reidel Press.

Grossberg, S. 1987. Competitive learning: from interactive activation to adaptive resonance. *Cognitive Science* **11**, 23–63.

Hebb, D. O. 1949. *The organization of behavior.* New York: John Wiley.

Hetherington, P. A. 1991. *The sequential learning problem in connectionist networks.* Unpublished Master's thesis, McGill University, Montreal.

Hopfield, J. J. 1982. Neural networks and physical systems with emergent collective computational abilities. *National Academy of Sciences of the USA, Proceedings* **79**, 2554–8.

Hornik, K., M. Stinchcombe, H. White 1989. Multilayer feedforward networks are universal approximators. *Neural Networks* **2**, 359–66.

Jacobs, R. A. 1988. Increased learning rates of convergence through learning rate adaption. *Neural Networks* **1**, 295–307.

Kohonen, T. 1972. Correlation matrix memories. *IEEE Transactions on Computers* **C–21**, 353–9.

Kohonen, T. 1982. Self-organized formation of topologically correct feature maps. *Biological Cybernetics* **43**, 59–69.

Kohonen, T. 1990. The self-organizing map. *IEEE, Proceedings* **78**, 1464–80.

Kruschke, J. K. 1990. ALCOVE: an exemplar-based connectionist model of category learning. *Psychological Review* **99**, 22–44.

Kruschke, J. K. & J. Rodriquez-Movellan 1991. Benefits of gains: speeded learning and minimal hidden layers in back-propagation networks. *IEEE Transactions on Systems, Man, and Cybernetics* **21**, 273–80.

Lewandowsky, S. 1991. Gradual unlearning and catastrophic interference: a comparison of distributed architectures. In *Relating theory and data: essays on human memory in honor of Bennet B. Murdock*, W. E. Hockley & S. Lewandowsky (eds), 445–76. Hillsdale, New Jersey: Lawrence Erlbaum.

Lockery, S. R., G. Wittenberg, W. B. Kristan, Jr, G. W. Cottrell 1989. Function of identified interneurons in the leech elucidated using neural networks trained by back-propagation. *Nature* **340**, 468–71.

McCloskey, M. & N. J. Cohen 1989. Catastrophic interference in connectionist networks: The sequential learning problem. In *The psychology of learning and motivation,* G. H. Bower (ed.), 109–65. New York: Academic Press.

McRae, K. & P. A. Hetherington 1993. *Catastrophic interference is eliminated in pretrained networks.* University of Rochester, Technical Report.

Mensink, G. J. & J. G. W. Raaijmakers 1988. A model for interference and forgetting. *Psychological Review* **95**, 434–55.

Mirchandani, G. & W. Cao 1989. On hidden nodes for neural nets. *IEEE Transactions on Circuits and Systems* **36**, 661–4.

Murre, J. M. J. 1992a. *Learning and categorization in modular neural networks.* Hemel Hempstead, England: Harvester Wheatsheaf.

Murre, J. M. J. 1992b. The effects of pattern presentation on interference in backpropagation networks. *14th Annual Conference of the Cognitive Science Society, Proceedings*, 54–9. Hillsdale, New Jersey: Lawrence Erlbaum.

Murre, J. M. J., 1995a. Neurosimulators. In *Handbook of Brain Research and Neural Networks*, M. A. Arbib (ed.), in press. Cambridge, Mass.: MIT Press.

Murre, J. M. J. 1995b. *A connectionist approach to transfer theory.* MRC Applied Psychology Unit, Technical Report, in preparation.

Nosofsky, R. M. 1985. Overall similarity and the identification of separable-dimension stimuli: a choice model analysis. *Perception & Psychophysics* **38**, 415–32.

Nosofsky, R. M. 1986. Attention, similarity, and the identification-categorization relationship. *Journal of Experimental Psychology: General* **115**, 39–57.

Nosofsky, R. M. 1992. Similarity scaling and cognitive process models. *Annual Review of Psychology* **43**, 25–53.

Nosofsky, R. M., J. K. Kruschke, S. C. McKinley 1992. Combining exemplar-based category representations and connectionist learning rules. *Journal of Experimental Psychology: Learning, Memory, and Cognition* **18,** 211–33.

Osgood, C. E. 1949. The similarity paradox in human learning: a resolution. *Psychological Review* **56**, 132–43.

Ratcliff, R. 1990. Connectionist models of recognition memory: constraints imposed by learning and forgetting functions. *Psychological Review* **97**, 285–308.

Rescorla, R. A. & A. R. Wagner 1972. A theory of Pavlovian conditioning: variations in the effectiveness of reinforcement and non-reinforcement. In *Classical conditioning II: current research and theory*, A. H. Black & W. F. Prokasy (eds), 64–99. New York: Appleton-Century-Crofts.

Rueckl, J. G., K. R. Cave, S. M. Kosslyn 1989. Why are "what" and "where" processed by separate cortical visual systems? A computational investigation. *Journal of Cognitive Neuroscience* **1**, 171–86.

Rumelhart, D. E. & J. L. McClelland (eds) 1986. *Parallel distributed processing: explorations in the microstructure of cognition*, vol. 2. *Psychological and biological models*. Cambridge, Mass.: MIT Press.

Rumelhart, D. E., G. E. Hinton, R. J. Williams 1986. Learning internal representations by error propagation. In *Parallel distributed processing: explorations in the microstructure of cognition*, vol. 1. *Foundations*, D. E. Rumelhart & J. L. McClelland (eds), 318–62. Cambridge, Mass.: MIT Press.

Shanks, D. R. & M. A. Gluck 1994. Tests of an adaptive network model for the identification and categorization of continuous-dimension stimuli. *Connection Science* **6**, 59–89.

Sharkey, N. E. & A. J. C. Sharkey 1993. Adaptive generalization. *Artificial Intelligence Review* **7**, 313–28.

Sharkey, N. E. & A. J. C. Sharkey 1995. An interference-discrimination tradeoff in connectionist models of human memory. Submitted.

Shepard, R. N. 1987. Toward a universal law of generalization for psychological science. *Science* **237**, 1317–23.

Singer, W. 1990. Ontogenetic self-organization and learning. In *Brain organization and memory: cells, systems, and circuits*, J. L. McGaugh, N. M. Weinberger, G. Lynch (eds), 211–33. Oxford: Oxford University Press.

Tulving, E. 1983. *Elements of episodic memory*. Oxford: Oxford University Press.

Wang, Y. F., J. B. Cruz, J. H. Mulligan, Jr 1993. Multiple training concepts for back-propagation neural networks for use in associative memories. *Neural Networks* **6**, 1169–75.

Werbos, P. J. 1974. *Beyond regression: new tools for prediction and analysis in the behavioral sciences*. Unpublished PhD thesis, Harvard University, Cambridge, Mass..

Widrow, B. & M. E. Hoff 1960. Adaptive switching circuits. *Western Electronic Show and Convention, Convention Record* **4**, 96–104.

Zipser, D. & R. A. Anderson 1988. A back-propagation programmed network that simulates response properties of a subset of posterior parietal neurons. *Nature* **331**, 679–84.

CHAPTER 6

Interactions between short- and long-term weights: applications for cognitive modelling

Joseph P. Levy & Dimitrios Bairaktaris

Introduction

In order to account for the variety of human cognitive phenomena, computational models of mental processing require processes that occur over different timescales. Human memory is often divided into three temporally defined categories: very short-term sensory memory (seconds), short-term memory (STM) (seconds to minutes) and long-term memory (LTM) (years).

Many cognitive processes appear to require short-term phenomena to be overlaid on a long-term learned competence. For example, in processing a sentence we need to be able to compute which entity a pronoun refers to and remember this for a short period. This information storage is motivated by our long-term knowledge of the grammar of language but does not need to be maintained permanently like our linguistic knowledge. From a computational viewpoint, similar short-term information needs to be stored in intimate contact with the processing system for phenomena like perceiving a complex stimulus and representing its conjunction of features (e.g. Treisman & Gelade 1980) or representing an individual in terms of its properties (Stenning et al. 1988, Levy & Stenning 1989). These binding phenomena require short-term representations for the correct groupings of features or properties that interact with long-term representations of the background knowledge that underlies the achievement of the overall task. In fact, most cognitive processes require the interaction between short- and long-term information whether they be perceptual processes requiring sensory information to be interpreted in the light of long-term experience or learning processes where new information is integrated into the organized structure of LTM.

Since the early 1980s a large body of work has accumulated that aims to apply connectionist systems to cognitive modelling. Not surprisingly, the problem of designing subsystems which interact on different timescales has not been ignored since it seems such an important aspect of cognitive phenomena. This chapter surveys some of this work, introduces a recently developed dual-store connectionist architecture and speculates on the variety of cognitive phenomena that might prove suitable for modelling by networks with weights that interact on different timescales.

It is helpful to introduce a simple taxonomy of connectionist architectures with short- and long-term components. The simplest is probably the *dual-trace* architecture as discussed by Hebb (1949) where long-term, learned knowledge is stored in the weights of a network and the short-term store is simply represented as the activations of the units. This mixture of knowledge held in weights and short-term information held as unit activation has been the (usually implicit) method of dealing with short- and long-term information in most connectionist cognitive modelling. A notable example of such a model where unit activation is explicitly described as representing a short-term trace of processing history is the TRACE model of speech perception (McClelland & Elman 1986). In TRACE, each level of representation (phonetic, phonological and word) is duplicated for over-lapping time slots. The time-course of processing is easily visualized by the relative unit activations across time slots.

A simple alternative to a dual-trace system is to have subcomponents with weights of differing timescales. The subcomponents might function as individual short- and long-term stores or use special short-term weights for a specific function within the system – e.g. binding of elements of long-term knowledge. We shall call this type of system a *dual-* (or *multi-*) *component system*, where distinct components have weights of different types.

The third main category of systems where short- and long-term capabilities interact is one where a single connection between a pair of units can carry more than one weight. These are systems with *dual-* or *multi-weight connections*. The effect of the weights can be additive (e.g. Hinton & Plaut 1987), multiplicative (e.g. Gardner-Medwin 1989) or independent (e.g. Levy & Bairaktaris 1991). The impetus behind multi-weight connections is the way that they allow processing at different timescales to be carried out on the same representations. This is possible because the representation is the pattern of activation of the units, while processing, the way in which the representations change, is controlled by interactions between the different types of weights. There is currently a great deal of neurophysiological research being done on mechanisms of synaptic plasticity over different timescales. Although the physiological details, let alone the functional significance, of phenomena such as post-tetanic potentiation, short-term potentiation, long-term potentiation, long-term depression and synaptic growth (arborization) are still unclear, the work on these mechanisms suggests that connectionist models using weights of differing timescales is plausible.

Dual-component and dual-weight systems

We shall review some of the recent work on dual-component systems and systems with dual-weight connections below. The review expands and extends the references cited by Plaut & Shallice (1993).

Dual-component systems

In a synthesis of psychological data, neuroanatomy and physiology, Crick (1984) suggested a use for short-term weights as an attentional mechanism – a way of implementing the internal attentional spotlight suggested by Treisman (e.g. Treisman & Gelade 1980) and others that serves to highlight a particular aspect of input information. The kind of data that Crick wanted to account for were the difficulty that people have in scanning visual scenes for objects with particular conjunctions of features (e.g. green Ts) when there are distractors in the array (e.g. brown Ts and green Xs). The time taken to achieve the task increases linearly with the number of distractors, suggesting a serial scanning process, the putative moving mental spotlight. Crick suggested that the spotlight was controlled by a sheet of cells called the thalamic reticular complex, through which pass all the axons that project from the thalamus to the neocortex and from the neocortex back to the thalamus. He proposed that the reticular complex could selectively increase the activity of certain groups of axons which project to or from specific cortical areas. He hypothesized that the spotlight was shone by a small area of the reticular complex, causing rapid firing in a subset of thalamic neurons, intensifying their input to particular cortical regions. He supposed that the bursts of firing cause rapid transient Hebbian facilitation of synapses in the thalamic–cortical circuits. The reticular spotlight then moves on to cause relatively increased excitation in another cortical area.

The way that the spotlight achieves attention to conjunctions of features is to cause simultaneous activation of units in the relevant cortical maps responsible for processing the particular features. This idea has since been developed in several papers that model the binding of features into objects as phase-locked activation of the units that represent the features (e.g. Shastri & Ajjanagadde 1989).

Goebel (1990) was concerned with using connectionist architectures to model symbol manipulation. He implemented a system consisting of mechanisms for STM, LTM, selective attention and variable binding. The system (PADSYMA) had three main components: an episodic STM, a "selection" layer and a context layer. The STM has fast autoassociative weights set by a Hebb rule. It also has one-to-one bidirectional weights to the selection level fixed at a value of 1. A phase-locking mechanism controlled by an "attention node" sets the focus of attention to be a group of corresponding units in the STM and selection units. Short-term serial order is represented by the degree of overlap between units or distributed representations – decay of short-term items means that temporal proximity is proportional to similarity in activation level.

97

LTM is stored in slow permanent recurrent weights between the selection and context units. There is an opportunity for long-term information to influence short-term processing via activation in the selection units, but this possibility is not explored by Goebel.

Goebel's system contained matching units in an episodic STM layer and a selection layer. Each unit was connected bidirectionally to its counterpart in the other layer and the weights were gated by an attention node that acted to synchronize the firing of groups of units to bind their activation into patterns representing individual objects. It would seem possible to implement this idea in a different way using independent multi-weighted connections. The short-term store could be implemented as short-term weights, and the attentional buffer as separate short-term weights. Phase locking could be used to bind units together in the attentional buffer, or individual objects could be represented as separate states in the buffer chained together using temporal associations. The long-term component of Goebel's system could be represented as a third, permanent weight type. Thus, Goebel's three layers could be collapsed into a single-layer system where each connection has three independently acting weights, one for a short-term buffer, one for an attentional buffer and one for long-term knowledge. It would be interesting to explore the possibilities of additive or multiplicative interactions between the weight types.

Additive dual-weight connections

Hinton & Plaut (1987) proposed a back-propagation network that used a fast weight (quickly decaying but fast learning) and a slow weight (non-decaying but slow learning) on each connection. The weights were simply *added* together to give an effective connection strength. The quickly decaying fast weights could be used to temporarily learn new associations without ultimately disturbing the slow weights. An interesting use for the new associations was to "revive" a set of old associations that had effectively been lost by interference by subsequent learning. The revival required only a subset of the old associations to be relearned using the fast weights. One hundred random 10-bit input vectors were associated with 100 random 10-bit output vectors using a standard version of the back-propagation algorithm on a network with 100 hidden units. This required 1300 sweeps through the training set and meant that the associations were held in the slow weights since the fast weights decayed by 1% after each weight update, and this dominated the tiny error derivatives that occur towards the end of training. The network was then trained on five new random associations without repeating any of the original 100. This took 400 epochs of training and effectively obliterated the ability of the network to make the original 100 associations. However, if the network is retrained on as few as ten of the original associations, there is a large (but not complete) transfer effect – a significant decrease in error on all of the 100 associations, not just the ones retrained. If the retraining occurs in the fast weights then the second set of five associations can recover after the fast weights have decayed away. The system is thus able to recover older memories without losing the newly acquired information.

In a discussion of connectionist models of sentence processing, McClelland & Kawamoto (1986) discuss an idea cited as a personal communication from Hinton for short-term weights to mediate recursive processing. In a system where sequential processing was implemented by associating one state with the next, a special set of units was set aside as a stack. Recursing down a level takes place by mapping the current state onto the first state of a subroutine while saving the current state in the stack by using fast weights to associate it with the current state of the stack. At the end of the subroutine, the stack can "pop" to the previous level using special permanent weights that associate stack levels with their previous state, and the state of the system returns to what it was when the subroutine was called because the fast weights can determine the next stage of processing when this would be indeterminate if left to the permanent weights.

Cleeremans & McClelland (1991) proposed a model of the implicit learning of temporal contingencies. They found that subjects gradually improved in a choice reaction time task where sequences of required responses were generated by a "noisy" finite-state grammar. The fact that subjects were picking up aspects of the underlying regularities was demonstrated by the faster reaction times for grammatical items than for ungrammatical noise over the very long sequence (60 000 items) of learning trials. Their first attempt at a computational model used a simple recurrent network (e.g. Elman 1990) to predict the next item given the history of previous ones. Their network learned some of the temporal regularities underlying the sequences but failed to fully account for the human data because it did not exhibit the priming that affected subjects' performance. Subjects were faster at responses that were repeated even after a different intervening item. There was also an advantage for items that would have been grammatical at the previous trial but had not occurred and for particular repeated sequential pairings of responses. The model was augmented with a decaying trace of previous activation on the output units as well as additive "fast" weights with a slightly higher learning constant than the permanent weights and a half-life of a single time step. The transient weights served to increase the tendency of the network to respond in the near future in a similar manner to the recent sequence of states. As Plaut & Shallice (1993) have noted, the weights are priming *associations* between items rather than just increasing the activation of an item or reducing its threshold as in a logogen model (Morton 1969). We shall return to the advantages of multi-weight connections for modelling priming in the final section of this chapter.

Cleeremans & McClelland's dual-weight model accounted for more of the variance in the data than their previous model because its responses were now influenced by a short-term sequential memory.

Plaut & Shallice (1993) describe a model of optic aphasia, a neuropsychological disorder of the naming of visually presented objects and pictures. Patients with this type of condition make characteristic errors. They make semantic errors (e.g. saying "hat" when shown a shoe or a picture of a shoe), visual errors (e.g. saying "hazel-nuts" when shown coffee beans) and mixed visual and semantic errors. In order to model how these errors might arise, Plaut and Shallice constructed a network that maps a high-level visual representation onto a semantic representation using recurrent back-propagation. The semantic level contains "clean-up units" which allow the learning of the semantics of objects as point attractors in a constructed semantic feature space. As well as making visual and semantic errors, patients with optic aphasia make "perseverative errors" – a tendency for recently presented objects to intrude into the errors made in attempts to name subsequent objects, e.g. a patient might correctly name a picture of a wristwatch but subsequently mistakenly name a picture of some scissors as "wristwatch". Plaut & Shallice used short-term correlational weights to slightly bias the system towards recently met items. Each short-term weight was an exponentially decaying average over the patterns in the training set of the activations of the units it connects. This has the consequence of giving only the most recent patterns any significant influence on the short-term weights. The correlational weights were additive and small in proportion to the long-term weights that supported the learning of the correct visual to semantic mapping. After lesioning, the network made similar patterns of errors to patients: mostly semantic with a significant pattern of perseveration caused by the short-term weights.

Multiplicative dual-weight connections

Gardner-Medwin (1989) proposed a dual-weight system where the short- and long-term weights combined in a *multiplicative* fashion. His computational aim was to build a system which had high-quality short- and long-term storage and could consolidate short-term traces into LTM. His building block was an autoassociative net with both STM and LTM traces at a single level of representation. His proposed system was justified biologically by the plausible suggestion that there are classes of synapse that act on different timescales and the assumption that the short- and long-term changes in synaptic efficacy combine *in series* to produce an overall dendritic current. In the network simulation this is achieved by *multiplying* short- and long-term weights together to calculate the effective overall weight on a connection between two units. This effectively means that the binary short-term weights gate the influence of the permanent weights.

Gardner-Medwin suggested that the weight modifications might arise from changes in presynaptic transmitter release or changes in dendritic spines. Temporary weights were changed from 0 to 1 if a Hebb rule was satisfied. They decay back to zero after a short time. The permanent weights increase gradually from a small initial value. Short-term recall is better for traces that are similar or identical to long-term traces since most or all of the non-zero temporary weights are enhanced by multiplication with a permanent weight. This property leads to a mechanism for recognition memory – the level of threshold for the constituent units of a trace that allows for the pattern to be self-sustaining is lower for ("familiar") patterns already held in LTM.

Gardner-Medwin's consolidation mechanism occurs by re-evoking the short-term traces and changing the permanent weights using a Hebb rule. Before the weight changes are made, an optimization mechanism reduces the overlap between new and old traces so as to decrease interference and increase the effective capacity of the long-term store.

Pure long-term recall requires a "booting" process to get around the fact that all connections have effective weights of zero in the absence of temporary traces. Gardner-Medwin suggests several methods of booting by setting some or all of the temporary weights to 1. He compares the booting procedure to the painstaking process of certain human long-term recall. Long-term recall is improved or "primed" by recent identical or related temporary traces.

Gardner-Medwin's dual-weight architecture has some psychologically interesting properties beyond those of a single-weight autoassociative network. Short- and long-term information interact in a simple and intimate manner. The dependence of short-term traces on LTM is achieved automatically. The work of Levy and Bairaktaris (see the next section) extends his ideas by using more powerful learning algorithms and exploring what happens when the temporary and permanent weights act *independently*.

Independent dual-weight connections

Some of our recent work (Levy & Bairaktaris 1991, Levy & Bairaktaris 1993) has explored the computational properties of a dual-weight system where short- and long-term weights act *independently* at different times. Interaction between short- and long-term information takes place through unit activation. The architecture uses a Hebbian learning algorithm for the short-term weights and a deterministic Boltzmann machine (BM) algorithm (mean field theory) for the long-term weights. The aim has been to construct an autoassociative memory with intimately coupled long- and short-term components. Autoassociators are networks where the patterns stored are point attractors, stable states in the dynamics of the system. This property makes them attractive modelling metaphors for aspects of human memory such as content addressability, cueing, noise resistance, etc. (e.g. see McClelland & Rumelhart 1986).

The short-term component is based on work by Bairaktaris (1990) using the two-layer BAM (bidirectional associative memory) architecture (Kosko 1988). This architecture consists of two layers, fully connected by bidirectional weights. An autoassociator is constructed by mapping each input vector to be stored onto another fixed vector on an output layer using a Hebb rule. Patterns are recalled by passing activation from the input layer to the output layer and then back again. The reverse pass is equivalent to multiplying by the transverse of the weight vector.

Instead of mapping between two predetermined representations, Bairaktaris increased the capacity of the system and decreased its sensitivity to correlation by mapping the representation on the input layer to a random internal representation (RIR) on the other layer. This has the effect of making the system an autoassociator. If the RIR is large, the dimensionality of the stored vector space is increased enough to lead to less overlap between vectors, allowing the network to store more patterns for the same dimensionality of input space. The architecture has a similar capacity for random vectors as a Hopfield net (McEliece et al. 1987) but is better able to tolerate non-orthogonal input patterns (Bairaktaris 1990). The Bairaktaris version of BAM shares the advantage of other Hebbian autoassociators for short-term storage of needing only one pass through the pattern set and lessens the disadvantage that this type of architecture has of small capacity and intolerance to non-orthogonality.

We train the long-term store component of our model using the mean field theory (MFT) algorithm discussed in Peterson & Hartman (1989). This is essentially the same algorithm discussed by Hinton and his colleagues which they call the deterministic BM (e.g. Hinton 1989). A convenient way to describe the algorithm is to compare it with the more familiar (stochastic) BM learning algorithm (Ackley et al. 1985).

Both types of network allow arbitrary connectivity between and within layers of units with or without hidden units, i.e. units whose values are not specified by the input and desired output of the training set. Units are typically not allowed self-connections, and each unit usually has a learnable bias.

Units in a BM network are binary – the *probability* of a given unit having an activation of 1 depends on the weighted sum of its input transformed by a sigmoidal "squashing" function. The shape of the sigmoidal curve depends on a global parameter called "temperature". At high temperatures the network effectively behaves randomly. As the temperature is lowered, the thermal noise involved helps avoid local minima, and the network settles into a good solution, ideally a global minimum.

In contrast, the units in an MFT network can vary continuously between 0 and 1 or between –1 and 1 using a similar but non-probabilistic sigmoidal output function. As the temperature is lowered, the shape of the output function approaches the form of a step function. Thus the output function of an MFT unit is simply the continuous deterministic form of the one used by the BM network.

The other main contrast between the two algorithms is the way statistics are collected during learning. The BM algorithm measures the frequency that both members of each pair of units are active during a *clamped phase* where both input and output units are forced to take on their desired values and a *free phase* where only the input units are clamped and the output units are free to vary. In both phases the activations of any hidden units are free to vary. Weight changes are made in proportion to the difference between the simultaneous activity statistics measured for each pair in the two phases. This form of measurement is necessarily slow because of the noisy stochastic nature of the system – reliable statistics require many measurements to be taken. The MFT algorithm speeds this process by using the product of the deterministic outputs of both members of each pair of units as the relevant co-occurrence statistic to determine weight change. This "mean field" approximation does not require as much computation as is needed in the BM algorithm and appears to perform well (Peterson & Hartman 1989, Hartman 1991), although there are certain problems with the approach (Galland 1993).

Peterson & Hartman (1989) describe how the MFT algorithm can be used to train an autoassociator with hidden units. This architecture contains a single layer that effectively represents both input and output, and a second hidden layer. The task for the learning algorithm is to find a set of weights that make the training patterns on the input/output layer attractors in the activation space of the network. This allows the pattern completion behaviour required to model content addressability and other properties of human memory. They demonstrate that such a network can have an impressive capacity (up to four times the number of visible units). They recommend that one-half of the visible units is chosen at random and clamped during the free phase when an autoassociator is trained. This leads to better performance than not clamping any of the visible units at all, which was the procedure suggested in the original BM literature (Ackley et al. 1985).

Unlike an architecture where the effectiveness of training can be tested by simply imposing an input pattern on the input units and comparing the output of the network to the desired output, the appropriate test for this kind of auto-associator is whether the training patterns are robust attractors. The ultimate test for this is giving incomplete cues and seeing whether the network completes the pattern correctly. During training, however, Petersen & Hartman used a less exacting test where the training pattern simply has to remain stable during the low-temperature part of the annealing schedule, i.e. the sequence of decreasing temperatures used to relax the network. Their procedure was as follows:

1. The visible units are clamped to their desired state or a given number of bits are changed if the noise resistance of the network is being measured.
2. Annealing begins and the hidden units are allowed to partially relax.
3. The visible units are released and the annealing schedule is completed.

Since the MFT algorithm allows the use of hidden units in an autoassociator, such a network can have a large capacity and can learn non-orthogonal patterns. In a similar way to networks trained using the back-propagation algorithm, the hidden units can be used to extract high-order statistics from a training set. These properties are all potentially useful for a model of long-term information storage.

The short- and long-term components of the dual-weight model share the same units and connections. Each connection can have either a short-term or a long-term weight or (as is usually the case) both. The combined system acts in either short- or long-term "mode", and information is passed by means of the activation levels of the units. In other words, information resulting from processing by one kind of weight is passed to the other component simply by means of the activations left on the units after the earlier processing.

An independent dual-weight architecture

The Hebbian BAM weights have a small capacity but can be trained in an incremental "one-shot" fashion. The long-term weights take a long time to train but have a large capacity and a powerful representational ability, being able to extract generalizations over a pattern set. These complementary properties mirror some of those that have been suggested for human short- and long-term memory.

The basic effectiveness of the architecture was demonstrated by pretraining the long-term component with a relatively complex training set and buffering several of these patterns in the short-term component. This is a simple way of showing how this kind of architecture can act as a way of boosting the availability of a limited number of LTM's by "caching" them in a short-term store. Something like this is suggested by some theories of short-term memory, whether the short-term store is simply an activated portion of the long-term store (e.g. Anderson 1976) or whether it is a distinct buffer.

Bitmap patterns corresponding to the letters of the Roman alphabet were stored in a long-term store trained using the MFT algorithm. The characters consisted of a simple 7×5 matrix. The network consisted of 35 visible units and 35 hidden units.

The result of the pilot simulation was that four or five of the letter patterns could be reliably buffered in a BAM-based short-term store. The procedure used was as follows:

1. A recall cue is imposed on the long-term store visible units.
2. The long-term store relaxes, giving a stable input and hidden vector.
3. The entire activation pattern evoked is stored in the short-term store using one-shot Hebbian learning.
4. Steps 1–3 are repeated for further items.

Catastrophic forgetting

Catastrophic forgetting is the unfortunate property that networks have of forgetting a previous set of learned patterns after being trained on a new set. The problem means that networks cannot be trained sequentially but must be presented with all the patterns they are likely to have to learn in one training set. There have been many recent papers on the problem and methods of reducing it (e.g. Hetherington 1990, McRae & Hetherington 1993). Most work on the problem has used back-propagation networks. We have demonstrated that the problem exists for mean field theory autoassociators, but can be made less serious by increasing the structural similarity within a pattern set (Levy & Bairaktaris 1992).

More recently, we have shown that using the hidden unit representations for old memories to help store new memories reduces catastrophic interference (Levy & Bairaktaris 1993). The method used is to train a set of patterns (*A*) using the normal mean field learning algorithm. The network evolves a set of weights and hidden unit activations that suffice to learn A. Then a second set of patterns (B) is trained in a somewhat different manner. Before training begins each pattern in B is clamped on the input layer and the hidden-unit activation patterns that result are saved. Then training begins with both input and hidden units clamped. The input unit patterns are thus augmented by their corresponding hidden-unit activations, and the net is trained as if it had no hidden units. Thus the hidden-unit representation from learning A is used to help train B. The metaphor behind this idea is that humans learn new information by relating it to past knowledge. In a trial simulation, two sets of patterns A and B were generated. Each set comprised ten binary vectors each comprising 20 input features. Structure was built into each set by means of an average correlation between the patterns within a set of 0.5. The interference error on set A caused by learning set B was reduced from 14% without using the hidden-unit representations to 4% when they are used. This method does not need to use the short-term component of our architecture, but, in a larger-scale demonstration, set B might be a small set buffered in STM and consolidated into LTM using the hidden-unit representations of a much larger A set representing past knowledge.

The dual-weight architecture described above differs from the additive and multiplicative systems described earlier by keeping the short- and long-term weights independent. This means that the mechanisms for STM and LTM can act separately but retains the advantage that they act on the same representations because the different kinds of weights share the same units. These properties extend the range of tools available to the cognitive modeller. It remains to be seen which type of multi-weight connection architecture is most suitable for modelling which cognitive phenomena. The following section speculates on areas where the use of this kind of modelling building block might prove fruitful.

Applications of multi-weight connections in cognitive modelling

This section speculates on the kind of cognitive phenomena that might be modelled using tools like our dual-store model. This includes both dual-weight mechanisms and dual-component mechanisms where the units comprising the different stores are distinct.

Priming and implicit memory

We have seen above that architectures with multi-weight connections are helpful in the modelling of priming phenomena. It is not hard to envisage how temporary weights can be used to bias a network into enhanced processing of more recent stimuli. Additive short-term weights captured the bias of subjects to use recent exemplars in the human grammar learning data modelled by Cleeremans & McClelland (1991). Plaut & Shallice (1993) used short-term weights to cause their network to "mistakenly" fall into a recent output attractor rather than the current target to model perseverative errors in optic aphasia.

Monsell (Monsell et al. 1992, Wheeldon & Monsell 1994) has recently suggested that priming effects in word recognition, where a prime interferes with the subsequent recognition of a lexical neighbour, might be modelled by changing the attractor structure in the recognition lexicon so that the successful recognition of the prime changes the shape of the basin of attraction for that lexical item in such a way that the network takes longer to recognize (fall into the attractor of) a lexical neighbour.

This might be achieved temporarily by the use of additive short-term weights. Alternatively, independent short-term weights might be used as a temporary short-term recognition buffer and account for this kind of data. This kind of buffer would act as a "working memory", not in the sense of the special-purpose speech or visuospatial stores described by Baddeley and his colleagues (e.g. Baddeley 1986) but a general mechanism for "processing memory", temporary storage within the computational machinery itself. This is strongly related to the procedural notion of short-term storage espoused by Crowder (1993) and others. Rather than using a separate short-term store, proceduralist models confound processing, learning and short-term storage.

An alternative general theoretical explanation for such priming phenomena might be preferred by those workers convinced of the power of the idea of implicit memory (e.g. Schacter 1987). They might prefer an explanation of priming that stressed the continual gradual alteration of *permanent* weights. There certainly is evidence of very long-term priming.

Other forms of priming, such as those in spoken word recognition mentioned above, appear to last much less long, but perhaps this is due to the continual flux of gradual permanent weight change that occurs in such a system.

One way of modelling this kind of explanation of priming would be to use dual-weight connections and include a consolidation mechanism for implicit memory. The computational properties of these alternative modelling frameworks need to be explored, and we intend to extend our model to allow this to be done.

Interactions between short- and long-term information

The most studied component in the working memory model proposed by Baddeley and his colleagues is the phonological loop. This buffer consists of a passive phonological store capable of accurately retaining about 2 seconds of speech and an articulatory rehearsal control process that refreshes the store, enabling it to store speech-based information for periods of several seconds. Since the capacity of the store is greater for words than non-words, it seems likely that the system does contact with long-term lexical information. Glasspool (see Ch. 1) describes an extension to Burgess & Hitch's connectionist model of the phonological loop that accounts for lexicality effects.

It seems likely that there are other influences on the phonological loop from long-term knowledge. Nelson (1992) reports effects of the semantic structure of word lists on subjects' working memory performance. Dual-component and dual-weight connectionist architectures offer the opportunity to build explicit computational models of this variety of interaction between short- and long-term stores.

Models of hippocampal function

A class of dual-store models not mentioned in our review is that of the interaction between putative hippocampal and cortical stores during learning in humans and animals.

We described above how our dual-store model can facilitate memory consolidation and partly alleviate the effects of catastrophic interference. Many human and animal studies (e.g. Cave & Squire 1992) support the idea that the hippocampal formation plays some role in new memory acquisition and consolidation. In this section we discuss possible ways in which our dual-store model can be used to simulate hippocampal function.

There are two major viewpoints in which hippocampal activity is related to memory consolidation processes. The first proposes that the hippocampus is a short-term store where novel memories are stored temporarily. At a later stage, possibly during REM sleep, these memories are transferred into LTM (McNaughton & Nadel 1990). The second viewpoint proposes that the hippocampus supports temporary associations between previously unassociated cortical areas of LTM, until permanent cortical-to-cortical associations can be established (Eichenbaum et al. 1992).

The architecture and dynamics of our dual-store system come to a significant degree in the realm of the first theory of hippocampal function. The BAM, fast-adapting component of our system is an autoassociative memory with limited storage capacity similar to the hippocampus (Gluck & Myers 1993). Furthermore, in Bairaktaris (1993) some anatomical correlates are established between the hippocampus and an optimally performing multilayer BAM network. The MFT component of our dual-store system functions as a slow-adapting LTM. Memories from the BAM component are transferred into the MFT component via a rehearsal process at a later stage. While a dual-store approach towards a model of the hippocampal function has undoubtedly the support of a major part of the neuroscience community there are still some unanswered questions.

It is fairly reasonable to assume that no experience is completely novel. On the one hand, any novel experience contains features which can be found in LTM. On the other hand, LTM traces always influence the way we perceive so-called novel experiences. In the unlikely situation where an event is completely novel we will most likely fail to perceive it. It is therefore reasonable to consider that a memory consolidation mechanism must be something more complicated than a mere device of rehearsal and storage of new memories in LTM. It must be a complex process where novel memories are truly integrated with existing LTM traces. Human LTM does not seem to be a collection of knowledge chunks completely dissociated from each other. Quite the opposite: it seems to be a continuum of knowledge with an extremely high degree of cross-reference and some form of hierarchical organization.

One may assume that STM traces stored in the hippocampus, include features from existing relevant LTM traces. Therefore, whatever complex integration processes are involved in memory consolidation happen before the memories reach the hippocampus. Later, when these hippocampal traces are to be transferred into LTM their representations contain all the necessary cues for the integration with the existing memories. Alternatively, one may claim that all the existing relevant LTM traces, together with the novel traces, are stored in the hippocampus and, of course, at consolidation time they completely overwrite their original LTM traces with the new material. Both ideas are worthy of further investigation.

In order to match the specifications of the second theory of hippocampal function our dual-store system can be modified to perform as a dual-weight system. In this case, both the MFT (LTM) and BAM (STM) components will share the same input and internal node representations. The fast adapting synapses of the BAM component will store temporary associations between the input and internal layers of the system until the slow-adapting synapses of the MFT component will establish permanent associations.

The single most convincing piece of evidence in favour of this approach comes not surprisingly from neurophysiology. Fast-induced long-term potentiation/depression (LTP/LTD) in the hippocampus (McNaughton & Nadel 1990) and in the cortex (Kirkwood & Bear 1994) can only last for a limited period of time and do not account for life-long lasting LTM effects. It is only new synaptic growth which could account for LTM at the cellular level. Since such growth takes a while to occur (Black et al. 1990), the hippocampus becomes an essential part of the mechanism for the formation of new memories, functioning as a temporary mapping device. In addition to synaptic growth, which can only account for LTM at a local level (synapses do not grow that much in length – rather, they become more dense in a particular area), gradual cortical remapping seems to be quite common in a number of areas in the brain (Pettet & Gilbert 1992, Recanzone et al. 1992), without any loss of functionality at any point in time.

If either or possibly both theories of hippocampal function prove correct the proposed dual-store system can be a useful simulation and analytical modelling tool.

Acknowledgements

We acknowledge the support of the Joint Council Initiative in HCI and Cognitive Science (grant no. SPG 9200496).

References

Ackley, D. H., G. E. Hinton, T. J. Sejnowski 1985. A learning algorithm for Boltzmann machines. *Cognitive Science* **9**, 147–69.

Anderson, J. R. 1976. *Language, memory and thought*. Hillsdale, New Jersey: Lawrence Erlbaum.

Baddeley, A. 1986. *Working memory*. Oxford: Oxford University Press.

Bairaktaris, D. 1990. A model of auto-associative memory that stores and retrieves, successfully, data regardless of their orthogonality, randomness or size. *Hawaii International Conference on System Sciences '90*. New York: IEEE Press.

Bairaktaris, D. 1993. Multi-layer bidirectional auto-associators. In *Computation and neural systems 1993*, F. H. Eekman & J. M. Bower (eds), 443–7. Dordrecht: Kluwer.

Black, I. B., E. DiCicco-Bloom, C. F. Dreyfus 1990. Nerve growth factor and the issue of mitosis in the nervous system. *Current Topics in Developmental Biology* **24**, 161–92.

Cave, C. B. & L. R. Squire 1992. Intact verbal and nonverbal short-term memory following damage to the human hippocampus. *Hippocampus* **2**, 151–64.

Cleeremans, A. & J. L. McClelland 1991. Learning the structure of event sequences. *Journal of Experimental Psychology: General* **120**, 235–53.

Crick, F. H. C. 1984. The function of the thalamic reticular complex: the search-light hypothesis. *National Academy of Sciences of the USA, Proceedings* **81**, 4586–90.

Crowder, R. G. 1993. Short-term memory: where do we stand? *Memory and Cognition* **21**, 142–5.

Eichenbaum, H., T. Otto, N. Cohen 1992. The hippocampus: what does it do? *Behavioral Neural Biology* **57**, 2–36.

Elman, J. L. 1990. Finding structure in time. *Cognitive Science* **14**, 179–211.

Galland, C. 1993. The limitations of deterministic Boltzmann machine learning. *Network* **4**, 355–79.

Gardner-Medwin, A. R. 1989. Doubly modifiable synapses: a model of short- and long-term auto-associative memory. *Royal Society B, Proceedings* **238**, 137–54.

Gluck, M. A. & C. E. Myers 1993. Hippocampal mediation of stimulus representation: a computational theory. *Hippocampus* **3**, 491–516.

Goebel, R. 1990. Binding, episodic short-term memory, and selective attention, or: why are PDP models poor at symbol manipulation. In *1990 Connectionist Summer School, Proceedings*, D. Touretsky, J. Elman, G. Hinton (eds), 253–64. Los Altos, Calif.: Morgan Kaufmann.

Hartman, E. 1991. A high storage capacity neural network content addressable memory. *Connection Science* **2**, 315–34.

Hebb, D. O. 1949. *The organization of behavior*. New York: John Wiley.

Hetherington, P. A. 1990. *The sequential learning problem in connectionist networks*. Master's thesis, Department of Psychology, McGill University, Montreal.

Hinton, G. E. 1989. *Deterministic Boltzmann learning performs steepest descent in weight-space*. Department of Computer Science, University of Toronto, Technical Report CRG-TR-89-1.

Hinton, G. E. & D. C. Plaut 1987. Using fast weights to deblur old memories. *9th Annual Conference of the Cognitive Science Society, Proceedings*, 177–86. Hillsdale, New Jersey: Lawrence Erlbaum.

Kirkwood, A. & M. F. Bear 1994. Hebbian synapses in the visual cortex. *Journal of Neuroscience* **14**, 1634–45.

Kosko, B. 1988. Bidirectional associative memories. *IEEE Transactions on Systems, Man and Cybernetics* **18**, 49–60.

Levy, J. & D. Bairaktaris 1991. A model of the interaction between long and short-term memory. *International Joint Conference on Neural Networks, Proceedings*, 1741–6. Piscataway, New Jersey: IEEE.

Levy, J. & D. Bairaktaris 1992. Sequential learning in mean field autoassociators. In *International Conference on Artificial Neural Networks*, I. Alexander & J. Taylor (eds), 1331–4. New York: Elsevier.

Levy, J. & D. Bairaktaris 1993. Memory and consolidation in mean field theory auto-associators. *1993 World Congress on Neural Networks, Proceedings*, vol. 2, 216–19. Hillsdale, New Jersey: Lawrence Erlbaum.

Levy, J. & K. Stenning 1989. Parallel distributed processing simulations of attribute binding in human memory. In *Neural networks: from models to applications,* L. Personnaz & G. Dreyfus (eds) 26–35. Paris: IDSET.

McClelland, J. L. & J. L. Elman 1986. The TRACE model of speech perception. *Cognitive Psychology* **18**, 1–86.

McClelland, J. L. & A. H. Kawamoto 1986. Mechanisms of sentence processing: Assigning roles to constituents. In *Parallel distributed processing: explorations in the microstructure of cognition*, vol. 2. *Psychological and biological models*, J. L. McClelland & D. E. Rumelhart (eds), 272–326. Cambridge, Mass.: MIT Press.

McClelland, J. L. & D. E. Rumelhart 1986. A distributed model of human learning and memory. In *Parallel distributed processing: explorations in the microstructure of cognition*, vol. 2. *Psychological and biological models*, J. L. McClelland & D. E. Rumelhart (eds), 170–215. Cambridge, Mass.: MIT Press.

McEliece, R. J., E. C. Posner, E. R. Rodemich, S. S. Venkatesh 1987. The capacity of the Hopfield associative memory. *IEEE Transactions on Information Theory* **33**, 461–82.

McNaughton, B. L. & L. Nadel 1990. Hebb–Marr networks and the neurobiological representation of action in space. In *Neuroscience and connectionist theory*, M. A. Gluck & D. E. Rumelhart (eds), 1–63. Hillsdale, New Jersey: Lawrence Erlbaum.

McRae, K. & P. A. Hetherington 1993. Catastrophic interference is eliminated in pretrained networks. *15th Annual Conference of the Cognitive Science Society, Proceedings*, 723–8. Hillsdale, New Jersey: Lawrence Erlbaum.

Monsell, S., G. H. Matthews, D. C. Miller 1992. Repetition of lexicalisation across languages – a further test of the locus of priming. *Quarterly Journal of Experimental Psychology: Human Experimental Psychology* **44**, 763–83.

Morton, J. 1969. The interaction of information in word recognition. *Psychological Review* **76**, 165–78.

Nelson, A. W. R. 1992. *The process of changing reference in simple texts*. PhD thesis, Centre for Cognitive Science, University of Edinburgh.

Peterson, C. & E. Hartman 1989. Explorations of the mean field theory learning algorithm. *Neural Networks* **2**, 475–94.

Pettet, M. W. & C. D. Gilbert 1992. Dynamic changes in receptive-field size in cat primary visual cortex. *National Academy of Sciences of the USA, Proceedings* **89**, 8366–70.

Plaut, D. C. & T. Shallice 1993. Perseverative and semantic influences on visual object naming errors in optic aphasia – a connectionist account. *Journal of Cognitive Neuroscience* **5**, 89–117.

Recanzone, G. H., M. M. Merzenich, C. E. Schreiner 1992. Changes in the distributed temporal response properties of SI cortical neurons reflect improvements in performance on a temporally based tactile discrimination task. *Journal of Neurophysiology* **67**, 1071–91.

Schacter, D. L. 1987. Implicit memory: history and current status. *Journal of Experimental Psychology: Learning, Memory and Cognition* **13**, 501–18.

Shastri, L. & V. Ajjanagadde 1989. *A connectionist system for rule based reasoning with multi-place predicates and variables*. Department of Computer and Information Science, School of Engineering and Applied Science, Philadelphia, Technical Report MS-CIS-89-06.

Stenning, K., M. Shepherd, J. Levy 1988. On the construction of representations for individuals from descriptions in text. *Language and Cognitive Processes* **2**, 129–64.

Treisman, A. M. & G. Gelade 1980. A feature-integration theory of attention. *Cognitive Psychology* **12**, 97–136.

Wheeldon, L. R. & S. Monsell 1994. Inhibition of spoken word production by priming a semantic competitor. *Journal of Memory and Language* **33**, 332–56.

READING

John A. Bullinaria

There is currently a lively debate concerning the nature of the mental processes underlying the act of reading aloud (i.e. text-to-phoneme conversion). The traditional position is that reading can only be described by a dual-route model with separate phonemic and semantic routes. The phonemic route, which consists of a series of grapheme-to-phoneme conversion (GPC) rules, appears necessary for the pronunciation of unfamiliar words or pronounceable non-words. The semantic/lexical route is thought to be necessary to produce the phonemes for irregular or exception words which do not follow the main GPC rules and to provide a contact point for traditional natural language processing and semantics. The standard "box and arrow" representation of this arrangement (adapted from Coltheart et al. 1993) is shown in Figure II.1. An alternative view is that a single-route system may be sufficient, and several explicit neural network models of reading have recently been constructed that are able to learn all the words (including the irregular words) in their training data and are also able to read new words or non-words with accuracies comparable to human subjects. Although it seems unlikely, at this stage, that a single-route model will be able to account for all aspects of human reading abilities (e.g. Coltheart et al. 1993), there is considerable evidence that the two routes of the traditional dual-route model cannot be totally independent (e.g. Humphreys & Evett 1985).

There are several directions from which the problem of modelling and understanding reading can be attacked. The approach agreed upon and adopted in the three chapters in this section is to construct explicit connectionist models of text-to-phoneme conversion and then examine how well these can fit in with more complete models of reading.

The first thing that has to be determined for any model of text-to-phoneme conversion is the choice of representation to use for the inputs (letters) and outputs (phonemes). It is clear that the same letters in the same positions in different words can be pronounced differently, resulting in a complicated hierarchy of

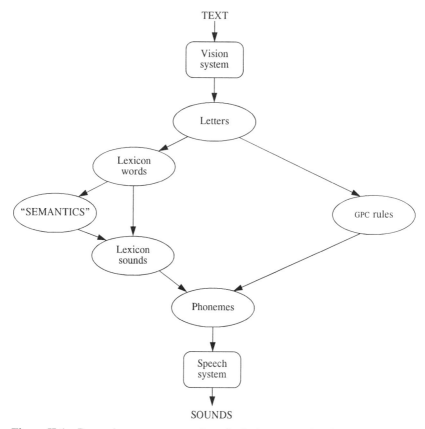

Figure II.1   Box and arrow representation of a dual-route model of reading.

rules, subrules and exceptions. This, of course, is precisely the kind of thing that neural networks are generally good at learning, but finding suitable input and output representations that will enable the networks to do this for reading has proven difficult. If there were a one-to-one correspondence between the letters and phonemes of every word, it would be fairly easy to set up a neural network to map from letter strings to phoneme strings. However, the relation is generally many-to-many: in English, up to four letters can map to one phoneme, e.g. "eigh" → /A/ in "eight", and one letter can map to two phonemes, e.g. "x" → /ks/ in "box". This leads to the so-called *alignment problem*: given a set of phonemes corresponding to a set of letters, it is not obvious how the letters and phonemes should line up, i.e. which subsets of the phonemes relate to which subsets of the letters. Another important issue, often referred to as the *recognition problem*, is that the same letters in different word positions should be recognized as being the same. If they are not, the result is not only a matter of computational inefficiency, but is also likely to result in a system with poor generalization ability.

One of the first successful neural network systems to get around these representational problems was NETtalk (Sejnowski & Rosenberg 1987). This used a moving window approach whereby a window was moved across the stream of input words and a single output phoneme then corresponded to the letter in the centre of the input window in the context of the other letters within that window. This proved to be a good solution to the recognition problem. Their solution of the alignment problem was simply to preprocess the training data by inserting special continuation (i.e. no output) characters into the phoneme strings to align the letters and phonemes. The overall performance of the system was good but, for many, the degree of preprocessing involved was considered unacceptable for cognitive modelling since it involved the system designer doing a significant amount of the work the network should be doing for itself.

This led Seidenberg & McClelland (1989) to develop a more sophisticated model that used a system of distributed Wickelfeatures in which each letter and phoneme string is split into sets of triples of characters (i.e. letters or phonemes) in the same way as for earlier models of past-tense acquisition (Rumelhart & McClelland 1986). This certainly bypassed the problem of aligning the letters and phonemes in the training data and solved the recognition problem (as will be discussed further in Ch. 8). Unfortunately, it also restricted their model to monosyllabic words, made the interpretation of the network's output difficult and presented enormous problems in understanding the nature of the internal representations. It has also been shown (Besner et al. 1990) that the model's generalization performance, i.e. its ability to pronounce new words or non-words, is unacceptably poor.

The three chapters in this section discuss a range of different approaches for improving on these original connectionist models of reading. Bob Damper, in Chapter 7, begins by discussing the traditional rule-based approach to text-to-phoneme conversion and then goes on to review the wide range of self-learning systems that have been developed for this task. These systems include generate-and-test rule induction, rule induction by clustering, decision tree induction, Markov modelling, synthesis by analogy and various neural networks. This chapter provides an excellent overview of the various techniques available for modelling reading and also considers the prospects for neural network implementations of the systems that are not explicitly connectionist (e.g. decision tree induction and Markov modelling).

David Plaut, Jay McClelland and Mark Seidenberg, in Chapter 8, discuss the underlying problems with the Seidenberg and McClelland model and formulate new input/output representations to solve the alignment and recognition problems based on the possible initial consonant, vowel and final consonant clusters in uninflected monosyllables. These representations are used, first in a standard feedforward network and then in a refixation network (with recurrent connections), to produce models of reading aloud that can handle both the regular and exception words in the training data and yet are still able to produce near human level generalization performance.

Finally, in Chapter 9, John Bullinaria argues that we should return to the conceptually simpler NETtalk system and shows how it can be modified to obviate the need to preprocess the training data. This model achieves perfect performance on the training data (including many exception words) and, like the models of Plaut et al., also exhibits near human level generalization performance. It gives simulated naming latencies with many (though not all) of the frequency and type effects found in humans, and under appropriate conditions it shows symptoms similar to developmental and acquired surface dyslexia.

Together, the three chapters in this section provide a complete survey of the state of the art in models of the direct (non-semantic) route to reading aloud.

References

Besner, D., L. Twilley, R. S. McCann, K. Seergobin 1990. On the connection between connectionism and data: are a few words necessary? *Psychological Review* **97**, 432–46.

Coltheart, M., B. Curtis, P. Atkins, M. Haller 1993. Models of reading aloud: dual-route and parallel-distributed-processing approaches. *Psychological Review* **100**, 589–608.

Humphreys, G. W. & L. J. Evett 1985. Are there independent lexical and nonlexical routes in word processing? An evaluation of dual-route theory of reading. *The Behavioral and Brain Sciences* **8**, 689–740.

Rumelhart, D. E. & J. L. McClelland 1986. On learning the past tenses of English verbs. In *Parallel distributed processing: explorations in the microstructure of cognition*, vol. 2. *Physiological and biological models*, D. E. Rumelhart & J. L. McClelland (eds), 216–271. Cambridge, Mass.: MIT Press.

Seidenberg, M. S. & J. L. McClelland 1989. A distributed, developmental model of word recognition and naming. *Psychological Review* **96**, 523–68.

Sejnowski, T. J. & C. R. Rosenberg 1987. Parallel networks that learn to pronounce English text. *Complex Systems* **1**, 145–68.

CHAPTER 7

Self-learning and connectionist approaches to text–phoneme conversion

Robert I. Damper

Introduction

The automatic derivation of the pronunciation of an English word from its spelling is a difficult problem of some practical significance. Most often, pronunciation will be specified as an idealized phonemic "baseform" (which may or may not include stress markers) so that we refer here to this process as *text–phoneme conversion*.

Interest in the conversion problem comes from two rather different points of view. First, interactive computer systems featuring speech output generally have a requirement for textual input. It makes sense to translate this input to some intermediate representation (e.g. phonemic) closer to the actual sounds to be synthesized. Secondly, the conversion process is a key component of many models of human language processing – particularly reading aloud. The earliest attempts at automated conversion (Ainsworth 1973) were made by speech scientists concerned with the technology of synthesis. These avoided dictionary matching (because of limitations of computer memory), and were based on a set of letter-to-phoneme rules, manually written by expert phoneticians, to capture the regularities which clearly exist in spelling–sound correspondence. This formalism remains popular in both fields of endeavour. For instance, present-day commercial text-to-speech (TTS) systems achieve acceptable performance by combining dictionary matching with the use of rules for the translation of words not present in the dictionary. Apart from this commonality in the use of a rule-based formalism, and given that essentially the same process is under study, it is perhaps surprising that the two fields of speech synthesis and psychological modelling have not informed one another more than they have so far done.

In both areas, however, there is currently much interest in inductive learning of the regularities of text–phoneme correspondence. If the automatic conversion process is also *self-learning*, then this should reduce the manual effort necessary to build a TTS system. Further, since the human ability to read is acquired rather than innate, a self-learning converter is better able to serve as a model of language processing.

Recently, connectionism has emerged as an influential paradigm in psychology, as well as figuring prominently in speech technology, largely stemming from the discovery of powerful self-learning algorithms such as error back-propagation (Rumelhart et al. 1986). From the psychological perspective, the parallel distributed nature of a connectionist model lends it a good deal more plausibility than a set of rules since, unlike the latter, we can readily see how the former might be implemented in "brainware". Quite apart from any concern with TTS systems or the modelling of human language processing, text–phoneme conversion has proved a popular application for workers interested in neural computation for its own sake. Because the mapping between text and phonemes is complex, the problem provides a good test for the power of neural solutions, and has become something of a standard in this respect.

The literature on connectionist and self-learning approaches to text–phoneme conversion is considerable but widely dispersed. In particular, as stated above, work in the area of speech synthesis has not always been informed by relevant work in psychological modelling, and vice versa. In this chapter, we present a critical, unifying review of this literature which we believe to be the first such to appear. As well as collecting the various pieces of work together so as to render them more accessible, we also categorise approaches and highlight relations between them. This is done to form a framework for the following two chapters, as well as in introduction to them. Not all the approaches dealt with are obviously connectionist in inspiration; they are included for completeness and because, in some cases, a rather obvious connectionist implementation is possible – if not necessarily parsimonious or efficient.

The chapter is structured as follows. The next section outlines the traditional, rule-based approach to text–phoneme conversion. Subsequently, a variety of automatic discovery techniques are treated, namely generate-and-test rule induction, rule induction by clustering, decision tree induction, Markov modelling, back-propagation networks in general, synthesis-by-analogy (including memory-based reasoning) and syntactic neural networks. Where the techniques are not neural in inspiration or in origin (e.g. decision tree induction and Markov modelling), prospects for neural implementation are considered. The final section summarizes.

Principles of rule-based translation

The synthesis of unrestricted-vocabulary speech from orthographic text is, in the words of Klatt (1987: 781), "a new technology with a rapidly changing set of capabilities and potential applications." Present-day TTS systems typically use a large dictionary of pronunciations in conjunction with a set of letter-to-sound rules to produce a phonemic transcription of the text (Allen 1976, Allen et al. 1987). The rules are invoked to transcribe words for which no dictionary match is found.

For English at least, the text–phoneme conversion process is far from trivial, reflecting the many complex historical influences on the spelling system (Venezky 1965, Scragg 1975). Indeed, Abercrombie (1981: 209) describes English orthography as ". . . one of the least successful applications of the Roman alphabet." Some of its well known vagaries are that letter combinations ("ch", "gh", "ll", "ea") frequently act as a unit (a "grapheme") signalling a single phoneme, a single letter occasionally corresponds to more than one phoneme (as in ("six", /sIks/)), pronunciation can depend upon word class (e.g. "convict", "subject") and there can be non-contiguous "markings" as with the final, mute "e" of ("make", /meIk/) (Wijk 1966, Venezky 1970).

Ever since the pioneering work of Ainsworth (1973), text–phoneme conversion for speech synthesis has (in the absence of a dictionary-derived pronunciation) traditionally used a context-dependent rewrite rule formalism of the sort favoured in generative phonology (e.g. Chomsky & Halle 1968: 14). Such rules can also be straightforwardly cast in the IF. . .THEN form commonly employed in expert systems technology. The notion underlying the conversion process is that it is possible to arrive at a translation for any "letter unit" (grapheme) – and thereby for a whole word – provided enough contextual information is available. Rules are of the form

$$[A] \, B \, [C] \rightarrow D \qquad (7.1)$$

which states that letter substring B with the left context A and right context C rewrites to phoneme substring D. Influential rule sets in this tradition are those of Elovitz et al. (1976) and Hunnicutt (1976).

Because of the complexities of English spelling-to-sound correspondence, more than one rule generally applies at each stage of transcription. The potential conflicts which arise are resolved by maintaining the rules in a set of sublists, grouped by (initial) letter and with each sublist ordered by specificity. Typically, the most specific rule is at the top and most general at the bottom. In the Elovitz et al. rules, transcription is a one-pass, left-to-right process. For the particular target letter (i.e. the initial letter of the substring currently under consideration), the appropriate sublist is searched from top to bottom until a match is found. This rule is then *fired*, the linear search terminated, and the next untranscribed letter taken as the target. The last rule in each sublist is a context-independent *default* for the target letter, which is fired in the case that no other, more specific rule applies. Transcription is rather more complex with the Hunnicutt rules, which use three passes; also, processing can be in either direction. First, affixes are stripped, then consonants are converted, and finally vowels and affixes are transcribed. This procedure allows converted (phonemic) strings which are highly dependable (i.e. for the consonants) to be used as context in the more complex rules for vowel and affix transcription. Again, rules are ordered to facilitate conflict resolution; however, the determination of an appropriate ordering is often problematic (Carlson et al. 1990: 275). After conversion of letters to phonemes, lexical stress can be added using a similar context-dependent formalism; see Church (1985) for a review.

In spite of obvious commonality between the computational process of TTS conversion and the psychological process of reading aloud, there has been very little interaction between techniques for the former and models of the latter. As we have just seen in the case of speech synthesis, the standard approach utilizes a pronouncing dictionary for known words and a set of general-purpose, context-dependent translation (CDT) rules which is invoked if the input word is not in the system's dictionary. In the case of reading aloud, the standard model also involves two routes to pronunciation. So-called *dual-route theory* (e.g. Forster & Chambers 1973, Coltheart 1978) posits a lexical route for the pronunciation of known words and a parallel, simultaneously-activated route utilizing abstract grapheme–phoneme conversion (GPC) rules for the pronunciation of unknown, or "novel", words (Fig. 7.1). Unlike the sequential application of dictionary matching followed by contingent rule-based translation in a TTS system, the two

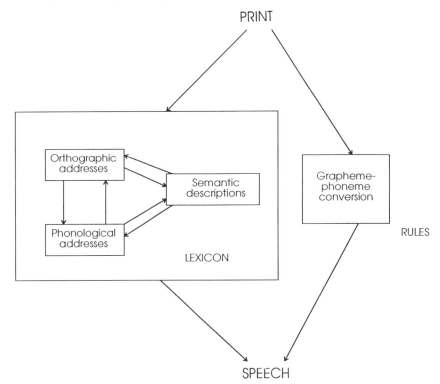

Figure 7.1 Schematic of the dual-route model for the production of pronunciation from print. (After Humphreys & Evett 1985.)

routes are usually conceived as operating essentially in parallel in the psychological model. Arguments for dual-route theory are based on the ability to pronounce pseudowords, latency difference effects between regular and exception words, and apparent double dissociation between the two routes in dyslexia (see Humphreys & Evett (1985) for an extensive review and discussion).

It follows from earlier discussion of the complexities of letter–sound correspondence in English that the task of writing an adequate set of CDT rules is labour-intensive and requires detailed, expert knowledge of the particular language. Once the rule set reaches a certain size, it becomes difficult to modify in the light of errors, because of the potential for interaction between rules. Problems such as this have led to the production of rule compilers (Carlson & Granström 1975, Hertz et al. 1985, Van Leeuwen 1989) which aim to give automatic assistance to rule developers for speech synthesis systems. Clearly, the necessity for such devices reduces the psychological plausibility of the rule-based approach greatly: it is difficult to imagine humans employing the mechanisms embodied in rule compilers for language acquisition. Recently, however, interesting advances have been made in applying self-learning – or "automatic discovery" – techniques to the problem of mapping text to phonemes. Not only does this offer the promise that much of the labour in traditional rule generation might be avoided, but the techniques are much more reasonable from a psychological point of view as they are well suited to implementation in "brainware". In this approach, transcribed texts or entries in a machine-readable pronouncing dictionary are treated as sources of training data from which the system learns generalizations useful for the transcription of seen and unseen words alike.

Review of self-learning systems

In this section, we review previous attempts to generate automatically systems capable of performing text–phoneme transcription.

Generate-and-test rule induction

Early work applying self-learning techniques to text–phoneme conversion (Oakey & Cawthorne 1981, Klatt & Shipman 1982) retained the CDT rule formalism and aimed to infer a rule set by a generate-and-test process. These attempts were not inspired by connectionist ideas: they are included here for completeness.

Oakey & Cawthorne started from a context-independent base set (the default rules) and worked through the dictionary, generating pronunciations for each word and comparing this with the known, correct version. Any difference between generated and correct pronunciation was then used to create a new special-purpose rule to cater for the mispronunciation. This creation process requires an *alignment* of text and phonemes, i.e. we need to know which letter(s) to associate with which phoneme. Oakey & Cawthorne's technique for this appears rather *ad hoc*, being based on a "look-ahead" heuristic. Subsequently, any such special-purpose rules which were sufficiently similar were generalized by combination to produce somewhat more general rules, which were themselves candidates for combination. Like the alignment technique, the combination methodology looks fraught with problems. Typical questions which arise during generate-and-test rule inferencing include: Are the problematic rules too general or too specific? Should the left context, the right context or both be adjusted? Should the target substring be increased or reduced in length?

Klatt & Shipman (1982), for instance, attempted to avoid such problems by selecting "only the most popular of each set of conflicting rules"; these were then formed into a decision tree to facilitate rapid translation. Although they do not give performance figures or make any subsequent report on their work, Klatt later states (1987) that an error rate of 7% was obtained. The rules were inferred from 10 000 words obtained by randomly dividing a 20 000-word pronouncing dictionary in two, and the system was tested on the unseen half of the dictionary. Such a high level of performance is remarkable in view of the gross way of treating conflicting rules.

Oakey & Cawthorne tested their set of rules on the training data after deleting all "basic" rules, i.e. the created rules which (by their very nature) were entirely specialized to the mispronunciations. Of course, if this had not been done, performance on the training data would have been 100%. In this way, the generalization power of the technique was under test (although not exactly on "unseen" words). On the Ladybird Key Word texts (477 words) for beginning readers, scores of 84% phonemes correct and 59% words correct were obtained. Corresponding figures for the Elovitz et al. rules, manually Anglicized, were 91% and 84%, respectively. Clearly, this is not a very demanding data set. On the 1015 words in their pronouncing dictionary with the initial letter *a*, however, the relevant figures were 74% phonemes correct and 21% words correct. This was actually superior to the respective figures of 64% and 16% obtained with the modified Elovitz et al. rules.

Van Coile (1990) describes a generate-and-test rule induction process which appears to be a significant improvement over that of Oakey & Cawthorne. Rules are generated for one letter at a time. Starting with aligned data, the training set is searched for occurrences of that letter, and each occurrence furnishes a pronunciation example. The left and right contexts are subject to an upper limit of three letters each. The phoneme that most frequently occurs in the examples is taken as the right-hand side of the (context-independent) default rule. Initially, pronunciation examples handled correctly by the default are marked *realized* and the remainder are marked *not yet realized*. The creation of new rules then proceeds iteratively. All possible different letter contexts are generated for the *not yet realized* examples. Next, one phoneme is associated with each generated context and the rule (i.e. context plus phoneme) that most improves the performance of the current rule set is added to the current set. The examples are then scanned again and either marked *realized* or *not yet realized*. (Van Coile notes that it is possible for *realized* examples to be re-marked *not yet realized*.)

The induction process is kept tractable by considering contexts of one fixed length, C, at a time in conjunction with two other parameters, M and T. M is the maximum context length that occurs in the previously determined rules and T is a performance-increase threshold which must be exceeded for a newly generated rule to be added to the current set. After learning on a random selection of 7000 of the 10 000 most common Dutch words, which generated an average of 14 rules per letter, the method was evaluated on the remaining 3000 words. Performance "at the grapheme level was better than 96%", which, ignoring stress, corresponds to roughly 82% correct at word level. Van Coile concedes that this is inferior to the performance obtained from traditionally developed rule sets (but see discussion above concerning the difficulty of assessing rules alone) yet believes the inductive technique is "very useful to obtain a good initial rule set for further manual development".

A question worth posing is: why does Van Coile's generate-and-test approach to rule induction produce so much better results than Oakey & Cawthorne's initial attempt? It seems likely that the key factors are the use of a principled alignment technique based on the Viterbi algorithm (see below), together with the strategy of processing one letter at a time across the whole training set, rather than processing one word at a time. It is difficult to say in quantitative terms how much easier the transcription task is for Dutch than for English.

Rule induction by clustering

Wolff (1984) describes a method of inducing CDT rules based entirely on clustering. While he calls his formalism "context-free" (Wolff 1984: 12) the end-result is a *concatenation* of context-independent rules, which amount to context-dependent rules.

The method first searches the dictionary (in multiple passes) looking for common clusters of contiguous symbols "using frequency and size of cluster as a guide to which amongst the manifold alternatives are best". Clustering occurs not only within the separate domains of letters (denoted L) and of phonemes (denoted P), but across domains also. This demands an extension of the concept of *contiguity* to the cross-domain (LP) situation. Wolff's solution is to parse the orthographic and phonemic representations "in step", and to consider as contiguous those elements (i.e. single letters, single phonemes or clusters) which are identified in sequence.

Figure 7.2 illustrates this for the example word ("this", /ðIs/), where all contiguities are shown as connecting lines. Initially then (parse 1), the letter "t" is considered contiguous not only with the letter "h" in the L domain but with phoneme /ð/ in the P domain also. Imagine that after the first pass through the dictionary, the L cluster "th" is identified. At parse 2, "th" is considered to be contiguous both with "i" and /ð/. Similarly, both the letter "i" and phoneme /I/

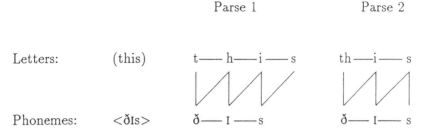

Figure 7.2 Contiguous symbol clusters in the letter (*L*) and phoneme (*P*) domains. Contiguities, within domains and between them, are shown with connecting lines. After (hypothetical) parse 1, the letter cluster "th" is recognized as a common substring, allowing the *LP* cluster (th)<ð> to be identified at parse 2. (After Wolff 1984.)

are considered to "follow" /ð/. In this way, Wolff aims to build up *LP* clusters such as (th)<ð>, which function as context-independent rules or, in Wolff's terminology, "basic" rules. Note how the clustering technique achieves an implicit alignment of text and phonemes, rather than requiring an explicit, prior alignment.

Clearly, the set of "basic rules" (being context-independent) will in general be ambiguous. For instance, both (th)<ð> and (th)<θ> are likely to exist, and there is no way of telling (without further analysis of the training data) which should be used in translating a given occurrence of "th". Wolff uses a method of disambiguation which is a variant of the clustering process. Having first identified ambiguous rules, contiguous pairs of such rules are clustered to produce new (context-dependent) "complex" rules. This should continue iteratively until no ambiguity remains in the rule set.

In the implementation reported, only one "rule" (cluster) is formed for each pass through the dictionary – clearly a very inefficient scheme. Unfortunately too, due to a software bug, Wolff was unable to run his program to completion. Nonetheless, interesting results were obtained. With a dictionary of 800 (most-frequent) words, the first *LP* cluster formed was (t)<t>. Other early *LP* clusters to emerge were (er)<ə> and (ng)<ŋ>. Since a complete set of "basic" rules could not be obtained automatically, the disambiguation process was tested on 252 hand-compiled "basic" rules. "Complex" rules were automatically formed up to an (arbitrary) limit of 200 in number, and their inclusion shown to reduce the size of the ambiguity set. Given the promise of this approach, it is a pity that a complete working system was never achieved.

Although Wolff's work was not obviously inspired by connectionism, clustering is a fundamental property of many unsupervised neural schemes so that a connectionist implementation should not be difficult.

Decision tree induction

Segre et al. (1983) describe a system using "two transcription rule sets which are machine generated over a set of sample transcriptions". The first operates on words and produces stress assignments; the second subsequently maps a letter "cluster" (with known stress value) onto its corresponding phoneme string. The approach of assigning lexical stress first is an interesting departure from the usual procedure. These authors base the rule inference process on machine-learning techniques like that of Quinlan (1979, 1990), which generate decision trees based on sets of examples and their classifications, in terms of an adequate set of *features*. Although they give a step-by-step description of the inference algorithm as applied to the generation of the stress rules, the learning of phonemic transcription rules is effectively ignored (beyond saying it is "similar"). In particular, the way the letter clusters are formed is not stated. Disappointingly, no performance data are given.

Lucassen & Mercer (1984), however, describe what was arguably the first self-learning technique to be theoretically well founded and to achieve a (quantified) measure of success. Accordingly, it has materially influenced later approaches, and so we describe it in some detail here.

Lucassen & Mercer view the transcription process as a transformation from a word's spelling, s, to its phonemic baseform, β, via a noisy channel. The term "phonemic baseform" describes the (idealized) word pronunciation as might be found in a pronouncing dictionary. It is used by Lucassen & Mercer in distinction to "pronunciation", which, in their terminology, is the phoneme or phonemes corresponding to a single letter. The current (target) letter has a context of letters to the left and right and of phonemes to the left; all these symbols together constitute the *channel context*.

The *features* of the model determine a partition either of the set of letters, L, or of the set of phonemes, P. The pronunciation π of the current letter is thus a string which is an element of the power set P^*. The authors point out the similarity of this formalism to a set of context-dependent rules (except that probabilities are involved here).

Lucassen & Mercer aim then to construct a vector, h, of binary features for channel contexts consisting of the current letter L_0, the four letters to the left and the four letters to the right of L_0 ($L_{-4}, \ldots, L_4$), and the three phonemes to the left of L_0 ($P_{-3}, \ldots, P_{-1}$). Note the requirement for text and phonemes to be aligned since the three phonemes to the left of L_0 must be known. They assume that the best binary feature for one of these symbols is that which maximizes the mutual information between features and pronunciations, and present a (suboptimal) algorithm to construct this. Given the best binary feature, another (presumed close to the next best) can be computed and included with the original to produce a two-element feature vector which effects a four-way partition of the relevant set. This feature inclusion proceeds until a complete partition is achieved. Altogether, for their lexical data, a six-element vector is required to give a complete partition of L, and an eight-element vector for P. The concatenation of the feature vectors for $L_{-4}, \ldots, L_4, P_{-3}, \ldots, P_{-1}$ corresponds to the 78-element feature vector h, for the entire channel context. At this stage, the automatically selected features are used to build a decision tree which determines the order in which the bits of h, should be examined, and how many bits should be examined, before selecting a pronunciation. To each leaf of the resulting decision tree is attached a probability distribution over the pronunciations, evaluated as the weighted sum of *a posteriori* maximum likelihood estimates at each node on the path from root to leaf.

After training on their lexicon, the system was tested on 194 words chosen at random from IBM's office correspondence database having a vocabulary of 5000 words. Of 1396 phonemes, 1308 (or 94%) were correctly transcribed. Interestingly, 47 of the 88 errors (53%) were reported to be errors of vowel stress placement. The authors do not state how many of the test words, if any, were absent from the training data. It is unfortunate that testing was on such a small data set.

The decision-tree induction approach is not connectionist in origin or inspiration. However, many workers have considered neural implementations of decision trees, or the combination of the two approaches (e.g. Stromberg et al. 1991, Rahim 1994).

Markov model techniques

Like many subsequent approaches, the Lucassen & Mercer methodology is dependent on having available aligned (s, β) word pairs, in order that the i index of P_i can be related to L_0. Alignment was achieved by recognizing that their translation formalism was a kind of hidden Markov model (HMM), so that standard algorithms could be exploited. In fact, they used for alignment a much simplified model of the spelling-to-baseform channel in which the channel "context" consisted solely of the current letter. The parameters of this model were optimized using the forward–backward algorithm (Baum 1972) and a dynamic-programming technique applied to find the most probable alignment of letters with phonemes.

Thus, Lucassen & Mercer faced a "bootstrapping" problem of alignment: to align text and phonemes effectively, as a first step in inferring correspondences between the two domains, one needs a set of correspondences (explicit or implicit) to start with. Indeed, the requirement for an explicit correspondence set is central to later alignment techniques, such as that of Lawrence & Kaye (1986) which has become a standard. In principle, for a technique to be truly self-learning, any necessity for a *prior* alignment phase (i.e. using explicit, precompiled correspondences) should be avoided – a point which is taken up later by Bullinaria (see Ch. 9).

The use by Lucassen & Mercer of the HMM techniques which were then starting to dominate speech recognition has inspired a small number of more recent attempts to employ an HMM formalism for text–phoneme conversion.

Van Coile (1990), whose rule induction work is described above, has used a hidden Markov phoneme model in conjunction with the Viterbi algorithm (Viterbi 1967, Forney 1973) to align the orthographic and phonemic representations of words prior to a rule induction phase. When trained on the 10 000 most-frequent Dutch words and using 1/72 (see below) as the initial output probability for all phonemic symbols, 98.5% correct alignment of words results (relative to a manual alignment), corresponding to better than 99% correct phoneme alignment.

Van Coile's program supports 72 different output (phoneme) symbols. The use of an equiprobable initial output distribution, while having the strong virtue of minimizing the assumptions made, might be expected to lead to problems with training. By contrast, initial transition probabilities appear to have been decided on the basis of intuition. They are detailed in the paper. The need to decide on some specific values for the initial probabilities is, of course, just another instance of the "bootstrapping" problem. This fact becomes plain when Van Coile also reports alignment results using output probabilities based on "some very simple observations about the correspondence between orthographic and phonetic representations in Dutch." In general, these output probabilities gave superior results to the equiprobable values for smaller training sets, but the advantage disappeared as the size of the training set increased.

Rather than using HMM techniques for prior alignment only, Parfitt & Sharman (1991) extend the formalism to give a complete, bidirectional (text–phoneme and phoneme–text) model. Thus, in the case of predicting pronunciation from spelling, the orthography is seen as the observed sequence of output symbols emitted by the model as it makes transitions between its hidden, phonemic states. In its present form, each state corresponds to a single phoneme; thus, the Markov property dictates a simple bigram model. Parfitt & Sharman point out, however, that the formalism is trivially expandable to higher-order n-grams. The problem then is to find the (phonemic) state sequence which accounts for the observed (orthographic) output with greatest probability; this problem is solved using the Viterbi algorithm. Once obtained, initial estimates of the model's parameters (phoneme transition probabilities and output probabilities) can be subsequently re-estimated using the forward–backward algorithm. While the initial transition probabilities were simply estimated by frequency counts using a (frequency-weighted) pronouncing dictionary, it is not so simple to estimate the initial output probabilities. Two possibilities are to assume equiprobability (as did Van Coile) or to make frequency counts of word pairs after orthographic–phonemic alignment using an algorithm like that of Lawrence & Kaye (1986). Parfitt & Sharman, however, adopt initially an intermediate position of assuming that each phoneme in a word's "phonetic" form has an equally likely chance of generating any of the letters in the orthographic form, although subsequently they do employ a dynamic-programming alignment algorithm as well.

The model was trained on a 50 000-word spoken English corpus, which had been previously transcribed into both orthographic and phonemic form. Parfitt &

Sharman do not specify the size of their training and test sets; the values which follow come from a personal communication. A 41 169-word section was selected for training, and a disjoint 1290-word section extracted for testing. There were 8149 distinct words in the training text, and 626 in the test text with an unknown number of these not present in the training data. Transcription accuracy is approximately 53% phonemes correct for initial estimates based on unaligned word pairs and with frequency counts normalized for *a priori* occurrence of phonemes. After four iterations of Baum–Welch re-estimation, this improves to approximately 70% phonemes correct. Using the dynamic-programming alignment improves the initial (normalized) estimates to 85% phonemes correct.

While forward–backward re-estimation improved performance in the case of the poorer initial estimates of the model's parameters, for the better initial estimates based on alignment, performance actually deteriorates (to around 76% after four iterations).

Luk & Damper (1991, 1994) describe an approach to text–phoneme translation that they call *stochastic transduction*. The approach is based on formal language theory and, in particular, the use of a stochastic "transduction" grammar to model the translation process. In this work, the terminal symbols of the transduction grammar are text–phoneme correspondences, and the translation of a word is modelled as sentential derivation producing a string of such correspondences. It is self-learning in that the terminals (correspondences) are inferred from training data, as are the probabilities of the rewrite rules of the stochastic grammar. Because the work in its current form embodies a regular grammar and the Markov assumption for the rule probabilities, it is essentially a form of Markov (but not hidden) modelling. However, these restrictions are not necessary so that the stochastic transduction formalism is in principle considerably more general than HMM approaches.

Luk & Damper (1994) have reported 100% alignment performance on 18 767 training words and 1667 unseen test words from the *Oxford advanced learners' dictionary of current English* (Oxford University Press 1989). The percentage of words correctly translated is 72% for the training set and also 72% for the test set, for a version of the model in which correspondences were inferred on the presumption that they should end just after a vowel or after at most two consonants in the phoneme domain.

Again, the HMM formalism developed in a way that was entirely divorced from notions of connectionism. Recently, however, a number of authors have considered relations between the two (Bridle 1990, Nádas 1994). HMMs are generative models based on statistical distributions, and can be trained by statistical estimation procedures. On the other hand, neural networks are not usually generative nor trained by statistical methods but by gradient descent, error minimization. These differences, however, are not irreconcilable. By considering neural models which can be trained on a maximum likelihood basis, connectionist implementations of HMMs can be realized.

Back-propagation networks

The best known example of a self-learning text-to-phoneme system is probably NETtalk, a back-propagation network created by Sejnowski & Rosenberg (1987). This is described in some detail by Bullinaria in Chapter 9, and so is only briefly dealt with here. NETtalk was primarily an attempt to model aspects of human learning. One of the most attractive characteristics of connectionist models is surely that their distributed, parallel nature lends them a degree of psychological plausibility. In what follows, we consider other work which has used multi-layer, feedforward nets trained by (supervised) error back-propagation for text–phoneme conversion.

McCulloch et al. (1987) describe NETspeak (Fig. 7.3), intended principally as a re-implementation of NETtalk. However, this work additionally explored the impact of different input and output codings, and examined the relative performance of separate networks for the transcription of common and uncommon words respectively.

Like NETtalk, NETspeak used a window of seven characters which was stepped across the input one character at a time. As does Klatt (1987), McCulloch et al. cite Lucassen & Mercer (1984) as having demonstrated that a seven-character window contains sufficient context to produce a reasonable transcription while avoiding the high computational expense of larger windows. However, Lucassen & Mercer actually used a *nine*-letter window. Further, the view of Church (1985: 246) is relevant: ". . . stress dependencies cannot be determined locally. It is impossible to determine the stress of a word by looking through a five or six character window."

The input and output codings were thought to be important in that "an appropriate coding can greatly assist learning whilst an inappropriate one can prevent it." Each input character was represented in NETtalk by a sparse (1 out of n) code of 29 bits – one bit for each of the 26 letters and three additional bits for punctuation marks. Thus, the number of input units, i, was $7 \times 29 = 203$. By contrast, NETspeak used a more compact (2 out of n) 11-bit coding, giving $i = 77$. The first five bits indicated which of five rough, "phonological sets" the letter belonged to, and the remaining six bits identified the particular character. In place of NETtalk's 21 "articulatory features" to represent the (single) phoneme output (plus five stress and syllable boundary units), NETspeak used $o = 25$ output features. Denoting the number of hidden units as h, NETtalk used $h = 120$ for most experiments, whereas NETspeak used 77 hidden units. The relative numbers of neurons are thus $(203 + 120 + 26) = 349$ for NETtalk versus $(77 + 77 + 25) = 179$ for NETspeak. Since, for a fully connected multi-layered perceptron with one hidden layer, the total number of interconnections is $(i \times h) + (h \times o)$, NETtalk had 27 480 adjustable weights (ignoring bias inputs to the neurons which are generally treated as adjustable weights also) compared to NETspeak's 7854.

NETspeak was trained on 15 080 of the 16 280 (previously aligned) words in a pronouncing dictionary. Given differences of detail between the networks

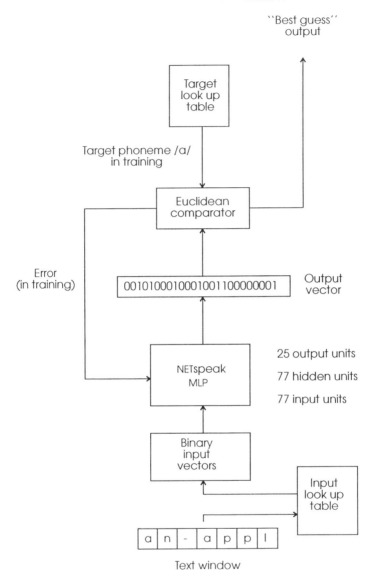

Figure 7.3   The architecture of NETspeak. (Redrawn from McCulloch et al. 1987.)

(e.g. in training material, in learning parameters and in evaluating the "best guess" output), a direct comparison with NETtalk is difficult. However, NETspeak achieved 87.9% "best guess" performance on these 15 080 words after some three passes through its training set. Further, performance on new (unseen) words was only slightly worse at 86%, indicating a high degree of generalization of the regularities of text–phoneme correspondence. The authors do not explicitly state the size of this test set. Presumably, it consisted of the (16 280 – 15 080 = 1200) held-out words.

The net was subsequently trained on a frequency-weighted version of the dictionary. The authors expected that the network would have more difficulty learning this training set, because of the higher proportion of more common (and presumably, therefore, irregularly pronounced) words. This turned out not to be the case, although performance on common words was reported to improve while that for "very regular words" deteriorated.

Further, McCulloch et al. trained two separate networks on common and on uncommon words. In the former case, the training set was 13 021 non-unique words representing in appropriate (frequency-weighted) proportions a randomly chosen 958 of the 1058 most common words. The test set consisted of 911 words representing the remaining 100 words in appropriate proportions. Relative figures in the latter case were a 6169-word training set obtained by scaling up 727 less common words, and a 531-word test set obtained by scaling up a disjoint 73 of the less common words. The number of (non-blank) characters in the two training sets, and in the two test sets, was the same. The results are reported to be "quite surprising" in that the network trained on the less common words (presumed to be more regular in their pronunciations) did not perform better than the network trained on the common words.

Additional light is thrown on this matter by Ainsworth & Pell (1989), who trained an MLP modelled on NETspeak on a 70 000-word dictionary divided into "regular" and "irregular" words according to whether pronunciation was correctly predicted by the Ainsworth (1973) rules or not. As one would expect, they found much better performance on the regular words than on the irregular words (asymptotic to 95.9% and 83.6% "best guess" phoneme scores on unseen words respectively) Thus, it seems that the assumption of McCulloch et al. that the more common words have less regular pronunciations may be incorrect.

Unlike McCulloch et al., whose concern was with the possible automatic creation of TTS systems, Seidenberg & McClelland (1989) studied the ability of a back-propagation network to simulate aspects of observed, human reading behaviour. Their network consisted of 400 input (orthographic) units, 200 hidden units and 460 output (phonological) units. Thus, there were 172 000 adjustable feedforward connections (excluding biases). There were also feedback connections (80 000) from the hidden to the orthographic units, but these do not appear to be essential to the operation of the model as implemented. Where NETtalk and NETspeak used a window of consecutive characters to provide context, Seidenberg & McClelland used input and output coding schemes which involve *triples* of consecutive "features". A consequence of this is that any input word is encoded into a single input vector, irrespective of its length.

The input units each code ten possible first characters (letters plus word boundary marker), ten possible second characters and ten possible third characters, so that each specifies 1000 possible character triples. An input string turns a unit on if it contains a three-character substring which is one of the 1000 triples allocated to that unit. The allocation is done entirely at random except that the word boundary marker cannot appear in the middle position. With this coding, the probability that two different input words would activate exactly the same set of units is effectively zero. Each of the output units represents a *single* triple of "phonetic features". To keep the number of units tractable (460), however, Seidenberg & McClelland discarded every coding for which the first and third features referred to a different phonetic dimension. This aspect of the coding scheme, and others, have been much criticized in the literature (Pinker & Prince 1988, Coltheart et al. 1993). In the following chapter, Plaut et al. take up the issue of input/output representation in this sort of connectionist model.

This treatment of input and output strings – by allocating units to code the occurrence of particular sequences (triples) – has many important implications. It is apparent that no prior alignment of text and phonemes is required since both input and output strings are encoded as single vectors. However, while it is straightforward to determine the output coding for a particular phoneme string, as required in training, the converse operation of converting an output coding to a phoneme string is not possible. This makes it difficult if not impossible to determine whether the net's output is correct or not. In our view, this inability to recover a phoneme sequence at the output is a virtually fatal flaw. It means that the model is incapable of simulating, for instance, the very basic human ability to name (say) the first phoneme of a printed word. How then did Seidenberg & McClelland overcome this profound objection?

In their model, a *phonological error score* is computed for any input as the sum of the squared differences between the obtained and desired output codings. By theorizing that this score would correlate with naming latency in reading experiments, the predictions of the model could be checked against empirical observations. In fact, Seidenberg & McClelland achieve an impressive degree of success in this. They write (Seidenberg & McClelland 1989: 540) that their "model simulates a broad range of empirical phenomena concerning the pronunciation of words and non-words". This error score also offered a possible way of deciding if the model's output was "correct". Seidenberg & McClelland computed all the output codings corresponding to the correct ("target") phoneme sequence plus all sequences differing by just one phoneme from the target. The output was then considered "correct" if the sum of the squared differences corresponding to the target string was lower than that corresponding to any of the out-by-one-phoneme strings. As Coltheart et al. (1993) point out, however, this procedure gives only a lower bound on the error rate of the model in pronouncing text strings.

Since psychological experimentation on reading traditionally favours the use of monosyllabic words, Seidenberg & McClelland used 2897 such words to train their net. Of this total, 2884 words were unique (e.g. the training set included ("wind", /wInd/) and ("wind", /waInd/) as separate "words"). Training words were presented a number of times proportional to the logarithm of their frequency of occurrence in English. After training (on 150 000 word presentations), and using the somewhat suspect scoring method described above, only 77 (2.7%) of the words in the training set produced outputs which were deemed incorrect. Generalization to unseen words was not tested systematically.

Before leaving the subject of back-propagation networks, it is worth pointing out a fundamental deficiency. Such nets, at least in their simple, feedforward form, are only suitable for mapping static input patterns to static output patterns. Because of the absence of feedback and/or memory, they are ill suited to processing dynamic input sequences. Yet the input (and output) strings encountered in text–phoneme translation are inherently sequential. This necessitates essentially *ad hoc* fixes like the use of a sliding context window (in NETtalk and NETspeak) or the wholly inappropriate input/output coding employed by Seidenberg & McClelland. A more principled way to proceed would be to employ dynamic, recurrent nets – as also suggested in the following two chapters.

Analogy-based methods

The use of letter-to-sound rules in conjunction with a dictionary of pronunciations to some extent mirrors dual-route, psychological models of reading aloud (see above). As stated above, arguments for dual-route theory are based on the ability to pronounce pseudowords, latency difference effects between regular and exception words, and apparent double dissociation between the two routes in dyslexia. However, it has been variously argued that all these observations can be explained by a *single* route. One pervasive idea in the literature is that pseudowords are pronounced by *analogy* with lexical words that they resemble (Baron 1977, Brooks 1977, Glushko 1979, Glushko 1981, Brown & Besner 1987). Glushko, for instance, showed that "exception pseudowords" like "tave" take longer to read than "regular pseudowords" such as "taze". Here, "taze" is considered as a "regular pseudoword" since all its orthographic "neighbours" ("raze", "gaze", "maze", etc.) have the regular vowel pronunciation /eI/. By contrast, "tave" is considered to be an "exception pseudoword" since it has the exception word ("have", /hav/) as an orthographic neighbour. Thus, in the words of Glushko (1979), the ". . . assignment of phonology to non-words is open to lexical influence" – a finding which is at variance with the notion of two separate, independent routes to pronunciation. Instead of this:

> . . . it appears that words and pseudowords are pronounced using similar kinds of orthographic and phonological knowledge: the pronunciation of words that share orthographic features with them, and specific spelling-to-sound rules for multiletter spelling patterns.

Thus, in place of *abstract* GPC rules in the dual-route model we have *specific* patterns of correspondence in the single-route, analogy model.

Pronunciation by analogy can be either *explicit* or *implicit*. The explicit form (e.g. Baron 1977) is a conscious strategy of recalling a similar word and modifying its pronunciation, whereas in implicit analogy (e.g. Brooks 1977) a pronunciation is derived from *generalized* phonographic knowledge about existing words. Implicit analogy has obvious commonalities with single-route, connectionist models (e.g. Sejnowski & Rosenberg 1987, Seidenberg & McClelland 1989). This commonality is highlighted by Glushko's arguments (1979, 1981) for the term *activation* in place of *analogy* since, for him, the process is naturally unconscious (i.e. implicit).

To test the computational feasibility of (explicit) analogy, Dedina & Nusbaum (1991) produced a prototype TTS system called PRONOUNCE. According to these authors, "pronunciation-by-analogy may provide the same pronunciation ability as a set of spelling-to-sound rules without requiring an explicit theory of rule induction . . . and may be relatively simple to automate." They identify the principal theoretical issue as "the degree to which orthographic consistency in the spelling patterns of words is related to phonographical consistency in pronouncing these words."

The lexical database of PRONOUNCE consists of a 20 000-word dictionary in which text and phonemes have been aligned. Dedina & Nusbaum acknowledge the crude nature of their alignment procedure, saying it "was carried out by a simple Lisp program that only uses knowledge about which phonemes are consonants and which are vowels." An incoming word is matched against orthographic entries in the lexicon by a process of registering the spelling patterns relative to one another and evaluating the number of contiguous, common letters for each registration index. Matched substrings, together with their phonemic mappings as stored in the lexical database, are used to build a pronunciation lattice which is then traversed to find a set of possible pronunciations for the input word. Pronunciations are rank ordered, first by length of path through the lattice and, second, by the sum of the arc frequencies, reflecting the number of matched substrings that produced that arc. (The system has no knowledge of specific word frequencies.) When tested on the 70 pseudowords employed by Glushko in his (1979) study, PRONOUNCE exhibited an error rate of 9%; here, a "correct" pronunciation is taken as one produced by any of Dedina & Nusbaum's seven human subjects. By contrast, the well known rule-based system DECtalk had an error rate of 3%. The authors interpret their results as showing "that pronunciation-by-analogy is computationally sufficient to generate reasonable pronunciations for short novel strings".

In explicit analogy, there is (apart from text–phoneme alignment) no prior training or inferencing phase. An input word is merely matched – as described above – with every orthographic entry in the lexicon and a pronunciation inferred from the matching substrings. This implies a good deal of computation to produce a pronunciation.

Subsequently, Sullivan & Damper (1992, 1993) extended this work in various ways. They employ an improved alignment procedure based on the Lawrence and Kaye (1986) algorithm. By pre-computing mappings and their statistics for use in the matching process, they have implemented a considerably more "implicit" form of synthesis-by-analogy than did Dedina & Nusbaum. This precomputation from the lexicon of possible mappings and their statistics amounts to a form of self-learning. They have also examined different ways of numerically ranking the candidate pronunciations. The analogy process is extended to the phonemic (in addition to the orthographic) domain. This latter extension has necessitated a reversion to some form of rules, in order that "plausible" pronunciations for an unknown word can be generated and matched against lexical entries. The "flexible" grapheme-to-phoneme rules of Brown & Besner (1987) are used for this purpose, where "flexible" means context-independent. Since the intention is to produce a *set* of plausible pronunciations, there is no necessity to resolve conflicts to ensure that only a single rule is fired. (Of course, the use of left and right contexts is an important strategy for conflict resolution in traditional rule-based transcription.)

Stanfill (1987) describes MBRtalk, a self-learning text–phoneme conversion system based on so-called *memory-based reasoning* (MBR). In our terms, however, this can be viewed as an analogy-based technique. The basic principle of MBR is that "best match recall from memory" can be regarded as a "primary inference mechanism". For every letter of every word in the dictionary of aligned orthographic–phonemic word pairs, a *frame* is created having five fields: the letter itself, the previous four letters, the succeeding four letters, the (single) phoneme aligned to that letter, and the stress assigned to it. Stanfill omits to say how many conflicting frames (having the same *letter*, *left-context* and *right-context* fields but different *phoneme* and *stress* fields), if any, ever arose. (Information on this point would give useful insight into the inherent difficulty of the transcription task.)

The complete set of frames is stored in memory for use during transcription. As with NETtalk and NETspeak, the assumption of a single-letter to single-phoneme alignment requires that a special character ("–") be used to denote a *null* phoneme, and that dipthongs, etc., be treated as single phonemes. In the case of a (possibly novel) input word, there will one frame for each letter of the word but the *phoneme* and *stress* fields of these frames will be empty; it is the task of MBRtalk to fill them. This is done by comparing each frame of the test word with every frame in memory and retrieving the best matching frame. The relevant contents of the retrieved memory frames are then transferred to the test frames to give a pronunciation.

This strategy of finding the best match to a lexically specified pronunciation stored in memory leads us to contend that MBRtalk is using a form of synthesis-by-analogy. The frames are effectively "analogy segments", i.e. fixed (nine character) substrings which, when matched, provide a single phoneme in the output pronunciation. Unlike PRONOUNCE, however, there is a significant pretraining phase consisting of extracting and storing in memory frames derived from the lexical database. Also, the simple mechanism of stepping the nine-character window through the input word a character at a time and concatenating output phonemes corresponding to the central letter (in the manner of NETtalk and NETspeak) avoids any need for a pronunciation lattice.

Clearly, the efficacy of MBRtalk depends critically upon having good similarity metrics. Dissimilarity between two frames is computed by assigning a

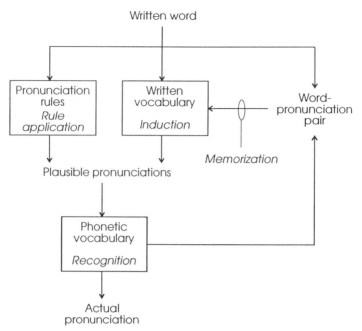

Figure 7.4 Schematic of JOHNNY, a memory-based model of the process of "learning to read". (After Stanfill 1988.)

(heuristically chosen) penalty for each field in which they conflict. With a 1024-frame (approximately 200-word) test set, Stanfill found a best accuracy of 88% frames correct with a 132 072-frame (about 25 000-word) training dictionary. This result was obtained with a penalty function which depended solely on the contents of the conflicting fields of the frames of the test word. Obviously, it would be more principled to score penalties according to the (conflicting) contents of fields for both test and training data. Indeed, the necessity to use (non-optimal) heuristics for scoring similarity seems currently to be a major weakness of synthesis-by-analogy.

In subsequent work, Stanfill (1988) presents a rule induction methodology within the framework of the MBR paradigm. The system, called JOHNNY, starts with "some knowledge of phonetic rules plus a phonetic lexicon". Unusually, there is no orthographic lexicon (or, at least, it is initially empty). As shown in Figure 7.4, the basic operations are rule application, "recognition", memorization and induction. All of these are implemented using MBR.

There are 104 rewrite rules, using a three-letter window as their left-hand side (LHS). The left and right contexts are a maximum of one character, which can be "don't care". An input word is broken into a series of such three-letter windows which are each matched against the LHSs of the rule set in memory. Mismatches are penalized such that the best matches will be the most specific rules which are applicable. The input word is assumed to be present in the phonetic vocabulary. The rules are ambiguous in that this process produces a number of plausible pronunciations.

"Recognition" scores these candidate pronunciations against the phonetic vocabulary, again using MBR. Testing with a text of 1024 dictionary-derived pronunciations of words spontaneously uttered by a child in the first grade, the error rate at this stage (rules plus recognition) was 10%. For 8192 words randomly selected from a dictionary, however, the error rate was lower at 7%. The implication is that randomly selected dictionary entries are more regular in their pronunciation than spontaneous utterances – at least, those of a child in the first grade.

Having obtained this pronunciation, JOHNNY assumes it is correct and commits it to its "learned" memory, using exactly the same nine-character frame representation and structure as MBRtalk. (Stanfill uses the terminology *frame* and *record* interchangeably in his 1987 paper; in his 1988 paper, he uses *record* exclusively, but we prefer to retain *frame*.) This memorization is seen as a process of acquiring spelling–pronunciation pairings in an unsupervised fashion. While the unsupervised nature of the learning is one of the system's most intriguing aspects, it seems likely that the success of the learning is critically dependent upon the 104 initial rules and on using as input only those words present in the "phonetic vocabulary".

Stanfill is not explicit on the mechanism of rule induction. It appears, however, that the rules and what we have called the "learned" memory are merely used in parallel to produce candidate pronunciations, as depicted in Figure 7.4. This impression is strengthened by the fact that rules and learned memory have different structures, i.e. a three-character window and nine-character frame, respectively. It is not clear why it is necessary to memorize a "sufficient number of words" before using the "learned" memory to supplement the rules.

When the rules are supplemented in this way, the error rates on the above test materials (after two passes) fall from 10% to approximately 7% for the "first grade" words, and from 7% to approximately 3% for the randomly selected dictionary words. JOHNNY also displays an impressive lack of sensitivity to the initial rule set – as demonstrated by, for example, deleting all multiletter rules and showing that the system is still capable of learned improvement. Finally, and not unexpectedly, JOHNNY is capable of further improvement when the learning is supervised (i.e. the system is told when it makes a mistake, and might also be given the correct pronunciation) rather than unsupervised.

Recently, Van den Bosch & Daelemans (1993) have described a very similar approach to memory-based reasoning that they describe as "a link between straightforward lexical lookup and similarity-based reasoning". The method takes a pronouncing dictionary as the training set but "solves the problem of lacking generalization power and efficiency by *compressing* it into a text-to-speech *lookup table*". Applied to Dutch, they say: "The most surprising result of our research is that the simplest method (based on tables and defaults) yields the best generalization results, suggesting that previous knowledge-based approaches were overkill".

Syntactic neural networks

Statistical translation models rely on associating "units" in one language, or domain, with "units" in another. Lucas & Damper (1992) describe a text–phoneme conversion model based on such statistical ideas and having a direct connectionist implementation in terms of so-called syntactic neural networks. They attempt to translate between commonly occurring substrings in the two domains. The assumption is that common substrings represent potentially useful abstractions, as in the case of "th".

The training process consists of three passes through the dictionary of orthographic–phonemic word pairs. First, the m most probable substrings (n-grams) in each domain are enumerated. The starting point is the alphabets of atomic symbols, Σ_o and Σ_p, where the subscripts denote orthography and phonemics, respectively. These atomic symbols are thought of as the terminals of a simple grammar for the generation of the observed words. Each word encountered (e.g. "the") is "exploded" into its constituent substrings (i.e. "t", "h", "e", "th", "he" and "the"). The cumulative count for each such substring is then incremented before considering the next word. Thus, a set (initially empty) of "non-terminal" n-grams is built up, to form new alphabets N_o and N_p such that $A_o = \Sigma_o \cup N_o$, an alphabet of orthographic symbols, and likewise an alphabet of phonemic symbols A_p.

At termination of the first pass, substrings are sorted into frequency order, and the first m (which include the atomic symbols) of each taken as the translation alphabets, i.e. $|A_o| = |A_p| = m$. The corresponding inferred grammar is trivial in the sense that each non-terminal can only rewrite to a single string of two "lower-order" symbols (i.e. productions are of the form $(th) \to (t)(h)$).

Following this initial pass, the second phase of learning calculates the bigram statistics of the symbols in A_o and A_p in their respective domains. For the example case of "the", this is effectively considered to be the four distinct words (t)(h)(e), (th)(e), (t)(he) and (the) in evaluating the bigram statistics. These are importantly different from "ordinary" bigram statistics, since many of the (non-terminal) symbols correspond to several atomic characters. In effect, we have an adaptive n-gram description of strings in each domain – in that the value of n varies so as to give the most compact representation for any given size of alphabet. The bigram statistics give the within-domain transition probabilities of going from one symbol to another.

Until this point the orthographic and phonemic domains have been considered separately. At the third pass, however, words are first considered as orthographic–phonemic pairs. This phase estimates the cross-domain (translation) probabilities $P(o \to p \,|\, o)$, i.e. given $o \in A_o$, the probability that o rewrites to a particular $p \in A_p$. This is done as follows. The possible segmentations of each word pair in the dictionary with respect to the alphabets A_o and A_p are identified, and used to build a pronunciation lattice. This is depicted in Figure 7.5 for the example word ("zoom", /zuːm/). (Note the treatment of the vowel lengthening symbol /ː/ as an atomic symbol although it only ever occurs in conjunction with a

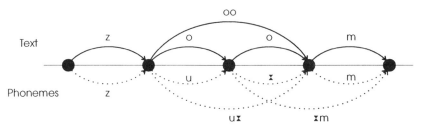

Figure 7.5 Pronunciation lattice for the word ("zoom", /zuːm/), given the non-terminals inferred by the system. A possible alignment corresponds to a pair of paths between word start and end, one in the orthographic domain and the other in the phonemic domain, and having an equal number of arcs. (After Lucas & Damper 1992.)

vowel. The system was left to infer this for itself.) Although not shown in the figure, transition probabilities are attached to each arc. A possible alignment then corresponds to a pair of paths between word start and end, one in the orthographic domain and the other in the phonemic domain, and having an equal number of arcs. The set of such possible alignments, $\{A_i\}$, is enumerated using a path algebra. For the example word, these are

$$A_1: \quad ((\text{"z"}, /z/) (\text{"o"}, /u/) (\text{"o"}, /ː/) (\text{"m"}, /m/) \; 1.6 \times 10^{-8})$$
$$A_2: \quad ((\text{"z"}, /z/) (\text{"oo"}, /u/) (\text{"m"}, /ːm/) \; 8.9 \times 10^{-7})$$
$$A_3: \quad ((\text{"z"}, /z/) (\text{"oo"}, /uː/) (\text{"m"}, /m/) \; 1.7 \times 10^{-6})$$

where the numbers represent the product of all arc (transition) probabilities on the respective o and p paths.

At the end of the pass, these figures are used to compute mapping probabilities for all the $(o \rightarrow p)$ correspondences seen. This is done, for a particular i and j, by simply counting the number of occurrences of $o_i \rightarrow p_j$ correspondences in the training data, and normalizing by the total number of correspondences involving o_i.

Although associations are (in principle) allowed between o and p substrings in any position within a word, the method enumerates contiguous paths only. Thus, non-contiguous markings (as with the final "e" of ("make", /meɪk/) indicating the dipthongization of the medial vowel) cannot be dealt with at all directly. In effect, the system has to learn that "ke" sometimes functions as an orthographic unit which associates with phoneme /k/ following /eɪ/. A related point is that there is no simple way to allow symbols in the source domain (letters here) to rewrite to *null* in the target domain.

As the enumeration of possible alignments is an intrinsic part of the learning process, the necessity for aligned training data is avoided. To this extent, the learning is unsupervised in that no target output is specified for each input symbol.

There are many ways the described model and its estimated probabilities could be used in a translation system. For instance, given a test word, its (orthographic)

139

segmentations could be determined and the translation (sequence of phoneme units) selected having the highest joint probability of the *o* path and associated rewrites for the arcs on that path. Another possibility would be to consider the *p* path probability additionally. However this is done, treating the probabilities as weights in a neural network allows a connectionist implementation to be conceived, as detailed in Lucas & Damper (1992). A similar connectionist model has recently been reported by Smith (1993), who employs an argument based on eye movements during reading to support the idea of subword units corresponding to Lucas & Damper's non-terminal *n*-grams.

While the performance obtained is not especially impressive (performance asymptotic to about 67% phonemes correct on largely monosyllabic words as test and training set increase in size), the authors point to "the system's immaturity" and describe their methods as "theoretically interesting from the point of view of . . . text–phonetics correspondence in general".

Summary

A critical review of self-learning approaches to text–phoneme conversion has been presented. To our knowledge, this is the first such review to appear. Work on this topic has been approached from two different perspectives: the automatic production of TTS (and possibly speech-to-text) systems, reducing the amount of manual work which must be done in implementing a new language as well as giving a potential means of improving existing rule-based systems, and the modelling of human reading processes – especially the acquisition of the skill of reading aloud. The chapter is intended to provide a framework and introduction for the following contributions which focus on connectionist models of reading.

The standard approach to text–phoneme conversion (in the absence of a dictionary containing all words ever encountered) in TTS systems uses a set of CDT rules. The operation of such rules has been described, and automatic-discovery techniques for inferring CDT rule sets reviewed. However, the main focus of this chapter is on self-learning, connectionist approaches. The rule-based approaches are included for completeness and as a basis of comparison.

Most commonly, connectionist models use the popular error back-propagation learning – as in NETtalk, NETspeak and the work of Seidenberg & McClelland. While the later model codes inputs and outputs in terms of a fixed set of features for any given word, NETtalk and NETspeak employ a sliding window which passes over the input character by character. Accordingly, they confront a "bootstrapping" problem in that they require text pre-aligned with corresponding phonemes for training. This implies prior knowledge of the correspondence between text and phonemes that we are actually trying to infer. Hence, interest centres on connectionist models which do not require pre-alignment, such as that of Bullinaria to be described later (see Ch. 9). Another problem is that, at least in their simple feedforward form, back-propagation networks are really only suitable for processing static patterns. The use of dynamic, recurrent nets appears highly appropriate, yet has not so far (to my knowledge) been reported in the literature.

HMMs have been popular in speech recognition, and there has been some interest in exploiting the trainable nature of HMMs in text–phoneme conversion. Connectionist implementations of HMMs have also been described in the literature.

Recently, synthesis-by-analogy has received renewed attention as a computational TTS model and as a single-route alternative to dual-route theory. In its explicit form, an unknown word is compared with lexical entries, and a pronunciation assembled from matching segments. In its implicit form, generalized knowledge concerning text–phoneme correspondences is pre-compiled and used in translation. Implicit analogy is close in spirit to connectionist approaches. Of course, much depends upon how matching is performed in the explicit case, and knowledge captured in the implicit case. Because of this, a number of approaches can be classified within the analogy framework. Examples are Stanfill's MBRtalk (founded on memory-based reasoning) and Van den Bosch & Daelemans's compressed table look-up method.

Ideas of statistical association and translation between features (letter or phoneme units) in the two different domains have been used by Lucas & Damper, and by Smith, to produce connectionist models of reading aloud that avoid the need for pre-alignment. As yet, these perform relatively poorly but they do represent an interesting direction for future research.

References

Abercrombie, D. 1981. Extending the Roman alphabet: some orthographic experiments of the past four centuries. In *Towards a history of phonetics*, R. E. Asher & E. Henderson (eds), 207–24. Edinburgh: Edinburgh University Press.

Ainsworth, W. A. 1973. A system for converting English text into speech. *IEEE Transactions on Audio Electroacoustics* **21**, 288–90.

Ainsworth, W. A. & B. Pell 1989. Connectionist architectures for a text-to-speech system. *Eurospeech '89, Proceedings*. Paris, 125–8.

Allen, J. 1976. Synthesis of speech from unrestricted text. *Proceedings of the IEEE* **64**, 422–33.

Allen, J., M. S. Hunnicutt, D. H. Klatt 1987. *From text to speech: the MITalk system*. Cambridge: Cambridge University Press.

Baron, J. 1977. Mechanisms for pronouncing printed words: use and acquisition. In *Basic processes in reading: perception and comprehension*, D. LaBerge and S. Samuels (eds), 175–216. Hillsdale, New Jersey: Lawrence Erlbaum.

Baum, L. E. 1972. An inequality and associated maximization technique in statistical estimation of probabilistic functions of Markov processes. *Inequalities* **3**, 1–8.

Bridle, J. S. 1990. Alpha nets: a recurrent neural network architecture with a hidden Markov model interpretation. *Speech Communication* **9**, 83–92.

Brooks, L. 1977. Non-analytic correspondences and pattern in word pronunciation. In *Attention and performance VII*, J. Renquin (ed.), 163–77. Hillsdale, New Jersey: Lawrence Erlbaum.

Brown, P. & D. Besner 1987. The assembly of phonology in oral reading: a new model. In *Attention and performance XII: the psychology of reading*, M. Coltheart (ed.), 471–89. London: Lawrence Erlbaum.

Carlson, R. & B. Granström 1975. A phonetically-oriented programming language for rule description of speech. In S*peech communication*, vol. 2, C. G. M. Fant (ed.), 245–53. Uppsala, Sweden: Almqvist and Wiksell.

Carlson, R., B. Granström, S. Hunnicutt 1990. Multilingual text-to-speech development and applications. In *Advances in speech, hearing and language processing*, vol. 1, W. A. Ainsworth (ed.), 269–96. London: JAI Press.

Chomsky, N. & M. Halle 1968. *The sound pattern of English*. New York: Harper and Row.

Church, K. W. 1985. Stress assignment in letter-to-sound rules for phonological rules for speech synthesis. *23rd Meeting of the Association of Computational Linguistics, Proceedings*. Chicago, 246–53.

Coltheart, M. 1978. Lexical access in simple reading tasks. In *Strategies of information processing*, G. Underwood (ed.), 151–216. London: Academic Press.

Coltheart, M., B. Curtis, P. Atkins, M. Haller 1993. Models of reading aloud: Dual-route and parallel-distributed-processing approaches. *Psychological Review* **100**, 589–608.

Dedina, M.J. & H.C. Nusbaum 1991. PRONOUNCE: a program for pronunciation by analogy. *Computer Speech and Language* **5**, 55–64.

Elovitz, H. S., R. Johnson, A. McHugh, J. E. Shore 1976. Letter–to–sound rules for automatic translation of English text to phonetics. *IEEE Transactions ASSP* **24**, 446–59.

Forney Jr, G. D. 1973. The Viterbi algorithm. *Proceedings of the IEEE* **61**, 268–78.

Forster, K. I. & S. M. Chambers 1973. Lexical access and naming time. *Journal of Verbal Learning and Verbal Behavior* **12**, 627–35.

Glushko, R. J. 1979. The organization and activation of orthographic knowledge in reading aloud. *Journal of Experimental Psychology: Human Perception and Performance* **5**, 674–91.

Glushko, R. J. 1981. Principles for pronouncing print: the psychology of phonography. In *Interactive processes in reading*, A. M. Lesgold and C. A. Perfetti (eds), 61–84. Hillsdale, New Jersey: Lawrence Erlbaum.

Hertz, S. R., J. Kadin, K. J. Karplus 1985. The delta rule development system for speech synthesis from text. *Proceedings of the IEEE* **73**, 1589–601.

Humphreys, G. W. & L. J. Evett 1985. Are there independent lexical and nonlexical routes in word processing? An evaluation of dual-route theory of reading. *Behavioral and Brain Sciences* **8**, 689–740.

Hunnicutt, S. 1976. Phonological rules for a text-to-speech system. *American Journal of Computational Linguistics* [microfiche] **57**, 1–72.

Klatt, D. H. 1987. Review of text-to-speech conversion for English. *Journal of the Acoustical Society of America* **82**, 737–93.

Klatt, D. H. & D. W. Shipman 1982. Letter-to-phoneme rules: a semi-automatic discovery procedure. *Journal of the Acoustical Society of America* **72**, supplement 1, S48.

Lawrence, S. G. C. & G. Kaye 1986. Alignment of phonemes with their corresponding orthography. *Computer Speech and Language* **1**, 153–65.

Lucas, S. M. & R. I. Damper 1992. Syntactic neural networks for bi-directional text-phonetics translation. In *Talking machines: theories, models and applications*, G. Bailly and C. Benoît (eds), 127–41. Amsterdam: Elsevier/North-Holland.

Lucassen, J. M. & R. L. Mercer 1984. An information theoretic approach to the automatic determination of phonemic baseforms, *ICASSP '84, Proceedings*. San Diego, 42.5.1–42.5.4.

Luk, R. W. P. & R. I. Damper 1991. Stochastic transduction for English text-to-phoneme conversion. *Eurospeech '91, 2nd European Conference on Speech Communication and Technology, Proceedings*. Genova, Italy, vol. 2, 779–82.

Luk, R. W. P. & R. I. Damper 1994. A review of stochastic transduction. *Joint European Speech Communication ESCA/IEEE Workshop on Speech Synthesis, Proceedings*. Lake Mohonk, New Paltz, New York, 248–51.

McCulloch, N., M. Bedworth, J. Bridle 1987. NETspeak – a re-implementation of NETtalk. *Computer Speech and Language* **2**, 289–301.

Nádas, A. 1994. Some relationships between ANNs and HMMs. In *Artificial neural networks for speech and vision*, R. J. Mammone (ed.), 240–67. London: Chapman and Hall.

Oakey, S. & R. C. Cawthorne 1981. Inductive learning of pronunciation rules by hypothesis testing and correction, *International Joint Conference on Artificial Intelligence, Proceedings*, IJCAI-81, Vancouver, Canada, 109–14.

Oxford University Press 1989. *Oxford advanced learner's dictionary of current English* 3rd edn, electronic handbook. Oxford: Oxford University Press.

Parfitt, S. H. & R. A. Sharman 1991. A bidirectional model of English pronunciation. *Eurospeech '91, 2nd European Conference on Speech Communication and Technology, Proceedings*, Genova, Italy, vol. 2, 800–4.

Pinker, S. & A. Prince 1988. On language and connectionism: analysis of a parallel distributed processing model of language acquisition. *Cognition* **28**, 73–193.

Quinlan, J. R. 1979. Discovering rules from large collections of examples: a case study. In *Expert systems in the micro electronic age*, D. Mitchie (ed.), 168–201. Edinburgh: Edinburgh University Press.

Quinlan, J. R. 1990. Decision trees and decision making. *IEEE Transactions Systems, Man and Cybernetics* **20**, 339–46.

Rahim, M. Z. 1994. A self-learning neural tree network for phone recognition. In *Artificial neural networks for speech and vision*, R.J. Mammone (ed.), 227–39. London: Chapman and Hall.

Rumelhart, D. E., G. E. Hinton, R. J. Williams 1986. Learning internal representations by error propagation. In *Parallel distributed processing: explorations in the microstructure of cognition*, vol. 2. *Physiological and Biological Models*, D. E. Rumelhart & J. L. McClelland (eds), 318–62. Cambridge, Mass.: MIT Press.

Scragg, D. G. 1975. *A history of English spelling*. Manchester: Manchester University Press.

Segre, A. M., B. A. Sherwood, W. B. Dickerson 1983. An expert system for the production of phoneme strings from unmarked English text using machine-induced rules. *1st Conference of the European Chapter of the Association of Computational Linguistics, Proceedings*. Pisa, Italy, 35–42.

Seidenberg, M. S. & J. L. McClelland 1989. A distributed, developmental model of word recognition and naming. *Psychological Review* **96**, 523–568.

Sejnowski, T. J. & C. R. Rosenberg 1987. Parallel networks that learn to pronounce English text. *Complex Systems* **1**, 145–68.

Smith, J. 1993. Using unsupervised feature detectors in a model of reading aloud. *Irish Journal of Psychology* **14**, 397–409.

Stanfill, C. W. 1987. Memory-based reasoning applied to English pronunciation. *6th National Conference on Artificial Intelligence, Proceedings*. Seattle, Wash., 577–81.

Stanfill, C. W. 1988. Learning to read: a memory-based model. *Case-Based Reasoning Workshop, Proceedings*, Clearwater Beach, Fla., 406–13.

Stromberg, J-E., J. Zrida, A. Isaksson 1991. Neural trees – using neural trees in a tree classifier structure. *ICASSP '91, Proceedings*, Toronto, Canada, vol. 1, 137–40.

Sullivan, K. P. H. & R. I. Damper 1992. Novel-word pronunciation within a text-to-speech system. In *Talking machines: theories, models and applications*, G. Bailly & C. Benoît (eds), 183–95. Amsterdam: Elsevier.

Sullivan, K. P. H. & R. I. Damper 1993. Novel-word pronunciation: a cross-language study. *Speech Communication* **13**, 441–52.

Van Coile, B. 1990. Inductive learning of grapheme-to-phoneme rules. *International Conference on Spoken Language Processes '90, Proceedings*. Kobe, Japan, vol. 2, 765–8.

Van den Bosch, A. & W. Daelemans 1993. Data-oriented methods for grapheme-to-phoneme conversion. *6th conference of the European Chapter of the Association of Computational Linguistics, Proceedings*, 45–93. Utrecht.

Van Leeuwen, H. C. 1989. A development tool for linguistic rules. *Computer Speech and Language* **3**, 83–104.

Venezky, R. L. 1965. *A study of English spelling-to-sound correspondences on historical principles*. Ann Arbor, Mich.: Ann Arbor Press.

Venezky, R. L. 1970. *The structure of English orthography*. The Hague: Mouton.

Viterbi, A. J. 1967. Error bounds for convolutional codes and an asymptotically optimum decoding algorithm. *IEEE Transactions on Information Theory* **13**, 260–9.

Wijk, A. 1966. *Rules of pronunciation of the English language*. Oxford: Oxford University Press.

Wolff, J. G. 1984. *Inductive learning of spelling-to-phoneme rules*. IBM UK Scientific Centre, Technical Report 128.

CHAPTER 8

Reading exception words and pseudowords: are two routes really necessary?

David C. Plaut, James L. McClelland, Mark S. Seidenberg

Introduction

This paper describes simulation experiments demonstrating that a unitary processing system in the form of a connectionist network is capable of learning to read exception words and pronounceable non-words aloud. We trained such a network on the 3000 word corpus used by Seidenberg & McClelland (1989). After training, the network was able to read over 99% of the training corpus correctly. When tested on the lists of pronounceable non-words used in several experiments, its accuracy was closely comparable to that displayed by human subjects.

The work addresses the ongoing debate about the nature of the mechanisms that are used in reading words aloud. One view, defended most recently by Coltheart et al. (1993), states that adequate performance on both pronounceable non-words and exception words depends on the use of two separate mechanisms, one that applies rules of grapheme–phoneme correspondence and another that retrieves pronunciations specific to particular familiar words. An alternative view, expressed by Seidenberg & McClelland, is that a single system may be capable of learning to read both kinds of letter strings.

The work relates more generally to the ongoing debate about the nature of the processing systems underlying human language use. The question is, should these systems be viewed as systems that learn and apply an explicit system of rules, augmented where necessary with an explicit enumeration of exceptions; or should these systems be viewed instead as connectionist systems that gradually develop sensitivity to the structure inherent in the mapping they are asked to learn, through a gradual learning procedure?

The connectionist approach has considerable appeal, because it accounts for the fact that regularity is a graded phenomenon and for the fact that human language users are sensitive to these gradations. In the domain of spelling-to-sound translation, Glushko (1979) was the first to emphasize that in fact the crucial variable for word pronunciation tasks is not regularity, but degree of consistency of the spelling-to-sound correspondences exhibited by one word to the correspondences exhibited by its neighbors. Since Glushko's work, it has been clearly established that the more consistent a word's spelling-to-sound correspondences are with those of its neighbours, the more rapid and accurate responses to that word will be. The connectionist approach, as exemplified by the model of Seidenberg & McClelland (hereafter the SM89 model), captures this graded consistency effect. It also captures the interaction of consistency and frequency that has been repeatedly found in experiments like those of Taraban & McClelland (1993) and Waters & Seidenberg (1985). As Seidenberg & McClelland showed, their connectionist model accounted in detail for the pattern of performance obtained in a large number of different experiments investigating frequency and consistency effects.

However, the model proposed by Seidenberg & McClelland did not perform adequately in reading pronounceable non-words. Attention has focused particularly on the non-words used by Glushko (1979) and by McCann & Besner (1987). Besner et al. (1990) pointed out that the SM89 model fell far short of human performance on either of these two word lists.

Two interpretations of these results have been proposed. Besner, Coltheart and their colleagues have suggested that the deficiencies of the SM89 model reflect the fact that no single system can actually read both exception words and pronounceable non-words adequately. Seidenberg & McClelland (1990) suggested instead that limitations in the SM89 simulation might have prevented the model from successfully capturing the non-word reading data. They suggested that the model's performance on non-words might be improved with either a larger training corpus or a different choice of input/output representations.

In the work we report here, we have focused our attention on the issue of representation. We will show that when a different representation is used, the networks' ability to read pronounceable non-words dramatically improves.

Representations

To understand the motivation for the new representation we have chosen, it is necessary to consider other possible representations that might be used. The most obvious thing to try is what we call a "slot-based" representation. On this approach, the first input letter goes to the first slot, the second to the second slot, etc. Similarly in the output, the first phoneme goes in the first slot, and so on. With enough slots, words up to any desired length can be represented.

In a connectionist system, in which the content of each slot is represented by a pattern of activation, and in which the knowledge that determines how the content of the slot is processed is stored in a set of connection weights, there is a problem with this scheme. The problem is that the same knowledge must be

Table 8.1 The "dispersion" problem.

(a) Position-specific letter units

Left justified					Vowel centred				
1	2	3	4	5	–3	–2	–1	0	1
L	O	G					S	U	N
G	L	A	D			S	W	A	M
S	P	L	I	T	S	P	L	I	T

(b) Context-sensitive triples ("Wickelgraphs")

LOG			*LO	LOG	OG*
GLAD		*GL	GLA	LAD	AD*
SPLIT	*SP	SPL	PLI	LIT	IT*

replicated several times. To see this, consider the words "log", "glad" and "split" (see Table 8.1). The fact that the letter "l" corresponds to the phoneme /l/ must be stored three separate times in this system. So the knowledge underlying this regularity is dispersed across positions, and there is no generalization of what is learned about letters in one position to the same letter in other positions. One can try to alleviate the problem a little by aligning the slots in various ways, but it does not go away completely. Adequate generalization still requires that the correspondences be learned separately in each of several slots.

Seidenberg & McClelland tried to get around this problem by using a scheme that avoided the specific limitations of the slot-based approach, but in the end it turns out to have a version of the same problem. They represented letters and phonemes, not in terms of their absolute spatial position, but in terms of the adjacent letters to the left or the right. This approach, which originated with Wickelgren (1969), makes the representation of each element context sensitive without being rigidly tied to position. Unfortunately, however, the knowledge of spelling-to-sound correspondences is still dispersed across a large number of different contexts, and adequate generalization still requires that the training effectively covers them all.

The hypothesis that guided the work we report here was the idea that this dispersion of knowledge is what prevented the network from exploiting the structure of the English spelling-to-sound system as fully as human readers do. We set out, therefore, to design a representation that minimizes this dispersion.

The limiting case of this approach would be to have a single set of letter units, one for each letter in the alphabet, and a single set of phoneme units, one for each phoneme. Unfortunately, this approach suffers from the problem that it cannot distinguish "top" from "pot" or "salt" from "slat". However, it turns out that a scheme that involves only a small amount of duplication is sufficient for unique representation of virtually all uninflected monosyllables. By definition, a monosyllable contains only a single vowel, so we only need one set of vowel

Phonology

Orthography

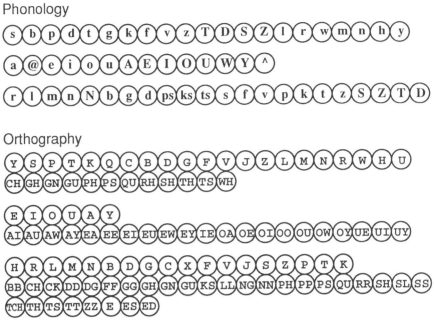

Figure 8.1 The phonological and orthographic representations. The meanings of the symbols in phonology are as follows: /a/ in "pot", /@/ in "cat", /e/ in "bed", /i/ in "hit", /o/ in "dog", /u/ in "good", /A/ in "make", /E/ in "keep", /I/ in "bike", /O/ in "hope", /U/ in "boot", /W/ in "now", /Y/ in "boy", /^/ in "cup", /N/ in "ring", /S/ in "she", /C/ in "chin", /Z/ in "beige", /T/ in "thin" and /D/ in "this". All other phonemes are represented in the conventional way (e.g. /b/ in "bat").

units. A monosyllable may contain both an initial and a final consonant cluster, and almost every consonant can occur in either the initial or the final cluster, so we need separate sets of consonant units for the initial and final consonant clusters.

The remarkable thing is that this is nearly all that we need. The reason is that within an initial or final consonant cluster there are strong phonotactic constraints. At both ends of the syllable, each phoneme can only occur once, and the order of phonemes is strongly constrained. Given this, if we know which phonemes are present in the initial consonant cluster, which phoneme is the vowel and which phonemes are present in the final cluster, then we can use the phonotactic constraints to uniquely determine the order in which these phonemes occur. These constraints can be captured by simply listing the units from left to right to correspond to the left-to-right ordering constraints that exist in initial and final consonant clusters. Once we do this, read-out occurs by simply reading out the phonemes that are active in sequence from left to right (Fig. 8.1).

There are a small number of cases in which two phonemes can occur in either order within a consonant cluster. One such pair is /s/ and /p/. The /s/ can precede the /p/ as in "clasp"; or the /s/ can follow the /p/ as in "lapse". To allow handling of such cases, it is necessary to add units disambiguating the order. To do this we include a unit for /ps/. The convention is that when /s/ and /p/ are both active the order is /s/ then /p/, unless the /ps/ unit is also active, in which case the order is taken to be /ps/. To cover the pronunciations of the items in the Seidenberg & McClelland corpus, we needed only three such units, for /ps/, /ts/ and /ks/. Interestingly, some of these combinations are sometimes treated as single phonemes called affricates and each is written with a single character in at least one Western orthography (Greek "ψ", German "Z", English "X").

We use a similar scheme to represent the spellings of words. Because English is an alphabetic language, this scheme works almost as well for spelling as it does for sound. There is one complication, however. As Venezky (1970) pointed out, the spelling units that correspond to sounds in English are not necessarily single letters. Rather, they are relational units sometimes called graphemes, consisting of one, two or, sometimes, even three or four letters. Because the spelling–sound regularities of English are grapheme–phoneme correspondences, the regularities in the system are most elegantly captured if the units represent the graphemes present in the string rather than simply the letters that make up the word. Unfortunately, it is not always clear what graphemes are present in a word. Consider the word "coltheart". In this case, there is a "t" next to an "h", so we might suppose that the word contains a "th" grapheme, but in fact it does not; if it did it would be pronounced "col-thart". Thus it is apparent that the input is ambiguous in such cases. Because of this, there is no simple procedure for translating letter strings into the correct sequence of graphemes. However, it is possible to translate a letter sequence into a pattern of activation representing all possible graphemes in the string, and this is what we did in our model. So in the case of "coltheart", we would activate units for "t", for "h" and for "th".

Simulation 1: feedforward network

The network we used consisted of 108 orthographic input units, one for each of the Venezky graphemes that occurs in the initial consonant cluster, the vowel cluster or the final consonant cluster. There were also 57 phonological output units, and 100 intermediate or hidden units. The grapheme units are fully connected to the hidden units which, in turn, are fully connected to the phoneme units. To encode any letter string forming an uninflected monosyllable into an input pattern, we parse the string into onset, vowel and coda, and activate all of the units corresponding to graphemes contained in each cluster. To translate the output into a pronunciation, we simply scan the output units from left to right, outputting in order the phonemes with activation greater than 0.5.

We trained the network on the entire list of 2897 monosyllabic words in the SM89 corpus, augmented with 101 words missing from the corpus but used as word stimuli in various experiments. We also included training patterns consisting of each single grapheme and the corresponding phoneme. These were included because children are taught these correspondences in the process of learning to read, although equivalent networks not trained on these correspondences exhibit equivalent behavior (see Plaut et al. 1994). Furthermore, rather than present each word with a probability proportional to its frequency of occurrence (Kucera & Francis 1967) and update the weights immediately, we accumulated the error derivatives for the training cases, each weighted by its frequency, before changing the weights. This enabled the learning rates on each connection to be adapted independently during training using the delta-bar-delta procedure (Jacobs (1988), with an additive increment of 0.1 and a multiplicative decrement of 0.9). Finally, phoneme units within 0.2 of their target values (0 or 1) for a particular training item did not receive any error for that presentation.

Two instances of the same network, differing only in the random initial connection weights, were trained using back-propagation for 300 epochs (with a global learning rate of 0.002 and momentum of 0.9). Performance steadily improved over time. Using the scoring procedure described above, the networks averaged 99.9% correct after training. One network made six and the other made two errors, all on very low-frequency exceptions.

Given this very high level of mastery of the training corpus, we are ready to consider our two competing hypotheses. One of these, due to Coltheart, Besner and their colleagues, is that no single network is capable of learning to read both exceptions and pronounceable non-words. According to this hypothesis, we should expect our networks to fare poorly on pronounceable non-words; either as poorly as the SM89 model did; or, if not that poorly, at least considerably more poorly than human subjects. The other hypothesis, the one that we have put forward above, is that in fact a single network can learn to read both exceptions and pronounceable non-words, as long as the regularities embodied in the spelling sound system are not too dispersed. On this hypothesis, we expect a substantial improvement in performance on non-words relative to the SM89 model.

First, let us begin with the results of experiments reporting non-word reading performance. We consider here two such experiments. Experiment 1 is by Glushko (1979), in which he tested subjects' reading of two sets of non-words: one set derived from regular words such as "dean" and another set derived from exception words such as "deaf". Corresponding example non-words would be "hean" and "heaf". The second experiment is one by McCann & Besner (1987), in which they tested a set of pseudohomophone pseudowords such as "brane" and a set of control pseudowords such as "frane". For present purposes, we concentrate on the control items since we believe pseudohomophone effects are mediated by aspects of the reading system that are not implemented in our simulation (e.g. semantics and/or articulatory processes; see Seidenberg et al. 1995).

Table 8.2 presents the percentage of times subjects used "regular" pronunciations, for both sets of Glushko stimuli and for the McCann & Besner non-words. Also included is the performance of the earlier, SM89 model, and the new model. We can see that the original SM89 model fared very poorly relative to human subjects. The new model, by contrast, produces a dramatic improvement.

Table 8.2   Percent of "regular" responses given to non-words.

	Glushko non-words		McCann–Besner non-words
	Regular	Exception	Control
Human subjects	93.8	78.3	88.6
SM89 network	60.5	53.5	51.0
Feedforward networks	94.2	69.8	88.8

It produces the correct, regular response as often as human subjects in two conditions: the Glushko regular non-words and the McCann & Besner control non-words. For the Glushko exception non-words, however, both the model and the human subjects fail to give the regular response on a substantial portion of items.

To delve more deeply into the data, it is important to consider how one determines whether a response is regular or not. This is not a simple matter, since it is far from clear *a priori* how to define what counts as a regular response. Consider the non-words "grook" and "wead". These items involve cases in which what counts as regular depends on whether we consider the specific context in which the vowel occurs. /U/ as in "boot" is the most common pronunciation of "oo", so we might be tempted to treat /grUk/ as regular. But the final "ook" is pronounced /uk/ as in "took" in 11 of the 12 words ending in "ook". Which one counts as regular depends on the scoring procedure we choose to employ. For the simulation results we have shown thus far, responses were considered regular only when they conformed to abstract spelling-to-sound correspondences. Thus, /grUk/ was considered to be the regular pronunciation for "grook" and /wEd/ as in "weed" was considered to be the correct pronunciation for "wead". We believe the criterion is at least as strict as the one used by Glushko, and is probably somewhat stricter than the one McCann & Besner used.

Let us now consider in more detail the performance of both Glushko's subjects and the model on his exception non-word stimuli (Fig. 8.2). Starting with the subjects, Glushko considered how many of the non-regular responses were consistent with any of the different pronunciations of the word's body that occur in the Kucera & Francis (1967) corpus. For example, given "grook", /gruk/ would now be considered correct, and for "wead", the response /wed/ as in "dead" would become acceptable. This criterion accounted for 80% of the non-regular responses, leaving only 4.1% of the responses unaccounted for. We applied the same criterion to the model's performance, and found that it accounted for 24 out of the 26 non-regular responses, leaving an error rate of only 2.3%. Both our model, and human subjects, read non-words derived from exceptions in a way that is not always consistent with abstract grapheme–phoneme conversion (GPC) rules, but is consistent with at least one of the possible pronunciations of the word's body, almost all of the time.

On the data considered thus far, the model appears a little more likely than

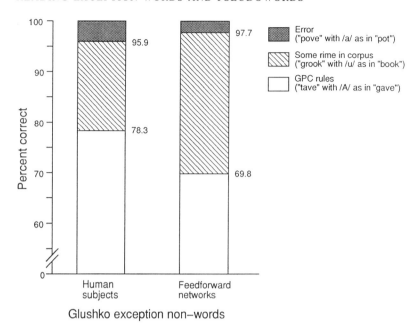

Figure 8.2 Percentage correct on non-word reading for human subjects and the feedforward networks as scored according to grapheme–phoneme conversion (GPC) rules or according to whether the pronunciation matches some rime for the same body in the corpus.

Glushko's subjects to use a pronunciation other than the one dictated by abstract GPC rules. What is it doing in these cases? It appears that it has a strong tendency to choose the most frequent vowel correspondence found among the words in the training corpus that share the same body. Thus, /uk/ is the most frequent pronunciation of "ook" in the corpus, and both runs of the network pronounced "grook" as /gruk/. Overall, in 14 of the 24 cases in which a correspondence other than the abstract GPC correspondence was used, the response actually agreed with the most frequent correspondence for that word body in the model's training corpus.

We are now in a position to consider whether the model's performance on the Glushko exception non-words represents a meaningful deviation from human performance. We believe it does not. If anything, it represents a slight relative difference in the tendency to rely on the pronunciation of word body neighbours, but this tendency is obviously present both in the model and in human subjects, and any difference is only a minor matter of degree. Furthermore, many low-frequency words in English are absent for the training corpus. These words tend to be regular by abstract GPC rules, and if they were included they would increase the extent to which the most frequent correspondence of a word body corresponds to the one that is regular by GPC rules. Therefore, we would expect an increase in the tendency of the model to adhere to GPC rules, if it were trained on a more complete sample of the actual words in the English language.

It must be acknowledged that the model is not absolutely perfect in its reading of non-words – it occasionally makes a frank mistake, such as leaving out a sound that should be pronounced, or pronouncing "ph" as /p/ instead of /f/. But in our view, the model comes close enough to mimicking human performance to seriously threaten the Coltheart–Besner hypothesis. This hypothesis claims that no single system can read both exception words and pronounceable non-words. Our model comes exceedingly close to doing exactly that.

There is one last escape for the Coltheart–Besner hypothesis. This is the possibility that the network has partitioned itself into two subnetworks, one that reads regular words, and another that reads exceptions. If this were the case, we would expect some of the hidden units to contribute to exception words and not non-words, and others to contribute to non-words but not exceptions. To investigate this possibility, we considered each hidden unit to make a significant contribution to pronouncing a given item if, when that unit is removed from the network, the overall error increases by some criterion (e.g. 0.025). We then counted the number of Taraban & McClelland's (1993) 48 exception words and the number of 48 body-matched non-words to which each hidden unit makes a significant contribution. Contrary to the hypothesis that hidden units had specialized for exception words versus non-words, almost all hidden units make substantial contributions to both exception words and non-words. In fact, there tends

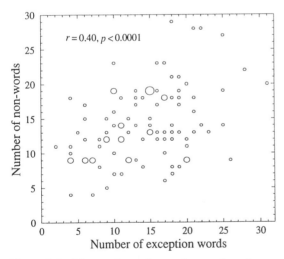

Figure 8.3 The numbers of exception words and non-words ($n = 48$ for each) to which each hidden unit makes a significant contribution, as indicated by an increase in error of at least 0.025 when the unit is removed from the network. Each circle represents one or more hidden units, with the size of the circle proportional to the number of hidden units making significant contributions to the numbers of exception words and non-words indicated by the co-ordinates of the circle.

to be a positive correlation between the number of exception words a hidden unit contributes to and the number of non-words ($r = 0.40$, $p < 0.0001$; see Fig. 8.3). Thus, it appears that the network has not in fact partitioned itself into a subset that learns the rules of English and a separate system that learns the exceptions.

Simulation 2: refixation network

The network developed in simulation 1 succeeds at reading non-words because its representations condense the regularities present within and between orthography and phonology. Although we designed these representations ourselves, based on phonotactic and orthotactic constraints, a network itself can learn to condense the regularities that are dispersed in a position-specific letter representation by refixating the input string. To demonstrate this, we trained a simple recurrent network (Elman 1990) to produce sequences of single phonemes as output when given position-specific letters as input.

The architecture of the network is shown in Figure 8.4. There are 26 letter units and a "blank" unit at each of ten positions. The third position from the left, indicated by the dark rectangle in the figure, corresponds to the point of fixation. These 270 letter units are fully connected to 100 hidden units which, in turn, are fully connected to 36 phoneme units. The hidden units also receive input from the previous states of phoneme units. In addition, there is a fourth group of *position* units that the network uses to keep track of where it is in the letter string as it is producing the appropriate sequence of phonemes, analogous to a focus of attention. Two copies of this group are shown in the figure simply to illustrate their behaviour over time. These position units have connections both to and from the hidden units.

In understanding how the network is trained, it will help to consider first its operation after it has achieved a reasonable level of proficiency at its task. A letter string is presented with its first letter at fixation, by activating the appropriate letter unit at each corresponding position, and the blank unit at all other positions. We assume that position information for internal letters is somewhat inaccurate, so that the same letter units at neighbouring internal positions are also activated slightly (to 0.3). In Figure 8.4, the gray regions for letter units are intended to indicate the activations for the word "bay" when fixating the "b". Initially, the position unit at fixation (numbered 0 by convention) is active and all others are inactive, and all phoneme units are inactive. (The figure shows the network after generating "b" → /b/ and attempting "ay" → /A/.) Hidden unit states are initialized to 0.2.

The network then computes new states for the hidden units, phoneme units and position units. The network has two tasks: (a) to activate the phoneme corresponding to the currently attended grapheme (as indicated by the position units), and (b) to activate the position of the *next* grapheme in the string (or, if the end

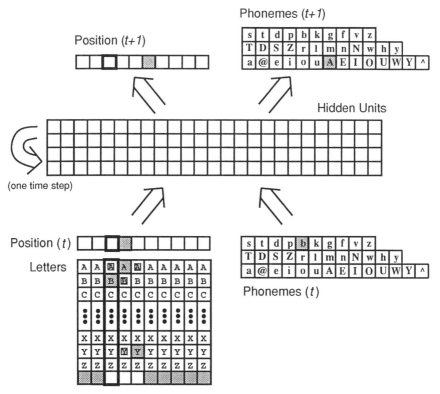

Figure 8.4 The network architecture for the refixation network. The arrows indicated full connectivity between groups of units. The recurrent connections among the hidden units only convey information about the last time step. The gray areas in the input and output units are intended to depict their activities at an intermediate point in processing the word "bay", after the "b" → /b/ has been pronounced (with no refixation) and the "ay" → /A/ is being attempted.

of the string is reached, the position of the adjacent blank). For example, when attending to the letter "b" at fixation in "bay", the network must activate the /b/ unit and position unit 1 (the position of "ay" in the input). To the extent that the activations over the phoneme and position units are inaccurate (i.e. not within 0.2 of their correct values), error is injected and back-propagated through the network.

Assuming that the network succeeds at generating the correct phoneme and position, this information is then used to guide the production of the next phoneme. As shown in Figure 8.4 for "bay", position unit 1 and the phoneme /b/ are now active, the letter input remains the same, and the network must activate /A/, the phoneme corresponding to the indicated grapheme "ay", as well as position unit 3 (corresponding to the blank following the string, thereby indicating a complete pronunciation). In general, when pronouncing a letter string of arbitrary length, the network activates the sequence of phonemes corresponding to its

155

pronunciation, while simultaneously keeping track of the position of the grapheme it is currently working on. If, throughout the process, every phoneme and position is generated correctly, the activations over the letter units remains fixed.

If, however, the network fails at generating the correct phoneme or position at some point in pronouncing a word, it *refixates* the input string so that the problematic grapheme falls at fixation. The network has information sufficient to do this because it generated the position of this grapheme (relative to fixation) on the previous time step. Although the network thus indicates *where* to fixate, the actual refixation is carried out by an external procedure that shifts the letter activations to the left (for a rightward saccade) the indicated number of positions and resets position unit 0 (at fixation) to be active. At this point the network tries again to pronounce the (now fixated) grapheme. Thus, for more difficult words and non-words, and when the network is only moderately trained, it pronounces as much of the input as it can until it runs into trouble, then refixates to that part of the input and continues.

The only remaining issue is what to do when the network fails on the first grapheme of a string, or immediately after refixating, because in these cases there is no accurate position information to drive refixation. This is solved during training by using the target for the position units as the location of the next fixation.

When the fully trained network is tested on an input string, its operation is slightly different. In this context, it cannot base its decision to refixate on whether its output is correct because it is not provided with this information. Rather, it refixates when its output is not *clean*. An output is defined to be clean when exactly one phoneme unit and one position unit are above 0.8 and the remaining phoneme and position units are all below 0.2. If it fails to generate clean output at fixation, it simply includes the most active phoneme unit in its pronunciation, and uses the most active position unit to guide refixation.

To summarize, as the network is trained to produce the appropriate sequence of phonemes for a letter string, it is also trained to maintain a representation of its current position within the string. The network uses this position signal to refixate a peripheral portion of the input when it finds that portion difficult to pronounce. This repositions the input string so that the peripheral portion falls right at the point of fixation, where the network has had more experience in generating pronunciations. In this way, the network can apply the knowledge tied to the units at the point of fixation to any portion of the string that is difficult for the network to read.

The network was given extensive training on the extended SM89 corpus (900 000 word presentations, sampled according to frequency, with a learning rate of 0.01 and momentum of 0.9). Early on in training, the network requires multiple fixations to read words, but as the network becomes more competent it eventually reads most words in a single fixation. At the end of training, the network reads correctly all but nine of the words (99.7%), with an average of 1.18 fixations per word.

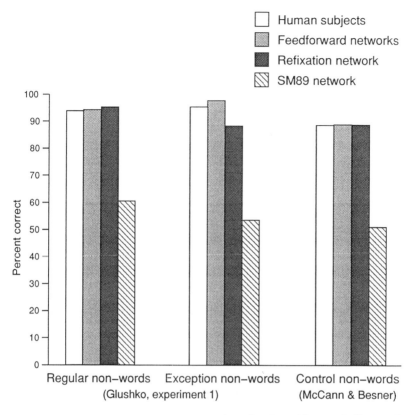

Figure 8.5 Percentage correct on non-word reading for subjects and all networks, where a correct response is one that is regular or is consistent with the pronunciation of a word with the same body.

When tested on non-words, the network gives regular pronunciations to 71/80 (88.8%) of McCann & Besner's control non-words, and 41/43 (95.3%) of Glushko's regular non-words. Although only 26/43 (60.5%) of Glushko's exception non-words are pronounced regularly, 12 of the 17 remaining pronunciations match some word in the training corpus, resulting in an overall level of performance of 38/43 (88.4%). Thus, while the refixation network shows a somewhat stronger tendency to give non-regular pronunciations to exception non-words than does the feedforward network or human subjects, nonetheless the network demonstrates that good non-word reading can coexist with good (exception) word reading in a network that must learn phonotactics itself. For comparison purposes, Figure 8.5 shows the percentage of acceptable pronunciations given to the different sets of non-words by subjects, by the feedforward and refixation networks, and by the SM89 network.

Conclusions

Coltheart et al. (1993) list several points beyond the ones considered here in support of their argument that two separate systems are needed to read pronounceable non-words and exception words. The present article addresses only one of their arguments, and so we cannot claim to have produced a single-route system that provides a complete account of reading of regular and exception words (although see Plaut et al. (1994) for additional simulations and discussion). However, we have addressed what we take to be the most central argument against a single-mechanism account. Coltheart and colleagues state in their article that dual-route theorists claim that there cannot be a single procedure that can correctly translate pseudowords and exception words from spelling-to-sound. We have shown that this claim is false.

References

Besner, D., L. Twilley, R. S. McCann, K. Seergobin 1990. On the connection between connectionism and data: are a few words necessary? *Psychological Review* **97**, 432–46.

Coltheart, M., B. Curtis, P. Atkins, M. Haller 1993. Models of reading aloud: Dual-route and parallel-distributed-processing approaches. *Psychological Review* **100**, 589–608.

Elman, J. L. 1990. Finding structure in time. *Cognitive Science* **14**, 179–211.

Glushko, R. J. 1979. The organization and activation of orthographic knowledge in reading aloud. *Journal of Experimental Psychology: Human Perception and Performance* **5**, 674–91.

Jacobs, R. A. 1988. Increased rates of convergence through learning rate adaptation. *Neural Networks* **1**, 295–307.

Kucera, H. & W. N. Francis 1967. *Computational analysis of present-day American English*. Providence, Rhode Island: Brown University Press.

McCann, R. S. & D. Besner 1987. Reading pseudohomophones: Implications for models of pronunciation assembly and the locus of word-frequency effects in naming. *Journal of Experimental Psychology: Human Perception and Performance* **13**, 14–24.

Plaut, D. C., J. L. McClelland, M. S. Seidenberg, K. E. Patterson 1994. *Understanding normal and impaired word reading: computational principles in quasi-regular domains*. Department of Psychology, Carnegie Mellon University, Technical Report PDP.CNS.94.5.

Seidenberg, M. S. & J. L. McClelland 1989. A distributed, developmental model of word recognition and naming. *Psychological Review* **96**, 523–68.

Seidenberg, M. S. & J. L. McClelland 1990. More words but still no lexicon: reply to Besner et al. 1990. *Psychological Review* **97**, 447–52.

Seidenberg, M. S., A. Petersen, M. C. MacDonald, D. C. Plaut 1995. Pseudohomophone effects and models of word recognition. *Journal of Experimental Psychology: Human Perception and Performance*, in press.

Taraban, R. & J. L. McClelland 1993. Conspiracy effects in word pronunciation, *Journal of Memory and Language* **26**, 608–31.

Venezky, R. L. 1970. *The structure of English orthography*. The Hague: Mouton.

Waters, G. S. & M. S. Seidenberg 1985. Spelling–sound effects in reading: time course and decision criteria. *Memory and Cognition* **13**, 557–72.

Wickelgren, W. A. 1969. Context-sensitive coding, associative memory, and serial order in (speech) behavior. *Psychological Review* **76**, 1–15.

CHAPTER 9

Neural network models of reading: solving the alignment problem without Wickelfeatures

John A. Bullinaria

Introduction

In this chapter it will be argued that many aspects of reading aloud (i.e. text-to-phoneme conversion) can be modelled successfully with a very simple feedforward network that does not require the use of any complicated input/output representations such as Wickelfeatures (Seidenberg & McClelland 1989) nor preprocessing of the training data. We will begin by considering the main constraints on models of reading and propose that the best way to proceed is to build upon the original and rather successful NETtalk model of Sejnowski & Rosenberg (1987). We then show how the problem of aligning the letters and phonemes in the training data (which required preprocessing by hand in NETtalk) can be solved by a straightforward modification of the learning algorithm. This is followed by a discussion of the performance of the resulting model under normal conditions and the possibility of it showing signs of developmental dyslexia under abnormal conditions as well as acquired dyslexia after network damage. The chapter ends with a brief discussion of some of the remaining problems with this and related models.

Constraints on reading models

Within the field of cognitive neuropsychology there is an important and common belief that double dissociation implies modularity of function (e.g. Shallice 1988), and with the advent of connectionist modelling this issue has become even more important (e.g. Bullinaria & Chater 1993). Indeed, it is the fact that acquired surface dyslexia (in which the ability to read exception words is lost) and acquired phonological dyslexia (in which the ability to pronounce non-words or words that have not been seen before is lost) constitute a double dissociation that leads many people to believe in the traditional dual-route model discussed in the section introduction. Similar dissociations between exception word and non-word reading are found in developmental dyslexics (e.g. Frith 1985) and provide further evidence for this dual-route model.

However, there is also evidence that the situation cannot be quite as simple as depicted by the box and arrow diagram in the section introduction. Glushko (1979) has shown that humans do not always use the simplest letter–phoneme rules when they pronounce non-words. They often work by analogy with larger segments of irregular words: for example, the non-word "tive" may be pronounced to rhyme with the irregular word "give" rather than the regular word "five". (Throughout this chapter we shall use quotes to denote word orthography and slashes to denote word phonology.) A related problem is the regularity effect on the naming latencies for non-words: non-words derived from regular words are found to be pronounced faster than non-words derived from irregular words (Glushko 1979). This is difficult to explain if all the non-words are pronounced using the same set of simple grapheme–phoneme conversion (GPC) rules. These two problems with non-words do not necessarily mean that the dual-route model is completely wrong. They may simply indicate that the GPC route must be expanded to accommodate correspondences between larger letter and phoneme strings. More problematic from this point of view is the well known pseudo-homophone effect (McCann & Besner 1987) whereby non-words such as "trax" which sound like real words (i.e. "tracks") are pronounced faster than matched non-words (e.g. "prax") which do not sound like real words. This seems to imply that there must (at least) be some interaction between the lexical/semantic and phonemic routes and is something that must clearly be investigated more closely in specific models.

In addition to this experimental evidence against the traditional dual-route model, the early connectionist reading models of Sejnowski & Rosenberg (1987) and Seidenberg & McClelland (1989) have demonstrated explicitly how the rules and exceptions can both be dealt with in a single system. With hindsight, this is not too surprising. Given that an exception word mapping can be thought of as a very low-frequency high-powered rule (i.e. a rule that is activated only for one specific word and overrides all other potentially useful rules), any rule-based system that has no restriction on the length of letter and phoneme strings it can work with should be able to handle exceptional words as well. Regular words will be pronounced according to simple rules, exception words will be pronounced according to complicated special purpose rules (effectively a lexicon) that must overrule the simpler rules. There will clearly be a continuous spectrum between these two classes of words and since there are very few (if any) "exception" words that do not contain any regular features at all, the need for true lexical entries will be minimal. The success of the model depends on the network maximizing its use of the simple rules whilst minimizing its use of the special-purpose rules. In this way, when presented with new words or non-words, none of the special-purpose rules will fire, and the network will output phonemes according to a full set of regular (GPC) rules, yet it will still be able to pronounce the exceptional words it has been trained on. In fact we do not really want the model to be totally successful in this way, it should still (like humans) pronounce some non-words irregularly (Glushko 1979).

The main constraints on any new reading model are therefore that, when functioning normally, it must produce appropriate pronunciations for non-words (including some which do not follow the main regularities) and it should give some indication of realistic naming latencies including a regularity effect and pseudohomophone effect for non-word naming latencies. When appropriate learning constraints are introduced it should exhibit both surface and phonological developmental dyslexia, and after appropriate forms of damage we expect both surface and phonological acquired dyslexia. Finally, for a preliminary investigation such as this, it is appropriate to keep our models as simple as possible and only introduce complications such as recurrent connections and inbuilt modularity when absolutely necessary.

The main aim of this chapter is to explore these issues for a particular neural network model. Of the existing models, the well known Wickelfeature-based models of Seidenberg & McClelland (1989) are restricted to monosyllabic words, have been shown to have an unacceptably poor non-word performance (Besner et al. 1990) and present many difficulties in analyzing their abnormal behaviour (Patterson et al. 1989). The proposal here is therefore to return to the earlier moving window (i.e. NETtalk) approach of Sejnowski & Rosenberg (1987) that does not suffer from these problems and (by modifying its learning algorithm) remedy its major defect, namely the need to preprocess the training data by hand.

Multitarget learning

The basic NETtalk architecture is a simple feedforward network with some representation of a single phoneme being produced on the output layer that corresponds to the central letter of a set of letters represented on the input layer. The input layer forms a "window" through which the training and testing words (i.e. letter strings) slide one letter at a time. This whole window of input letters is needed, rather than just a single input letter, to provide the necessary context information. Clearly, for this arrangement to work, the length of each output string must be equal in length to the corresponding input string. This means that the set of phonemes corresponding to each word in the training data needs to be padded out with blanks (i.e. phonemic nulls) to the same number of phonemes as there are letters in the word. If there are nl letters and np phonemes in the word, then there are $ntarg = nl!/np!(nl - np)!$ ways that the $nl - np$ blanks can be inserted, but, clearly, we only want the network to train on one of these $ntarg$ possible targets. If we attempt to train on more than one of these targets we will run into the well known over-fitting problems of learning from noisy training data. From this point of view, the alignment problem is simply a statement of the fact that we do not know which of the aligned targets should be used.

We can understand most easily how we need to develop a modified learning process to choose the most appropriate targets by considering exactly what is happening for a typical word. Let us consider, for example, the word "bade" which is pronounced /bAd/. (We shall use the phonemic notation and terminology of Seidenberg & McClelland (1989) throughout.) The basic problem is for our system to decide which of the targets /bAd–/, /bA–d/, /b–Ad/ or /–bAd/ is the appropriate one to learn from. The original NETtalk system had the most appropriate one (namely /bAd–/) picked by hand, but this can be regarded as doing too much of the work the network should be doing. Here we ask if it is possible for a system, such as a neural network with an appropriate learning algorithm, to choose the most suitable target for itself. The natural idea would be to consider all the possible targets and use the one that best fits in with the existing knowledge. This would seem to tie in nicely with how humans normally tend to proceed with new learning instances. Suppose that our system already had a grasp of the basic rules of English pronunciation and was then asked to learn to pronounce the irregular word "give" → /giv/ that it had never seen before. It would probably be expecting it to be pronounced /gIv–/ to rhyme with the regular word "five" and so, by any reasonable definition of similarity, the aligned target that best fits in with that is /giv–/ (which is the correct one) rather than /gi–v/, /g–iv/ or /–giv/. This seems like a natural learning process: first learn the basic rules from small regular words that only have a single aligned target and then use these as a framework for learning longer, less regular and multiple target words.

It would be preferable, however, if our system could learn the whole training set from scratch without the need for some externally imposed incremental learning scheme. The above discussion suggests that if it could just pick up the main regularities from scratch, then the rest should follow fairly easily. Suppose that the system began with no prior knowledge at all and hence the targets for each word were chosen at random. Table 9.1 shows the expected number of letter-to-phoneme correspondences for our word "bade". These are calculated by counting the number of each letter-to-phoneme correspondence over all possible targets for the word and then dividing by the number of targets (e.g. "b" → /b/ three times out of four targets giving an expectation of 0.75 if we chose the target at random a large number of times). For the single word "bade", all the targets are equally good, and we arrive at the somewhat worrying conclusion that the most likely pronunciation to be learnt for the letter "e" is /d/. However, as we consider the total expected correspondences for larger sets of words, such unlikely correspondences tend to get lost as noise. With a reasonably representative set of only five words (namely "bade", "ate", "bed", "bad" and "feel") we are already seeing the most likely correspondences becoming more realistic. By the time we average over the full training set (namely the 2998 word Seidenberg & McClelland corpus) the erroneous correspondences like "a" → /b/

Table 9.1 The expected letter–phoneme correspondence counts.

Letter	Phoneme	Just "bade"	Five words	All words
b	b	0.75	2.75	262.89
	–	0.25	0.25	51.19
a	a	0.00	1.00	247.53
	–	0.25	0.58	190.75
	A	0.50	1.17	132.96
	b	0.25	0.25	7.23
d	d	0.25	2.25	339.15
	–	0.25	0.25	68.08
	A	0.50	0.50	6.90
e	–	0.25	1.08	348.01
	e	0.00	1.00	203.22
	E	0.00	1.00	174.27
	d	0.75	0.75	38.15

and "d" → /A/ are totally overpowered by the more regular correspondences and this turns out to be true for all the possible letters. (For each letter in the table, the two most common correspondences are shown, plus any others more than half as likely as the most likely, plus those actually occurring for the word "bade".) Thus, even if we start off choosing from the potential targets at random, there is still a strong general trend towards learning the regular letter-to-phoneme correspondences. It follows that any learning algorithm, which learns (sufficiently slowly) only from the targets that already best fit in with its existing knowledge, can automatically learn from scratch which are the most appropriate targets to use. Moreover, as the learning proceeds, the target selection becomes increasingly less random, thus improving the chances of learning the sensible letter–phoneme correspondences even further.

The neural network models

It would appear that all that remains to be done is to incorporate our multitarget learning process into NETtalk and we should obtain a successful model of reading without recourse to preprocessing of the training data. No fundamental changes of the network architecture should be required. Our basic model thus consists of a standard fully connected feedforward network with sigmoidal activation functions and one hidden layer set up in a similar manner to the original NETtalk model of Sejnowski & Rosenberg (1987) shown in Figure 9.1. The input layer consists of a window of *nchar* sets of units, each set consisting of one unit for each letter occurring in the training data (i.e. 26 for English). The output layer consists of one unit for each phoneme occurring in the training data (i.e. about 40 units). In due course we may want to replace the moving window approach by a more biologically and psychologically plausible system of recurrent

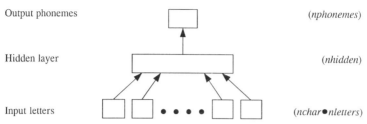

Output phonemes   (*nphonemes*)

Hidden layer   (*nhidden*)

Input letters   (*nchar●nletters*)

Figure 9.1   The basic NETtalk style network architecture.

connections (e.g. as in Jordan 1986) or some other form of input/output buffer system, but for the present study we shall simply assume that this will not significantly affect any of the main results.

The input words slide through the input window which is *nchar* letters wide, starting with the first letter of the word at the central position of the window and ending with the final letter of the word at the central position. Each letter activates a single input unit. Since we are using aligned (i.e. padded) targets there is a one-to-one correspondence between the letters and phonemes, and the activated output phoneme corresponds to the letter occurring in the centre of the window in the context of the other letters occurring in the window.

The networks were trained using the back-propagation gradient descent learning algorithm (Rumelhart et al. 1986) with the extended Seidenberg & McClelland training corpus of 2998 monosyllabic words consisting of the original Seidenberg & McClelland (1989) set plus 101 other words missing from that set (Plaut & McClelland 1993). To allow a study of word frequency effects, about 25% of the training data was presented in each epoch in random order with the same logarithmic frequency distribution as in Seidenberg & McClelland (1989).

The multitarget learning is implemented in the network by calculating (with a suitable error measure) for each input word the total error corresponding to each of the possible targets and only propagating back the error from the target with the least error. The network is then (with a suitably diverse set of training words and a sufficiently small learning rate) able to *learn* which is the appropriate target for each word. For example, consider our word "bade" again. This training example will be presented $nl = 4$ times, each with $ntarg = 4$ possible target outputs, as shown in Table 9.2. For each of the four input presentations the error is calculated for each of the four target outputs. The sum of the errors for each target over the four input presentations is then computed, and the target with the minimum total error is used to update the weights in the appropriate manner. Given small random initial weights and a representative set of training words, common potential rules such as "d" → /d/ will dominate the weight changes over others such as "d" → /A/, and the network will automatically settle into a useful set of targets including "bade" → /bAd–/ rather than /bA–d/, /b–Ad/ or /–bAd/. Once this basic alignment has been learnt, the network can then proceed to fit efficiently the less regular correspondences into this framework.

Table 9.2 The four training data presentations for the word "bade".

Presentation	Inputs	Target outputs			
			or	*or*	*or*
1	– – – – b a d e –	b	b	b	–
2	– – – b a d e – –	A	A	–	b
3	– – b a d e – – –	d	–	A	A
4	– b a d e – – – –	–	d	d	d

One problem that needs to be overcome with this model and training data is that some letters (namely "j", "g", "x") can correspond to more than one phoneme (namely /dZ/, /dZ/, /ks/) and hence some words can have fewer letters than phonemes (e.g. "box" → /boks/). In order to solve this problem without complicating the model, in these cases the combinations /dZ/ and /ks/ were replaced by the additional "phonemes" /J/ and /X/, respectively, bringing the total number of phonemes up to 40. This recoding, however, is not necessary in more sophisticated versions of this model that allow more than one output phoneme per word presentation (Bullinaria 1994).

The small initial weights were chosen randomly with a rectangular distribution in the range –0.1 to +0.1 . After some experimentation, the back-propagation learning rate was fixed at 0.05 and the momentum factor at 0.9. In order to prevent activations getting stuck hard wrong (where the error propagated back is zero) targets of 0.1 and 0.9 are often used rather than 0 and 1 (as in Seidenberg & McClelland 1989). However, a few initial trial runs suggested that slightly better results were achieved using a sigmoid prime offset (Fahlman 1988) of 0.1 instead, and so this approach was used throughout. Over-learning was controlled by not propagating back the error signal for words that already had the correct phoneme outputs and a total error less than some threshold *errcrit*.

Simulation results

For each network simulation we plotted the percentages learnt correctly and the mean square output activation errors for the full set of training data plus various interesting subsets (high/low-frequency regular words, exception words, homographs, etc.) against the number of training epochs. Also, to test the networks' generalization ability (i.e. success at learning the GPC rules), the percentages of three standard sets of non-words that were pronounced "acceptably" and the corresponding error scores were also plotted. These plots were then used to determine the best values for the various parameters used in the model.

For ease of comparison with the results of various experiments on humans and with other reading models we used exactly the same word subsets as Taraban & McClelland (1993). The three sets of non-words used were the regular non-words and exception non-words of Glushko (1979, experiment 1) and the control non-words of McCann & Besner (1987, experiment 1). The allowable pronunciations of these non-words were derived from the training data by matching word segments (particularly rimes) in the non-words with the same segments in the training data and constructing possible non-word pronunciations by concatenating the pronunciations of the segments from the training data. For the regular non-words this typically led to a single allowable pronunciation (e.g. "wote" → /wOt/ by analogy with "note" → /nOt/), but for the exceptional non-words there were often several allowable pronunciations (e.g. "mone" → /mOn/ as in "cone", → /m^n/ as in "done", → /m*n/ as in "gone"). Finally, for comparison with Glushko's (1979) experimental results and other models we also distinguished between the strictly regular pronunciation (e.g. "mone" → /mOn/) and the other pronunciations that were deemed acceptable.

Due to the large amount of processing power required for these simulations, only 25 fairly small runs have so far been carried out, and it is often difficult to distinguish real improvements caused by parameter or architecture changes from statistical fluctuations. The following will be presented in a necessarily longer paper elsewhere (Bullinaria 1994): a full analysis of how the models' performance varies with parameters such as the window size *nchar*, the number of hidden units, *errcrit*, etc.; what structures are actually being represented in the hidden units; the detailed relationship between the output activation errors and naming latencies; possible causes of developmental dyslexia; and exactly how the models respond to different types of damage. This paper will also discuss a number of possible variations on the basic model described above. Here we will restrict ourselves to making a few fairly general observations and present some results from one typical run in which all the training words were learnt correctly.

Since the Seidenberg & McClelland corpus contains 13 pairs of homographs it is clear that the network can never be totally successful in learning this training data. Humans can use various contextual information to deal with homographs, so some experiments were carried out on the use of context flags to resolve these ambiguities in our networks. As a preliminary investigation, this was implemented by introducing a single extra character into the input alphabet and appending that character to the least regular input word of each pair of homographs. This not only allowed the network to achieve 100% success rate on the homographs (compared with a maximum of 50% before) but also seemed to improve its performance on certain non-homographs as well.

Networks with a window size of nine characters and as few as 40 hidden units were able to learn all but one of the training examples, namely "though" → /DO/. The reason the network fails on this word is that the training data also include the word "thought" → /T*t/, in which the subword "though" has to be pronounced as /T*/, and unless the input window is large enough to have the final "t" in the window while the initial "t" is in the centre of the window the network has no way of resolving the ambiguity. By increasing the window size to 13 this long-range dependency can be handled and the network achieves 100% success rate on its training data. To confirm the networks' capability of handling long-range dependencies and also their ability to deal with more complex multisyllabic words, some runs with the words "photographic" → /fotOgrafik/ and "photography" → /fot*grafE/ incorporated into the training data were carried out. Each of these words contain the letter "o" pronounced in two different ways, and the pronunciation of the second "o" is determined only by letters at least six characters away. With a window size of 13 the network was able to learn both words without any difficulty. With a window size of 11 (for which the crucial "i" and "y" fall outside the window while the problematic second "o" is in the central position) the network failed to learn the two words.

The alignments learnt by the networks are easily checked and are generally found to be good, though not always the same for each run (e.g. sometimes "ph" → /–f/ instead of /f–/). We have also determined that the final generalization performance of our networks trained with the multitarget approach is not significantly different from that of the same networks trained with preprocessed training data.

Figure 9.2 shows the learning curves for a typical network with 120 hidden units, a window size of 13 characters, *errcrit* = 0.0001 and a context flag to resolve homograph ambiguities. The generalization performance peaks at about 40 epochs and then falls slightly as the exception words are learnt. The network eventually achieved 100 % performance on the training data, and for non-words plateaued at 95.3 % for regulars, 95.3 % for exceptions and 92.5 % for controls. Comparison with other models is complicated by different authors using different non-word sets and scoring criteria, so bearing this in mind: the original Seidenberg & McClelland (1989) model achieved 97.3 % on the training data and about 65 % on the Glushko non-words; the best models of Plaut et al. described

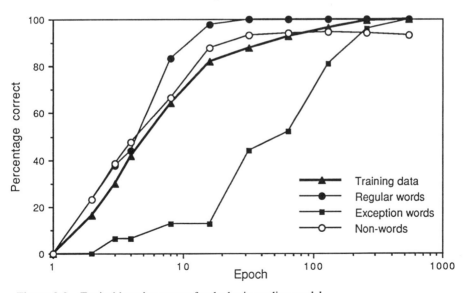

Figure 9.2 Typical learning curves for the basic reading model.

169

in the previous chapter achieved 99.9% and 97.7%; the non-connectionist GPC procedure of Coltheart et al. (1993) also achieved 97.7% on the Glushko non-words but, by construction, did not learn the exception words in the training data. Human subjects would typically achieve virtually 100% for monosyllabic words under unconstrained conditions and about 95% on the Glushko non-words in time-constrained tests (Glushko 1979).

For each word (or subword) presentation, the output phoneme of the network has thus far simply been defined to be that phoneme corresponding to the output unit with the highest activation. An important design choice that needs to be reconsidered at this stage is whether we really should have a separate output unit corresponding to the blank (i.e. the phonemic null). It would seem more realistic to simply have some threshold such that if no output activation exceeds that threshold then the network output is deemed to be a blank. In order to check the feasibility of doing this, networks adopting this new threshold approach were constructed and their results compared with existing networks that do have explicit "blanks" in the output units. Figure 9.3 shows the learning curves for such a network with the same parameters as used for the network of Figure 9.2. We see that as far as the training data are concerned there is very little difference in performance between the two networks. However, there is a significant difference in their generalization performance, with the threshold approach having many more errors.

Table 9.3 lists the actual non-word errors produced by the two networks. Both the explicit blanks and the subthreshold blanks are denoted by "–". For the network with explicit blanks, all the errors are due to incorrect or extra phonemes. None are due to the presence of extra explicit blanks. For the network without explicit blanks a large proportion of the errors are caused by phonemes

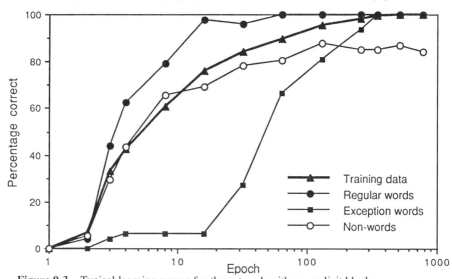

Figure 9.3 Typical learning curves for the network with no explicit blanks.

Table 9.3 Typical non-word errors for the two network types. In each case we have the input, closest correct output and actual output.

Explicit blanks (10 errors)

Missing phonemes
  None

Other problems

fues	fyUz	fyez
nazz	naz-	nazz
tolph	tOlf-	t^lf-
wosh	woS-	w*S-
blaie	blA--	blAA-
mone	mOn-	mw^-
mune	myUn	myn-
phoyce	f--Ys-	f--Yz-
wone	w^n-	w^nn
zute	zyUt	zyt-

No explicit blanks (26 errors)

Missing phonemes

bood	bu-d	b--d
brove	brOv-	br-v-
doad	dO-d	d--d
dold	dOld	d-ld
drean	drE-n	dr--n
drood	drU-d	dr--d
gome	gOm-	g-m-
goph	gof-	go--
heaf	hE-f	h--f
lokes	lOk-s	lOk--
lome	lOm-	l-m-
phoyce	-fY-s-	-fY---
plove	plOv-	pl-v-
praid	prA-d	pr--d
steat	stE-t	st--t
sweak	swE-k	sw--k
sweal	swE-l	sw--l
zope	zOp-	-Op-
zupe	zUp-	-yp-
zute	zUt-	-yt-

Other problems

fues	fyUz	fyez
nazz	naz-	nazz
tolph	tOl-f	tUl-f
wosh	wo-S	w*-S
faije	fA-J-	fAIJ-
wush	w^-S	w^US

missing because they failed to reach the required threshold. This additional source of errors accounts for the worsened generalization performance and means that we will, for the present, have to abandon our no explicit blank models. More sophisticated versions of our model in the future will undoubtedly benefit from the imposition of more complicated decision criteria or basins of attraction in the output unit activation space (e.g. as in Hinton & Shallice 1991), and this may eventually allow us to reconsider this design choice once again.

We noted earlier Glushko's (1979) observation that humans often pronounce non-words by analogy with irregular words rather than by using the main GPC rules, and listed this as being one of the minimum requirements of our model. It actually turns out that our networks tend to pronounce too many non-words by analogy. For the Glushko exception non-words, humans were found to pronounce 18.4% by analogy whereas the model corresponding to Figure 9.2 pronounces 34.9% by analogy. Moreover, many of the generalization errors listed in Table 9.3 can also be recognized as being due to the over-use of analogy (e.g. "mone" → /mw^-/ instead of /mOn-/ is clearly caused by confusion with "one" → /w^n/). Whether this excessive use of incorrect analogies can be resolved by the introduction of recurrent connections, clean up units or different learning algorithms/parameters is not clear at present. However, for small networks it was found that increasing the number of hidden units improved generalization. It was also noticed that if smaller training sets were used, then obviously incorrect grapheme–phoneme correspondences could be learnt (e.g. "bade" → /-bAd/ instead of /bAd-/), without any noticeable effect on the output performance on that training set. It is likely, therefore, that simply by using larger networks and training sets we could further improve our performance on non-words.

Another requirement of our network's non-word performance was that it exhibit some form of regularity effect for its simulated naming latencies. Seidenberg & McClelland (1989) gave convincing arguments that the experimental naming latencies in humans should be monotonically related to the output activation errors in the corresponding neural network model. To check this, we first performed an analysis of variance on the error scores for the 48 regular words and 48 exception words of Taraban & McClelland (1993). We found a significant regularity effect ($F_{1,92} = 153.1$, $p < 0.0001$) and frequency effect ($F_{1,92} = 6.6$, $p = 0.012$) and lack of interaction effect ($F_{1,92} = 0.23$, $p = 0.63$), as is observed in humans. Applying the same analysis to the 43 regular non-words, the 43 exception non-words and the matched control words of Glushko (1979), we found a significant word versus non-word difference ($F_{1,168} = 74.3$, $p < 0.0001$), a significant regularity effect ($F_{1,168} = 13.0$, $p < 0.0005$) and a not quite significant interaction effect ($F_{1,168} = 3.2$, $p = 0.076$). Our simulated naming latency results thus appear to be in broad agreement with the experiments on humans. However, there are still some problems that need to be overcome. First, separate t tests of the word versus non-word differences reveal that, while there is a significant difference for the exceptions ($t_{1,84} = 4.7$, $p < 0.0005$), there is no significant difference for the regulars ($t_{1,84} = 1.1$,

$p = 0.27$). This is not only in disagreement with the experimental naming latency results for humans but also means that we cannot use the error scores as a basis for lexical decision. The second problem concerns the pseudohomophone effect, which is always likely to be a problem for single-route models (e.g. Besner et al. 1990). Performing t tests on the error scores of the 80 pseudo-homophones, the 80 control non-words and 80 control words of McCann & Besner (1987) shows that we again have significant differences between the pseudohomophones and the control words ($t_{1,158} = 4.0$, $p < 0.0002$) and be-tween the control non-words and the control words ($t_{1,158} = 4.2$, $p < 0.0001$), but we have no significant difference between the pseudohomophones and the control non-words ($t_{1,158} = 0.20$, $p = 0.84$); that is, our model does not exhibit the pseudohomophone effect. We shall discuss the implication of these difficul-ties once we have determined what other problems the model has.

Developmental and acquired dyslexia

The gross features of the learning curves shown in Figure 9.2, in particular the clear differences we see in the rates of learning of the regular and exception words, are in broad agreement with human development (e.g. Backman et al. 1984). Moreover, anything that slows down the learning process or prevents it from proceeding much past epoch 20 is liable to reproduce the effects of one common form of developmental dyslexia in which there is a particularly large regularity effect but no unusual problem with non-word reading (e.g. Backman et al. 1984). However, our model appears to be unable to offer any explanation of the other main form of developmental dyslexia in which we have a less than normal regularity effect and increased problems in pronouncing non-words (e.g. Frith 1985).

As discussed previously, an important method of constraining cognitive models is to examine their performance after damage (e.g. Coltheart et al. 1993). It has already been mentioned that, for models of reading, we are particu-larly interested in the two forms of acquired dyslexia that constitute a double dissociation: patients with phonological dyslexia exhibit a dissociation between word and non-word reading – for example, there can be complete failure to read non-words whilst maintaining around 90% success on words (Funnell 1983); patients with surface dyslexia exhibit a dissociation between regular and excep-tion word reading – for example, 90% success on low-frequency regular words against 40% on low-frequency exceptions (Bub et al. 1985) and 80% on regular words against 35% on very irregular words (Shallice et al. 1983).

Connectionist models that can deal with both the regular and exception words in a single system have previously cast doubt on the traditional dual-route explanation of the double dissociation (e.g. Seidenberg & McClelland 1989, Bullinaria & Chater 1993). For this reason, it was important for us to examine the effect of a wide range of forms of simulated brain damage on the performance of our model. There are numerous ways we can interfere with either whole hidden units (which correspond to neurons in real brains) or individual network connection weights (which correspond to synaptic strengths). We actually inflicted five representative types of damage on our networks, namely the global reduction of all weights by constant amounts, the global scaling of all weights by constant factors, the addition of Gaussian random noise to all weights, the random removal of connections, and the random removal of hidden units. We found that none of these forms of damage were able to produce the loss of non-words (i.e. the GPC rules) to a significantly greater extent than the words, let alone produce the enormous dissociations found in some human patients. We have to conclude that the network cannot be considered to constitute a realistic single-route model on its own.

The most obvious explanation is that phonological dyslexia must still be explained by the loss of the GPC route, but not the lexical–semantic route, of some form of dual-route model (Coltheart et al. 1993). Some other possibilities have been suggested by Patterson & Marcel (1992), but we await explicit computational models that test their feasibility. It would seem appropriate, therefore, to consider our model to be an implementation of the GPC route of a dual-route model. However, given our model's inherent success with the exception words, losing the lexical–semantic route but not the GPC route is no longer enough to explain surface dyslexia. We must, at the same time, lose the exceptions to a greater extent than the rules in our GPC route. Fortunately, all five forms of damage described above had this effect on our networks. For networks with large numbers of hidden units, globally reducing all the weights by successive factors of 0.9 is a convenient deterministic procedure that has a very similar effect in

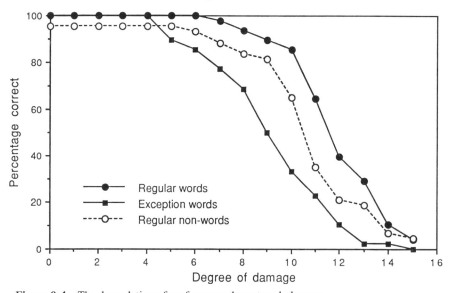

Figure 9.4 The degradation of performance by network damage.

Table 9.4   Initial regularization errors due to network damage.

Input	Normal output	Post-damage output
bush	buS	b^S
pear	pAr	pEr
pint	pInt	pint
shoe	SU	SO
spook	spUk	spOk
worm	werm	wOrm
both	bOT	boT
do	dU	d^
move	mUv	mOv
says	sez	sAz
word	werd	wOrd

large networks to adding random noise or removing appropriate fractions of the hidden units and/or connections at random. Figure 9.4 shows that the effect of doing this (for the same network as Fig. 9.2) is the required dissociation. We see that the exception words are preferentially lost over the regular words, so that at the point where the weights have been reduced by a total of ten factors we have regular word performance at 85% compared with exception word performance at 33%. The exact percentages seem to vary somewhat with different network parameters, in particular the number of hidden units, but they are generally not far from those found in human patients. The networks also show the same remarkable resilience to small amounts of damage as humans.

Most of the exception word errors made by surface dyslexics who have near-normal regular word performance are found to be regularizations, whereas in patients that also have impaired regular word reading the errors are not necessarily perfect regularizations (e.g. Shallice & McCarthy 1985). We find a similar pattern of errors with our networks. Table 9.4 shows the exception word errors at damage point 6 on Figure 9.4, which is the last point before we begin to get errors on the regular words. We see that there is a clear tendency to regularize. As we increase the damage from here, the regular word errors begin to appear along with the non-regularization errors on the exception words. It seems that our model can provide a fairly accurate account of surface dyslexia but not phonological dyslexia.

Conclusions

We have shown in detail how the NETtalk model of Sejnowski & Rosenberg (1987) can be modified to produce a simple neural network model of reading aloud with an output performance on the training data and on non-words that is already comparable to that of humans in time-constrained experiments.

Moreover, unlike earlier models, multisyllabic words and long-range dependencies can be handled without the need for preprocessing of the training data, and we do not require any complicated input/output representations such as Wickelfeatures. We have also shown how the use of very simple context flags can solve the problems of homograph ambiguity. The networks' performance can almost certainly be improved further simply by using more hidden units, more hidden layers, more realistic training sets and/or introducing recurrent connections. The inclusion of stress markers and continuous speech in the training data should be possible in just slightly more complicated versions of the model, and the "inverse problem" of spelling should be equally tractable.

At the beginning of this chapter we identified a number of features which models of reading such as ours should exhibit. The first of these was that the model should be able to pronounce non-words with a realistic proportion by analogy with exception words rather than using the main GPC rules. If anything, our model has exceeded our expectations in this respect. We were next concerned with naming latencies, in particular the regularity effect for non-words and the pseudohomophone effect (which are both difficult to understand in terms of traditional dual-route models). We found that our models' simulated naming latencies were in broad agreement with experiments on humans, including the regularity effect for non-words, but they failed to exhibit the pseudohomophone effect and did not show the required differences between regular words and non-words derived from those regular words. We have also noted that the model provides no basis for lexical decision. These problems all seem to indicate that our model still needs to be supplemented by an additional lexical–semantic system.

The main problem we were expecting to have (as with any proposed single-route model of reading) was in reproducing the double dissociation between exception word and non-word reading that is found in dyslexics. After several different representative types of network damage, our models were found to exhibit symptoms similar to acquired surface dyslexia, but there appears to be no way for them to experience acquired phonological dyslexia. Similarly, it seems likely that we will be able to understand developmental surface dyslexia but not developmental phonological dyslexia. These problems seem to confirm the indications from the naming latencies that we still need another route. Clearly, the dual-route model is not dead yet!

It would seem fair to conclude, however, that a class of neural network models has been presented which, given their simplicity, performance and room for improvement, provides a promising basis for the rule-based route of a realistic dual-route model of reading. Moreover, now that we have a "GPC route" that can deal with all the exceptional words as well as the regular words, we no longer have to posit a mechanism to resolve the conflicting outputs of the two routes for exceptional words. Work is now in progress to build on the current model to produce a complete connectionist dual-route model of reading.

Acknowledgements

The author acknowledges the support of the Joint Council Initiative in HCI and Cognitive Science.

References

Backman, J., M. Bruck, M. Hébert, M. Seidenberg 1984. Acquisition and use of spelling–sound information in reading. *Journal of Experimental Child Psychology* **38**, 114–33.

Besner, D., L. Twilley, R. S. McCann, K. Seergobin 1990. On the connection between connectionism and data: are a few words necessary? *Psychological Review* **97**, 432–46.

Bub, D., A. Cancelliere, A. Kertesz 1985. Whole-word and analytic translation of spelling-to-sound in a non–semantic reader. In *Surface dyslexia: neuropsychological and cognitive studies of phonological reading*, K. E. Patterson, J. C. Marshall, M. Coltheart (eds), 15–34. London: Lawrence Erlbaum.

Bullinaria, J. A. 1994. Representation, learning, generalization and damage in neural network models of reading aloud. Submitted.

Bullinaria, J. A. & N. Chater 1993. Double dissociation in artificial neural networks: implications for neuropsychology, *15th Annual Conference of the Cognitive Science Society, Proceedings*, 283–8. Hillsdale, New Jersey: Lawrence Erlbaum.

Coltheart, M., B. Curtis, P. Atkins, M. Haller 1993. Models of reading aloud: dual-route and parallel-distributed-processing approaches, *Psychological Review* **100**, 589–608.

Fahlman, S. E. 1988. An empirical study of learning speed in back-propagation networks. A technical report from Carneige Mellon, CMU-CS-88-162.

Frith, U. 1985. Beneath the surface of developmental dyslexia. In *Surface dyslexia: neuropsychological and cognitive studies of phonological reading*, K. E. Patterson, J. C. Marshall, M. Coltheart (eds), 301–30. London: Lawrence Erlbaum.

Funnell, E. 1983. Phonological processes in reading: new evidence from acquired dyslexia. *British Journal of Psychology* **74**, 159–80.

Glushko, R. J. 1979. The organization and activation of orthographic knowledge in reading aloud. *Journal of Experimental Psychology: Human Perception and Performance* **5**, 674–91.

Hinton, G. E. & T. Shallice 1991. Lesioning an attractor network: Investigations of acquired dyslexia. *Psychological Review* **98**, 74–95.

Jordan, M. I. 1986. Attractor dynamics and parallelism in a connectionist sequential machine. *8th Annual Conference of the Cognitive Science Society, Proceedings*, 531–6. Hillsdale, New Jersey: Lawrence Erlbaum.

McCann, R. S. & D. Besner 1987. Reading pseudohomophones: implications for models of pronunciation assembly and the locus of word–frequency effects in naming. *Journal of Experimental Psychology: Human Perception and Performance* **13**, 14–24.

Patterson, K. & A. Marcel 1992. Phonological ALEXIA or PHONOLOGICAL alexia? In *Analytic approaches to human cognition*, J. Alegria, D. Holender, J. Junça de Morais, M. Radeau (eds), 259–74. Amsterdam: Elsevier.

Patterson, K., M. S. Seidenberg, J. L. McClelland 1989. Connections and disconnections: acquired dyslexia in a computational model of reading processes. In

Parallel distributed processing implications for psychology and neurobiology, R. G. M. Morris (ed.). Oxford: Oxford University Press.

Plaut, D. C. & J. L. McClelland 1993. Generalization with componential attractors: word and non-word reading in an attractor network, *15th Annual Conference of the Cognitive Science Society, Proceedings*, 824–9. Hillsdale, New Jersey: Lawrence Erlbaum.

Rumelhart, D. E., G. E. Hinton, R. J. Williams 1986. Learning internal representations by error propagation. In *Parallel distributed processing: explorations in the microstructure of cognition*, vol. 2. *Physiological and biological models*, D. E. Rumelhart & J. L. McClelland (eds), 318–62. Cambridge, Mass.: MIT Press.

Seidenberg, M. S. & J. L. McClelland 1989. A distributed, developmental model of word recognition and naming. *Psychological Review* **96**, 523–68.

Sejnowski, T. J. & C. R. Rosenberg 1987. Parallel networks that learn to pronounce English text. *Complex Systems* **1**, 145–68.

Shallice, T. 1988. *From neuropsychology to mental structure*. Cambridge: Cambridge University Press.

Shallice, T. & R. McCarthy 1985. Phonological reading: from patterns of impairment to possible procedures. In *Surface dyslexia: neuropsychological and cognitive studies of phonological reading,* K. E. Patterson, J. C. Marshall, M. Coltheart (eds), 361–97. London: Lawrence Erlbaum.

Shallice, T., E. K. Warrington, R. McCarthy 1983. Reading without semantics. *Quarterly Journal of Experimental Psychology* **35A**, 111–38.

Taraban, R. & J. L. McClelland 1993. Conspiracy effects in word pronunciation. *Journal of Memory and Language* **26**, 608–31.

COMPUTATION AND STATISTICS

Dimitrios Bairaktaris

The recent dramatic developments in connectionist models of human behaviour are mainly due to the revival of neural computation, and in particular the simplicity and effectiveness of one specific algorithm, back-propagation. As the number of variations and improvements made to the standard back-propagation algorithm increases, more psychologically plausible models of human behaviour are emerging. A number of these models can be found in the other sections of this book. The widespread use of neural algorithms and architectures demands in-depth understanding of their computational properties and relative strengths and weaknesses.

We can enhance our understanding of the computational qualities underpinning neural algorithms by attempting to discover the level of "computing" equivalence with respect to other calculating machines. One such early attempt, using simple pretrained networks, can be found in Foster (1990). Foster demonstrates how, by means of abstraction, the final network layout can be gradually reduced to a number of internal states which are directly comparable to those of other symbolic algorithms performing the same input/output function. In Chapter 10, Bob Kentridge presents an exciting new perspective on internal machine state equivalence, by attempting to classify a particular set of highly complex biologically plausible network architectures in terms of the standard Chomsky hierarchy of grammars. Using the Crutchfield & Young (1990) classification algorithm, a number of networks with different dynamical properties are shown to be equivalent to a set of varying complexity finite-state automata, thus revealing to its full extent the structure and computational power of the true computational engine concealed under the standard node and connection neural formalisms. What makes his approach particularly timely with respect to connectionist models of cognitive function is its immediate relevance to biologically motivated neural architectures. His work will be especially useful to anyone with a limited computer science background who is interested in discovering the "true" computational abilities of neural networks.

While the majority of analytical attempts at understanding connectionist algorithmic complexity remain firmly within the realm of pretrained network configurations, the majority of cognitive modellers are attracted to connectionism because of the availability of a wide range of useful learning tools. Learning techniques are not exclusive to connectionist models. "Traditional" symbolic artificial intelligence learning techniques have been used with success for some time now. However, a direct comparison between artificial intelligence and connectionist learning seems to be an unprofitable path to explore, as these two areas are attempting to address rather different problems. What emerges as a more useful relationship to investigate is the relation between connectionist and statistical models of cognitive function. In Chapter 11, Nick Chater makes a successful attempt to relate these two approaches. His analysis quite clearly indicates that connectionist models should be viewed as an integral part of a long-standing psychological tradition in statistical modelling. This new perspective on connectionist modelling is taken on board in by Steve Finch, Nick Chater and Martin Redington in Chapter 12. They present a model of syntax acquisition using distributional statistics. Their work provides a helpful insight into how connectionist modelling ideas about machine learning can be integrated into standard statistical techniques.

References

Crutchfield, J. P. & Young, K. 1990. Computation at the onset of chaos. In *Complexity, entropy and the physics of information*, W. H. Zurek (ed.), 105–7. Redwood City, Calif.: Addison-Wesley.

Foster, C. L. 1990. *Algorithms, abstraction and implementation: a massively multilevel theory of strong equivalence of complex systems*. PhD thesis, Centre for Cognitive Science, University of Edinburgh.

CHAPTER 10

Cortical neurocomputation, language and cognition

Robert W. Kentridge

Introduction

Understanding or producing natural language requires considerable computational power. Although we cannot specify precisely what these requirements are, it is clear that they differ qualitatively from the simplest formal grammars. I argue that the computation taking place in most artificial neural networks as they process stimuli "on-line" cannot exceed the power of these simple regular grammars so alternative models of linguistic neurocomputation must be sought. One alternative framework for neurocomputation, pattern formation in networks with critical dynamics, is investigated. The results of reconstructions of the symbolic dynamics of one of these networks suggests that they may indeed be capable of implementing qualitatively more powerful forms of computation that natural language processing requires.

Any process can, in principle, be described in formal computational terms. Chomsky (1963) defined a hierarchical classification of different classes of formal computation which qualitatively distinguishes processes in terms of the types of grammars necessary to generate (or parse) different behaviours. This hierarchy can also be defined in terms of the automata equivalent to these grammars (Table 10.1).

Table 10.1 The Chomsky hierarchy of grammatical classes and their equivalent formal machine types.

	Grammar	Machine	Machine description
Type 0	Unrestricted	Turing machine	FSA with infinite tape
Type 1	Context-sensitive	Linear-bounded automaton	FSA with finite tape
Type 2	Context-free	Pushdown stack automaton	FSA with stack
Type 3	Regular	Finite-state automaton (FSA)	Bare FSA

This is a strict hierarchy, no grammar in it can accept (or produce) behaviour only describable by a higher-level grammar. Because the hierarchy is equally applicable to the generation and the parsing of behaviours we can use it to infer the properties of unknown systems which generate particular behaviours from the class of computation required to parse those behaviours. In other words, if we can only meaningfully interpret a code produced by an unknown machine with a context-free grammar then we know that the machine cannot be described by a regular grammar (or finite-state automaton), it must be performing computation of at least context-free power in order to generate this behaviour (unless, of course we are interpreting "information" which was never encoded in the first place!).

Although it is not currently a popular area of research, one of the central questions addressed by cognitive scientists in the 1960s and early 1970s was the position of natural language in the Chomsky hierarchy of formal grammars. It is reasonably certain that natural language generation requires computational power above the lowest level of the hierarchy, regular grammar. It is less clear how the generative power of natural language is restricted relative to the highest class of the hierarchy (e.g. see Winograd 1983: 174–5). Nevertheless, we might assert that since humans generate and understand natural language, and presumably use their brains to do it, then one behaviour which we can assume a model of human brain function should exhibit is the ability to perform the computation necessary to parse or generate natural language. In other words, its intrinsic computational power should exceed that of regular grammars.

Neural networks

The dominant approach to modelling linguistic and cognitive processes used to be based around systems of symbolic propositions. Such "traditional" artificial intelligence systems could clearly implement computation above the power of regular grammars. They did, however, encounter many other problems when dealing with ambiguous or incomplete information. Artificial neural networks, on the other hand, are well suited to dealing with such imperfect stimuli and have the added appeal that they may provide us with some insights into the biological implementation of cognition.

Artificial neural networks consist of a large number of "neuron-like" elements connected together by links through which the activity of one unit influences the activities of those it is connected to. Entities are represented in neural networks as patterns of activation in these units. The strength of the connections between units determines how the pattern of activity in the network changes in response to an input stimulus. Depending on the arrangement of connections the effect of an external input can either propagate through the network in a single pass, or, if some of the connections feed back into the net, then its effects can develop over time (perhaps forever). In this way the pattern of activity produced in some of the units by an externally applied stimulus, together, perhaps, with the initial state of the network, evolve to produce a new pattern of activity which represents the network's "interpretation" of the stimulus. Given an appropriate method of adjusting the strengths of the connections between units a neural net can be made to classify stimuli and produce useful outputs even if the stimulus is incomplete or quite novel.

Many methods of adjusting connection strengths in neural networks have been devised, and have been used to solve real problems. This appears to be a very promising situation; however, some more detailed consideration of the way in which knowledge is represented in neural networks reveals potentially serious problems.

The problem of representing rules and relations in neural networks

There are two broad classes of neural nets, those in which a unit in the net "stands for" some quality (e.g. "redness", "having feathers"), and those in which some "input" units stand for qualities of the stimulus (perhaps in terms of a sensory map, or perhaps in terms of more abstract qualities assumed to have been analyzed at an earlier stage) and some "output" units stand for qualities of the system's interpretation of the stimulus (e.g. "is a fire engine", "can fly", "is a connected figure", etc.), but most units are "hidden" and, although the relationship between input and output depends on them, they have no explicit individual meaning – their response to particular stimuli is determined by the algorithm used to construct appropriate connections in the net. In both cases, however, the representation of a stimulus in the network, unlike its representation in a symbolic system, is not arbitrary – it is dependent, albeit indirectly, on some qualities of the stimulus itself reflected in the coding of the stimulus in the input to the network. As representations are not arbitrary and are distributed over the whole network it is not possible to distinguish between a state of the network which represents the presence of two separate objects and one which represents a single complex object. The consequence of this is very serious – there is not only a problem representing multiple objects, it is also follows that it is equally hard to represent any relation between objects, and hence any rule. As Fodor & Pylyshyn (1988) argue, models of thinking require systematic representation of knowledge at a very minimum – that is, knowledge represented in a manner which allows generalizable relations to be represented. This does not appear to be the case for most neural networks.

Representing multiple objects in networks

One obvious method of allowing multiple representation to exist in a net, which is the first step towards encoding relations and rules, is to separate representations in time. This principle can be applied either sequentially, so that representations of items and relations follow one another, or in parallel, so that representations of multiple items and relations exist together in the network but are distinguished on the basis of the temporal characteristics of their encoding (for example in terms of the relative phase or oscillation frequency of units involved in particular representations). In general, rule representation has been the explicit aim of networks using parallel temporal coding schemes (e.g. von der Malsburg & Bienenstock 1986, Shastri & Ajjanagadde 1989, Shastri & Ajjanagadde 1993), while it emerges and may not even be recognized in networks which identify structures over time. These networks are, however, capable, in principle, of encoding relations between items as well.

Computational power and network dynamics

There are then a number of methods which, either in principle or in practice, are capable of allowing neural networks to represent relations between objects. Before considering what types of computation these methods can, in principle, implement we must avoid a possible confusion. It has been shown that purely feedforward networks with at least one "hidden" layer of units are, in principle, capable of approximating any function in their mapping of input onto output states (Hornik et al. 1990). This corresponds to the highest level of the Chomsky hierarchy – unrestricted grammar. I would like, however, to draw a distinction between the computation performed by a neural network as it responds to stimuli ("on-line" computation) and the computation performed by the combined processes of training the network and preprocessing the stimuli combined with the network's response to those stimuli. Although a feedforward network may behave as a universal function approximation, the network's response to stimuli is essentially just a matter of table look-up (which can be conceived of as a very simple regular grammar in which a single transition is made from each starting state to a corresponding terminal state). The computation which makes this table look-up interpretable as an unrestricted grammar actually takes place during the training of the network. Given sufficient training and enough units, a feedforward network could be trained to decode messages produced by a stack automaton for example. The network is not, however, learning an algorithm, it is learning a set of input/output mappings. If we present the network with a coded message which it has not been exposed to during training, the extrapolation or interpolation of the function it has learned may not produce an appropriate output. Consider the extremes of representation which a feedforward net could apply to the problem. One possibility is that each starting state could be mapped onto an output consisting of the complete decoding of the message. At the other extreme, each symbol in the message could be mapped onto a single decoded symbol and a new "context" added to the next symbol of code to be processed. The first scheme is likely to fail to produce generalization as there is no way of simply partitioning input patterns (which is all a feedforward network can do) in order to uncover the regularities salient to a correct decoding. On the other hand, the single-symbol approach is likely to fail because, while the output required in response to a novel input state is related to generalizable features in the pattern of the input, this relationship is ambiguous unless it is set in the context of more than one prior code symbol.

Having put "computation" in networks with no feedback, and hence no dynamics, to one side, let us now consider the behaviour of dynamic networks in a computational framework. The behaviour of networks with dynamics is not just dependent on the stimuli to which they are exposed at any instant, but also on their prior state. If we wish to make inferences about the computational power of networks it is important to establish how far this dependence of responses on the processing of prior stimuli extends back in time. There is a clear computational distinction between systems in which the transitions between states depend on current state and input – finite-state automata, and those in which state transitions depend on current input and a potentially infinite history of prior states – pushdown stack automata, linear-bounded automata, Turing machines, etc. (the distinctions between these latter types derive from constraints on the construction of the state histories they can use). What implication does this argument have for information processing with dynamic neural networks?

Information is often represented in dynamic neural networks as point attractors in the network's dynamics. From any given starting state, the state of the network evolves towards the attractor state whose basin of attraction the starting state happens to be in. The rules used to determine where attractors are formed and the addition of noise and other constraints allow this general scheme to be used in a wide variety of problems (e.g. pattern recognition or content-addressable memory (Hopfield 1982), optimization (Hopfield & Tank 1985) and constraint satisfaction (Ackley et al. 1985)). In computational terms the "normal" way of using these networks is of minimal computational power. The transition from an arbitrary starting state to an encoded attractor state is simply a mapping, just as the behaviour of the feedforward network was. If, however, we ask the question of how a subsequent stimulus might affect the network the situation becomes a little more interesting. If this new stimulus perturbs the states of some, but not all, of the neurons in the network then the basin of attraction which the network will consequentially find itself in depends on both the new stimulus and the prior state of the network. The transitions from basin of attraction to basin of attraction (which are the informationally significant subdivisions of the network's states) depend on the history of the network. This dependence is, however, only the "single-step" dependence of a finite-state machine, not the arbitrarily long time dependence of more powerful stack or tape machines. This becomes clear when we consider the nature of attractors. The reason that attractors are useful information-processing tools is that their use allows a system to selectively throw away "irrelevant" information. All of the different states of a system within one basin of attraction are reduced to a single attracting state. Once the network has reached one of these attracting states no information remains which can distinguish where in that state's basin of attraction the network had previously been. When a system makes a transition to a new attractor, all information of its prior states is therefore lost. The direct dependence of the response of an attractor network to a stimulus cannot therefore extend beyond its current state. Exactly the same argument can be applied to networks which support multiple limit cycle or chaotic attractors as opposed to point attractors (as do various recurrent error-propagating networks (Pearlmutter 1988, Williams & Zipser 1989)) – the point is that once the system has converged onto one of a set of attractors it cannot directly recover information about which other members of the set it has visited or when it did so.

Any system using attractors to represent the informationally significant subdivisions of its state space does not appear capable of supporting computation above the level of finite-state automata. Nevertheless, physical dynamic systems which have greater computational power clearly exist, and some natural examples appear to have neural network processors. How can their dynamics be harnessed for computation if not in terms of a sequence of attractor transitions?

Much has been made in recent work on information processing in complex dynamical systems of the importance of "the edge of chaos" (Packard 1988, Crutchfield & Young 1989, Crutchfield & Young 1990, Mitchell et al. 1993). Phenomenologically, the interesting thing about physical systems undergoing "slow", "second-order" or "critical" phase transitions (one of which involves being at "the edge of chaos") is that their behaviour is dependent on interactions between their components which extend over arbitrary distances of space and time (e.g. see de Gennes 1975, Uzunov 1993). In terms of the computational consequences of dynamics we have been discussing this is very interesting since it implies that under these conditions dynamical systems may implement computation of greater power than finite-state automata. In some macroscopic physical systems these long-range interactions manifest themselves in the formation of coherent spatiotemporal patterns (e.g. see Nicolis & Prigogine 1989). The formation of these patterns in systems composed of billions of interacting components indicates that long-range interactions between components at phase transitions can greatly simplify the dynamics of complex systems – some relatively low-dimensional dynamics are governing the behaviour of the system, as opposed to those in which the system's components behave more or less independently. There are two quite different ways in which fluctuations (which, for our purposes, may be the stimuli we occasionally inject into a network) can effect pattern-forming systems – multistability and defects (see Gaponov-Grekhov & Rabinovich (1992: ch. 4) for an excellent concise review). Effects which operate via multistability essentially move the system from a regime in which one type of pattern is stable into a regime where a different type of pattern is stable. Although it can be difficult to conceptualize complex pattern-forming systems in terms of simple attractors, these different pattern-forming states are clear analogues of attractors and, just as we argued in the case of multiple-attractor systems, information processing acting through fluctuations which move a system *between* stable pattern-forming modes cannot be more powerful than that of finite-state automata. In contrast to effects operating through multistability, fluctuations which produce defects do not result in the wholesale replacement of one mode of behaviour by another. Defects are boundaries between areas whose dynamics differ in some way. These areas may be forming different types of patterns, or they may be forming different variants on the same pattern (just as defects in crystals separate regions of the crystal lattice which are out of register with one another). Once formed, defects can move and may either spread or decay. Most importantly, an existing defect can influence the effect that a fluctuation has on a system and, although it may be changed in some way if the fluctuation induces a new defect, the original defect is not inevitably destroyed. If we wish to think about information processing using defects in terms of attractors it is simplest to imagine a system governed by a single attractor with many different lobes (the Lorenz butterfly attractor has two lobes for example). The dynamics of the system lies on a transient converging towards the attractor. The system may move from orbiting around one lobe to another lobe spontaneously; however, externally applied fluctuations can force such changes to occur. In contrast to the information processing schemes discussed earlier which move *between* attractors, we can conceive of fluctuations in systems with critical phase transition dynamics as moving the state of the system *within* a single complex attractor (or at least on a transient very near to it).

Experiments on network dynamics and computational power

Introducing the model

The arguments presented above suggest that it would be interesting to examine pattern formation in neural networks with critical dynamics, and, if possible, to assess their computational power in this regime. As I mentioned earlier in the chapter, current artificial neural networks are not constructed with these dynamics in mind. On the other hand, both the argument above and some evidence from electrophysiological recordings (Freeman & van Dijk 1987, Kelso et al. 1992, Young et al. 1992) suggest that real brains may well support such dynamics. I have therefore studied these phenomena in network models which are more closely based on the anatomy and physiology of the cerebral cortex than most artificial neural network models. In the First Neural Computation and Psychology Workshop I discussed a biologically plausible neural network model which could be maintained in a critical dynamic regime over a wide range of conditions (Kentridge 1994). Very briefly, the model consists of simple spiking neurons which fire when their membrane potentials exceed a threshold. They are leaky integrators of their input stimulation, their membrane potential decaying exponentially with time: if $v_j(t) \geq th_j$ then

$$\forall \{t_r : t_r < t + r\}, v_j(t_r) = 0, x_j(t_r) = 0$$
$$\{t_r : t_r = t + r\}\, v_j(t_r) = 0, x_j(t_r) = ap_j$$

(10.1a)

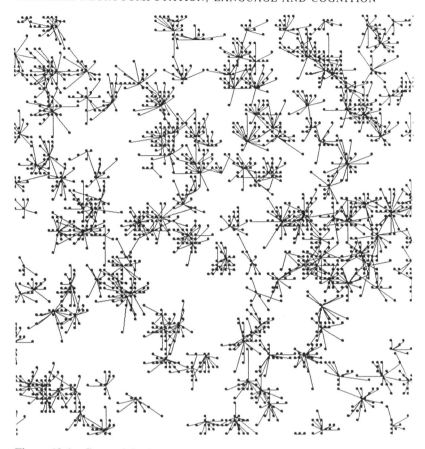

Figure 10.1 Connectivity in the network. Solid circles represent neurons, lines the connections between them. For clarity, only 5% of the connections in the network and their associated neurons are shown.

if $v_j(t) < th_j$ then

$$\frac{\mathrm{d}v_j(t)}{\mathrm{d}t} = -kv_j(t) + \sum_{\forall \{i \neq j\}} w_{ij}x_i \qquad (10.1\mathrm{b})$$

where $v_j(t)$ is the membrane potential of neuron j at time t, $th_j(t)$ is its threshold, $x_j(t)$ is its action potential, ap_j is the sign of action potential for that neuron ($+1$ for excitatory cells, -1 for inhibitory ones), r is the refractory period of the neuron, k_j is the membrane decay time-constant of the neuron and w_{ij} is the strength of synapse from neuron i to neuron j.

The network consists of a two-dimensional sheet of these neurons. Connectivity between neurons in this sheet is spatially localized – the probability of a synapse from one neuron to another being present being inversely proportional to a Gaussian of the distance between them (Fig. 10.1). The strengths of all such

188

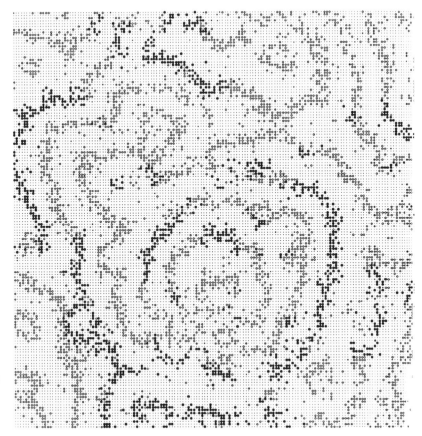

Figure 10.2 A two-dimensional network in which a small amount of potential is uniformly and constantly added to each neuron. The waves of activity produced are partially desynchronized allowing more complex interactions than those occurring in very weakly driven networks with completely synchronized "target wave" activity patterns. Firing units are shown as solid circles. The membrane potential of other units is proportional to the diameter of the open circles corresponding to them.

synapses are set randomly. All neurons are excitatory in the simulations discussed here. When such a network is driven by uniform low-intensity stimulation of all neurons, spatiotemporally organized patterns of activity develop (Fig. 10.2). The network exhibits a range of behaviours typical of systems undergoing critical phase transitions, including pattern formation.

In the rest of this chapter I analyze the behaviour of this model in computational terms.

Measuring computation in dynamical systems

The general problem we face is to characterize the computational power of a system without prior knowledge of the way in which invariances in the system's behaviour might correspond to "symbols" or "rules" which could be interpreted as forming that computation. This problem has been faced by a number of other authors. One approach is to attempt to adapt the system to solve computational problems of some known complexity. Attempts to use genetic algorithms to adapt cellular automata to solve particular computational problems in this manner (Packard 1988) have been beset by difficulties both in characterizing the dynamics of these discrete space and time systems (Li et al. 1990) and by determining whether it is those dynamics themselves, rather than their interaction with the genetic algorithms used to search the systems' parameter space, which determines the capabilities of the systems found (Mitchell et al. 1993). The problem of searching the parameter space of the model neural network would be much more severe than those encountered in one-dimensional cellular automata space. This approach does not therefore seem appropriate.

An alternative to studying the solution of explicit computational problems is to characterize the implicit computational complexity of systems, that is, to determine the computational power of the simplest machine which can describe the behaviour of the system irrespective of any interpretation of that behaviour (Crutchfield & Young 1989). The aim is not, therefore, to find a way of achieving a particular computation, but rather of finding what class of computation the system would be capable of if a suitable interpretation could be found. Application of this method to very simple dynamical systems shows a qualitative difference between the implicit finite-state computation found in both ordered and chaotic regimes and the computational power of the type of stack machine required to describe the dynamics of the system at the transition between ordered and chaotic regimes (Crutchfield & Young 1990). Our hypothesis regarding the network is that it too should show this type of behaviour in the critical regime.

Before describing Crutchfield & Young's algorithm for producing symbolic description of dynamical systems I will pause briefly to consider some general issues regarding the computational description of dynamical systems. In general we expect computational systems to be composed of discrete symbols and rules, whereas dynamical systems are much more general, covering systems with both continuous and discrete space and time evolution. The symbolic dynamics of a system with continuous space or time variables can be investigated by discretely sampling or averaging those variables to produce a version of the system with completely discrete dynamics. This is one of the features of Crutchfield & Young's algorithm. Once we have produced a discrete picture of a system's dynamics we then need to abstract a set of rules which describe those discrete dynamics. Rule abstraction can be based upon two separate views of the fundamental processes governing a system's dynamics: deterministic or probabilistic. Rule abstraction from discrete sequences is, perhaps, most usually encountered in the concepts of algorithmic complexity (Chaitin 1975); here the aim is to find the most concise description of a discrete sequence of symbols – the rate of growth of the size of this description with the length of string to be described is an index of the complexity of the string. The most concise descriptions will often be in terms of sets of mappings between symbol sequences in the string and new sets of meta-symbols and of rules governing the allowable transitions between meta-symbols. These rules are deterministic, the outcome of this is that a symbol sequence generated from a chaotic data stream will always have maximal complexity (because no prior sequence of symbols is guaranteed to predict the next symbol deterministically, so the only effective description of the string is the string itself) even though we know the sequence was produced by a mathematically simple process and we know that only a limited set of states can ever be encountered. The induction of purely deterministic rules is therefore unlikely to capture a true picture of a dynamical system in symbolic computational terms. The alternative is the production of a probabilistic symbolic description of a system's dynamics. Such a description can capture important features of dynamical systems symbolically, for example the two distinct orbits of the Lorenz "butterfly" attractor and the relative probabilities of orbiting within and switching between them. Crutchfield & Young's algorithm produces probabilistic descriptions of this type, although, if presented with data which are described by a simple deterministic process, then an appropriate deterministic reconstruction will be produced.

The Crutchfield & Young algorithm

The aim of the algorithm is to produce the minimum complexity symbolic description of a dynamical system relative to a random-register Turing machine. The output of the system is the formal machine equivalent (referred to by Crutchfield & Young as an "ε-machine") of a stochastic grammar. Essentially the algorithm depends on finding the set of time invariances in a system's dynamics which have the simplest (minimum entropy) set of transitions between them. In outline the algorithm produces an ε-machine from a system's dynamics according to the following procedure.

Produce a quantized time series from the system's dynamics by dividing its state space into regions and sampling its position in state space in terms of these regions at fixed time intervals. From this time series construct a probabilistically labelled tree of all paths through the series up to a given length. From this tree produce the set of all unique types of subtrees (that is, subtree equivalence classes) of some smaller depth. These subtree equivalence classes correspond to time translation invariances in the original data stream. Subtree uniqueness is defined both by node labels, topology and by transition probabilities within the subtrees. The accuracy of the match in transition probabilities required for subtrees to be considered identical is a parameter δ. From the original tree find all of the transitions which transform one subtree into another (i.e. the transitions from the head of one subtree to the head of another subtree). Using these transitions produce a probabilistically labelled directed graph of the allowable transitions between subtrees (subtree equivalence classes correspond to nodes in this digraph). Repeat the above procedure with different tree and subtree depths and values of the matching parameter δ until the graph indeterminacy I_G is minimized. The graph indeterminacy I_G is defined as

$$I_G = \sum_{v \in V} p(v) \sum_{s \in A} p(s \,|v) \sum_{v' \in V} p(v' \,|v;s) \log\big(p(v' \,|v;s)\big) \tag{10.2}$$

where V is the set of vertices in which $v \to v'$ is a particular transition whose edge is labelled s, $p(v'|v; s)$ is the probability of that transition and $p(s|v)$ is the probability that s is emitted on leaving v. This is simply the weighted conditional entropy of the graph. It sums the entropies of all individual transitions between nodes $(p(v'|v; s))\log(p(v'|v; s))$ weighted by the probability of encountering the output node $(p(v))$ times the probability that the symbol labelling an edge from that node will occur at that node $(p(s|v))$.

Symbolic reconstructions of cyclic and chaotic dynamics

As the machine reconstructed by this algorithm is stochastic, the algorithm can produce concise symbolic descriptions of both chaotic and limit cycle systems. Figure 10.3 shows some ε-machine reconstructions from limit cycles produced by an iterated logistic map $(X_{t+1} = rX_t(1 - X_t))$. Figure 10.4 shows ε-machine reconstructions from chaotic regions with single, multiple and merging chaotic bands. In these cases the ε-machine reflects the different probability densities of value ranges of X in the map and the transitions between those regions.

Note that all of these machines are finite – they correspond to regular languages – the lowest level of the Chomsky hierarchy. ε-machines corresponding to higher-level grammars can, however, be produced by the Crutchfield & Young algorithm. If a system's dynamics is optimally described by a higher-level grammar the algorithm produces truncations of an infinite machine corresponding to that grammar. Regularities in the structure of this machine may allow us to infer the data structures (stacks or tapes) which would allow the machine to be represented in finite terms and the grammar to which the machine corresponds. We can distinguish truncations of infinite machines from finite machines by examining the machines produced from a single data set using successively larger tree depths in the reconstruction algorithm. If the optimal ε-machine is finite then this machine will eventually be produced using a particular tree depth, and when reconstructions are performed using deeper trees the same machine will be produced. On the other hand, if the optimal ε-machine is

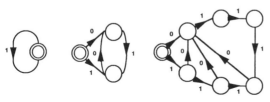

Figure 10.3 ε-machine reconstructions of periods 1, 2 and 4 limit cycles from the iterated logistic map $X_{t+1} = r X_t$ $(1 - X_t)$ with $r = 2.5, 3.4$ and 3.5 respectively.

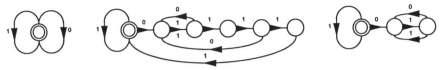

Figure 10.4 ε-machine reconstructions of single-band chaos, typical multiple-band chaos and the merger from double to single chaotic bands in the iterated logistic map X_{t+1} $= r X_t (1 - X_t)$ with $r = 4.0$, 3.7 and 3.67859 respectively.

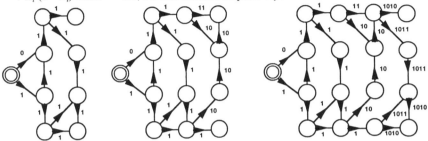

Figure 10.5 ε-machine reconstructions of the transition to chaos in the iterated logistic map $X_{t+1} = r X_t (1 - X_t)$ with $r = 3.570$, at reconstruction depths of 5, 9 and 17.

infinite then successively larger and larger truncations of the infinite machine will be produced as deeper trees are used in the reconstruction. An example of this can be seen in reconstructions of the ε-machine produced from the logistic map at the transition between periodic and chaotic behaviour shown in Figure 10.5. In order to make regularities in the growth of this machine apparent the digraph is "de-decorated" by collapsing together all deterministic transitions between nodes.

We can see from these examples that in order to identify whether a system such as a neural network is performing intrinsic computation in the context-free grammar class or above then our criterion must be that the minimal indeterminacy machine size does not reach a limit as reconstruction depth is increased.

Applying the Crutchfield & Young algorithm to the neural network

It may appear impractical to apply the Crutchfield & Young algorithm to a network consisting of thousands of neurons (although it is in principle possible) when it can take considerable computational resources to reconstruct an ε-machines from one-dimensional maps in some parameter ranges. If, however, there is some systematic collective behaviour occurring in the network then this should be reflected in the behaviour of single neurons if sampled over a sufficiently long time. I now present results of machine reconstructions from time series of membrane potentials of a single unit from a network showing critical dynamics sampled over 100 000 time steps. As a control, the same neuron was sampled under identical driving conditions but disconnected from all other units. First of all I present some data which give a general impression of the behaviour of the network and of a single neuron in isolation.

193

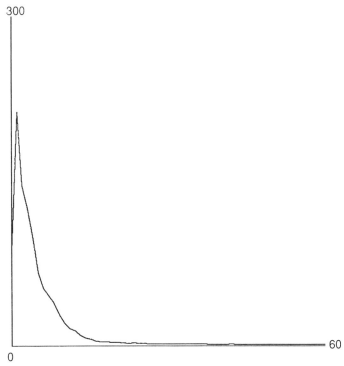

Figure 10.6 Power spectrum (frequency on the abscissa, power on the ordinate) of the 5625-neuron network from one neuron of which ε-machine reconstructions were subsequently made.

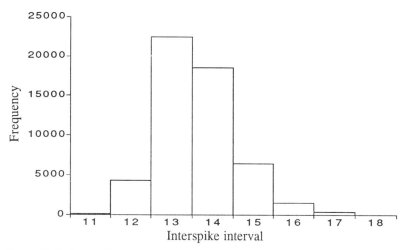

Figure 10.7 Interspike interval histogram collected from an individual unit in a network over approximately 50 000 time steps.

The network's power spectrum shown in Figure 10.6, measured in terms of the number of neurons firing in the network per time step, gives some indication of the global dynamics of the network. The spectrum shows the $1/f$ pattern typical of critically organized systems but superimposed with a single peak indicating the presence of some collective periodicity in the network. This spectrum is remarkably similar to those found by Young et al. (1992) in the visual and medial–temporal cortices of macaque monkeys and is quite similar to those found in the visual cortex of rhesus monkeys by Freeman & Van Dijk (1987) (who found $1/f$ power overlaid with multiple, rather than single, periodic peaks). Work is currently in progress on a more accurate simulation of EEG recording from the network for comparison with these results.

The power spectrum of individual units does not convey much useful information due to the response of the Fourier transform to the instantaneous drops of membrane potential as the unit fires. An interspike interval histogram, showing the distribution of times between the generation of successive action potentials in a single neuron (Fig. 10.7), however, gives a good conception of an individual unit's behaviour when embedded in the network.

The same neuron when isolated produces a constant interspike interval of 18, as would be expected from its simple evolution equation and constant driving current. It is also obvious that the effect of being embedded in an excitatory network can only reduce a unit's interspike interval.

Initially, I would like to explore the application of the ε-machine reconstruction algorithm to the isolated neuron. The relative simplicity of its behaviour allows the meaning of the machine reconstructions to be understood easily. At the maximum possible sampling rate we can use for reconstruction (corresponding to single time steps in the simulation which generated the data) we can choose a potential value at which to code 0s and 1s in the initial reconstruction string which discriminates perfectly between the refractory and active phases of the neuron. This discretization value falls at a potential between 0 and 0.4 for the current simulation. The machine reconstructed at this sampling rate and using this discretization value is shown in Figure 10.8.

This machine was reconstructed to a subtree depth of 19 in order to capture the full 18-step periodicity of the unit's behaviour. The indeterminacy I_G is zero. The first two layers of reconstruction establish the initial point in the time series

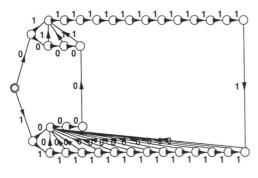

Figure 10.8 Machine reconstructed from the membrane potential time series of an isolated model neuron. The sampling rate and potential discretization were chosen to highlight the refractory and active phases of the neuron's behaviour. A subtree depth of 19 was used in the reconstruction.

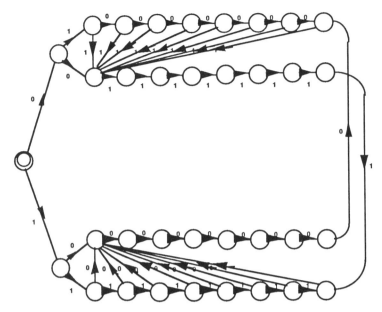

Figure 10.9 Machine reconstructed from the membrane potential time series of an isolated model neuron. The sampling rate and potential discretization were chosen to minimize the reconstruction depth required to describe the neuron's behaviour. A subtree depth of 10 was used in the reconstruction.

at which a tree is encountered. Beyond that the paths by which four refractory steps are always followed by 14 active ones can clearly be traced around the machine reconstruction.

Although this reconstruction is easy to relate to the neuron's membrane potential time series, we can produce simpler reconstructions. If, instead of discretizing in order to discriminate refractory and active phases of the neuron's behaviour, we discretize so that half of the labels in our initial string are 0s and half are 1s by choosing a value near to half of the neuron's firing threshold (in this case about 2.5) then much shallower reconstructions can capture the full dynamics of the neuron, as shown in Figure 10.9.

Instead of needing to reconstruct to a subtree depth of 19, an invariant zero indeterminacy machine is produced at depths of 10 and above.

In addition to changing the discretization value we can also change our sampling rate in order to further simplify the reconstructed machine. The machine shown in Figure 10.10 was produced using a sampling rate four times as fast as that used in the previous examples.

This appears to be the most concise symbolic description of the neuron's behaviour possible. Higher sampling rates can fail to discriminate the refractory period of the neuron (which also represents the strongest nonlinearity in its otherwise almost linear behaviour) and hence tend to produce non-zero indeterminacy reconstructions.

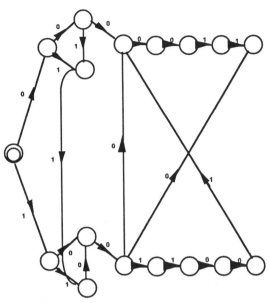

Figure 10.10 Machine reconstructed from the membrane potential time series of an isolated model neuron. Potential discretization is the same as that used in Figure 10.8; however, data are only sampled every fourth time step. A subtree depth of 8 was used in the reconstruction.

We now turn to reconstruction of the behaviour of the same neuron when it is embedded in the network. A reconstruction using exactly the same parameters as used in Figure 10.10 is shown in Figure 10.11.

The difference between Figures 10.10 and 10.11 is striking. The machine reconstructed from the isolated neuron is clearly finite whereas the neuron embedded in the network produces a machine which appears to grow as reconstruction depth increases, which gives no sign of imminent closure and yet which still has zero indeterminacy. Note, however, that reconstruction was only possible to a subtree depth of 6, as the number of possible paths through the data becomes extremely large for the embedded neuron, and computing deeper reconstructions becomes very time- and space-intensive.

It is also possible to produce machine reconstruction by producing strings labelled 1 whenever an action potential has just occurred and 0 otherwise. This initial step does not follow Crutchfield & Young's algorithm; however, the rest of the reconstruction is standard. This method produces far fewer initial paths through the data, so a depth 12 reconstruction of the embedded neuron's behaviour was possible. This is particularly useful since, at a sampling rate of four time steps, machine closure due to the maximum periodicity of 18 time steps would be expected at a depth of no more than 8. The depth 10 machine produced is shown in Figure 10.12.

This reconstruction also appears to be a truncation of an infinite machine; moreover, it shows the lengthening sequences of deterministic transitions which are typical signs of data structures such as stacks or tapes. Its indeterminacy was zero. These findings strongly suggest that the behaviour of this single neuron, when embedded in a network, is optimally described by, and is capable of implementing, computation at the level of context-free grammars or above.

197

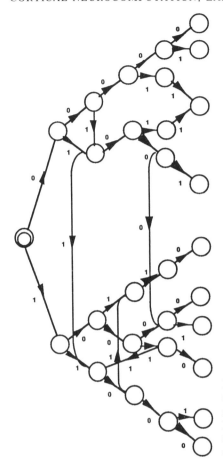

Figure 10.11 Machine reconstructed from the membrane potential time series of a model neuron embedded in a network. Reconstruction parameters were the same as those used in Figure 10.10. A subtree depth of 6 was used in the reconstruction.

Before accepting this conclusion, however, an alternative explanation for this complex behaviour must be investigated. Would the simple presence of multiple interspike interval periodicities in the data stream produce such behaviour? This alternative explanation can easily be tested using surrogate data.

Figure 10.13 shows the invariant machine reconstruction from surrogate data in which three periodicities, 4, 5 and 6, were mixed with different probabilities. Each choice of period was independent of previous ones. It can clearly be seen that the reconstructed machine is finite. In fact, it does not differ in form from a period 6 reconstruction; the only influence of the shorter periodicities is to change the transition probabilities in the graph.

It might still be argued that the strong nonlinearity of refractoriness combined with multiple independent periodicities might produce complex machines such as that reconstructed from the neuron embedded in the network. Figure 10.14, which shows a machine reconstructed from surrogate data with three independent periodicities of 4, 5 and 6 each followed by a fixed two-step "refractory" period, indicates that this is not the case.

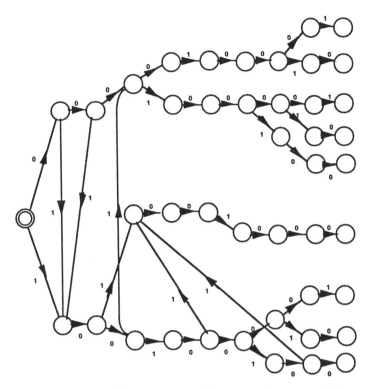

Figure 10.12 Machine reconstructed from an action potential time series of a model neuron embedded in a network. A subtree depth of 12 was used in the reconstruction.

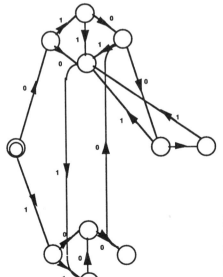

Figure 10.13 Machine reconstructed from surrogate data in which periodicities of 4, 5 and 6 follow one another randomly with different occurrence probabilities (0.3, 0.4 and 0.3, respectively).

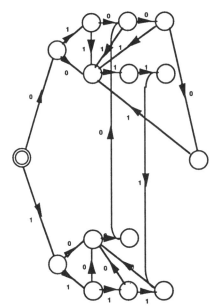

Figure 10.14 Machine reconstructed from surrogate data in which periodicities of 4, 5 and 6 follow one another randomly with different occurrence probabilities (0.3, 0.4 and 0.3, respectively). Each period is followed by a fixed two-step "refractory" period.

The conclusions we can reach from these ε-machine reconstructions are that the collective effect of the network on an individual neuron's dynamics is to raise its intrinsic computation to the level of context-free grammars or above. This effect is due to the production of multiple periodicities; moreover, the length of each interspike interval must depend on the previous behaviour of the neuron. This dependence continues to outweigh a simple random prediction of all of a neuron's subsequent behaviour in minimizing graph indeterminacy even at large reconstruction depths. Such long-range influences are typical of systems with critical dynamics, lending credence to the hypothesis that neural networks in a critical regime might perform computation beyond regular grammars.

Discussion

Reconstruction issues

The work presented here represents a considerable advance on preliminary work I reported in Kentridge (1995a). In that paper the network used produced interspike intervals in the range of 30–50 time steps at the maximum sampling rate. Although the reconstructions presented could identify the refractory and active phases of an isolated neuron, and could differentiate between the relatively simple isolated neuron behaviour and more complex embedded behaviour, it was quite impractical to reconstruct to a depth sufficient to reveal the true nature of the neuron's dynamics. Higher sampling rates could not significantly improve the situation, probably because the refractory period of the units was relatively small and its active behaviour almost linear. These results, together with those presented in this chapter, suggest that adding an explicit nonlinearity in addition to refractoriness to the model neuron's behaviour may have a significant effect on the practicalities of machine reconstruction.

Explicit models of "neural" symbol processing

The recent work aimed at producing "neural" implementations of explicitly symbolic processes mentioned in the introduction (e.g. von der Malsburg & Bienenstock 1986, Shastri & Ajjanagadde 1989, Shastri & Ajjanagadde 1993) has taken a very different approach from the one I have described here. These explicit approaches have centred on the problem of variable binding. When the features of objects are represented as distributed patterns of activity in a network, the representation of more than one object at a time in a network becomes problematic – how can different aspects of the overall pattern of activity in the network be associated with distinct objects? The ability to temporarily bind sets of representations or characteristics together has far more general uses than the avoidance of perceptual chimeras. Any use of syntax in information processing depends upon binding – how, otherwise, can "the man" sometimes be the subject and sometimes be the object of a sentence? The most common solution to the binding problem in recent years has been to link the aspects of a network's activity to be associated with a distinct entity on a finer timescale than that at which the co-occurrence of those entities is represented. In practice this separation of timescales is achieved by using "neurons" whose activity oscillates quickly relative to the rate of appearance of the entities being represented. Bindings are made through the relative phases of these oscillations. Such a system allows the patterns of activity (defined by the amplitudes of oscillations) corresponding to long-term representations (e.g. "happy", "sad", "man", "dog", "subject", "object") to be separated from specific instances in which sets of these patterns are temporarily bound together (defined by the phase of oscillations). We can, for example, begin to represent a sentence like "The sad man hit the happy dog" by arranging for neurons in the patterns corresponding to "sad", "man" and "subject" to oscillate in phase with one another while those representing "happy", "dog" and "subject" also oscillate together but in a different phase from the first set. This solution to the binding problem is especially attractive in the light of recent evidence of stimulus-dependent oscillations of neural activity in the visual cortex. In the framework I have been discussing, variable binding is only part of the more general problem of implementing symbolic processes neurally. Taken in isolation, the oscillation through binding story implies that bound symbols are represented as limit cycles; however, the necessarily temporary nature of binding implies that, at another level, these limit cycles are only transiently stable. Although I will argue that the neural computation underlying cognitive *processes* is best viewed in terms of transient, rather than attractor, dynamics, the difference between the approach I have described in this chapter and work on oscillation and binding is really one of emphasis. Whether we choose to concentrate on stability or the organization of systematic transitions in neural dynamics depends on our interests – on one hand the nature of neural implementation of representations, on the other hand the nature of the processes which transform those representations. On short timescales, when concentrating on mechanisms of representation, it may be appropriate to treat oscillations, both in model networks and in real brains, as attractors. In this chapter, however, I have attempted to begin addressing a problem which is set in a longer timescale where the transient qualities of a network's dynamics determine its functional properties.

Implications for cognitive neurocomputation

The work presented in this chapter can be seen as a hypothesis of brain function, and one can indeed derive propositions from it which may be tested electrophysiologically. Our purpose here, however, is to examine the implications of this work for our understanding of the nature of neural computation and its relationship to cognitive processes in language and memory.

The rest of this section rests on the premise that computation at the level of context-free grammars or above in dynamical systems requires nonlinear feedback (see also Kentridge 1993, Kentridge 1995b). This should be quite uncontroversial. If this computation is to proceed effectively in the presence of noise then the system must be in a critical regime, as the network model described earlier was. Given these assumptions, how are the equivalents of symbols and rules instantiated in dynamical systems?

Neural computation is transient, memories and experience are not identical

When we think about dynamical systems we tend to envisage them in terms of attractors. Computation at the level of context-free grammars or above in noisy dynamical systems cannot, in principle, occur if an attractor is reached. Computation at this level requires the system to follow a potentially infinite transient, perturbed occasionally by external stimuli. The path of this transient is, of course, influenced by topography of the system's state space, which in turn is determined by the form of attractors in the system. There is a reason for this beyond the fact that critical dynamics imply infinite transients *per se* (the phenomenon of "critical slowing" in physical systems near criticality is evidence of the "reluctance" of such systems to reach attractors). In a system performing computation at or above the level of context-free grammars the prior history of the system at some arbitrary point in the past must be able to influence the system's evolution. Once a system has reached an attractor it is impossible to determine the path taken into the attractor. When an external influence moves the system out of the attractor, all information about the system's history prior to reaching that basin of attraction is lost. In order to implement computation more powerful than regular grammars, then, the dynamics of the system must be forever transient.

What does this mean psychologically? In this model, information is stored by altering the form of the attractor which governs the system's dynamics. The attractor is, however, never reached. Stimuli therefore do not recreate memories. The effect a stimulus has on the state of the system is *influenced* by memories (the attractor) but the state of the system is not *the same* as those memories. This, of course, makes perfect psychological sense!

Symbols are non-atomic

Symbols in a dynamical system are invariances in the system's dynamics. However, in chaotic or nearly chaotic regimes the system will never exactly retrace its path, so the invariances we can make use of as symbols are only approximate. Although one particular level of approximation may produce some desired characteristics, such as minimal indeterminacy in the system's symbolic description, there is no reason to assume that invariances at finer or coarser levels of approximation might not also be regarded as symbols. As these invariances are approximate, predictions of the transitions between them will also often have to be approximate. More accurate predictions can be made for the transitions between finer-grained approximate invariances than between coarse ones. The upshot of this is that a whole range of different ways of coding a segment of dynamics in symbolic terms is possible. Transitions which appear random at one level of coding are less so at other levels. When viewed in these terms, "symbols" in neurocomputation are not truly symbolic – rather than being atomic and content-free they consist of hierarchies of more detailed contents. These hierarchies may mix semantic and syntactic information in a similar way to the structures revealed by analyses of invariances in text corpora (e.g. see Chater & Conkey 1994, Finch & Chater 1994). This conclusion says slightly more than "representations are distributed". Their distribution is also meaningfully structured. It is particularly interesting to note that Wuensche (1993) arrives at a very similar notion from a quite different direction in his cellular automaton model of memory.

Acknowledgements

This research was supported by DDRA Fort Halstead, UK. Contract number 2051/047/RARDE.

References

Ackley, D. H., G. E. Hinton, T. J. Sejnowski 1985. A learning algorithm for Boltzmann machines. *Cognitive Science* **9**, 147–69.
Chaitin, G. 1975. Randomness and mathematical proof. *Scientific American* **233**(5), 47–52.

Chater, N. & P. Conkey 1994. Sequence processing with recurrent neural networks. See Oaksford & Brown (1994), 269–94.

Chomsky, N. 1963. Formal properties of grammars. In *Handbook of mathematical psychology*, vol. 2, R. D. Luce, R. B. Bush, E. Galanter (eds), 323–418. New York: Wiley.

Crutchfield, J. P. & K. Young 1989. Inferring statistical complexity. *Physical Review Letters* **63**, 105–8.

Crutchfield, J. P. & K. Young 1990. Computation at the onset of chaos. In *Complexity, entropy and the physics of information*, W. H. Zurek (ed.), 105–8. Redwood City, Calif.: Addison Wesley.

de Gennes, P. G. 1975. Phase transitions and turbulence: an introduction. In *Fluctuations, instabilities and phase transitions*, T. Riste (ed.), 1–14. New York: Plenum Press.

Finch, S. & N. Chater 1994. Learning syntactic categories: a statistical approach. See Oaksford & Brown (1994), 295–321.

Fodor, J. A. & Z. W. Pylyshyn 1988. Connectionism and cognitive architecture: a critical analysis. *Cognition* **28**, 3–71.

Freeman, W. J. & B. W. van Dijk 1987. Spatial patterns of visual cortical fast EEG during conditioned reflex in a rhesus monkey. *Brain Research* **422**, 267–76.

Gaponov-Grekhov, A. V. & M. I. Rabinovich 1992. *Nonlinearities in action: oscillations, chaos, order, fractals*. Berlin: Springer-Verlag.

Hopfield, J. J. 1982. Neural networks and physical systems with emergent collective computational properties. *National Academy of Science of the USA, Proceedings*, **79**, 2554–8.

Hopfield, J. J. & D. W. Tank 1985. "Neural" computation of decisions in optimization problems. *Biological Cybernetics* **52**, 141–52.

Hornik, K., M. Stinchcombe, H. White 1990. Universal approximation of an unknown mapping and its derivatives using multilayer feedforward networks. *Neural Networks* **3**, 551–60.

Kelso, J. A. S., S. L. Bressler, S. Buchanan, G. C. DeGuzman, M. Ding, A. Fuchs, T. Holroyd 1992. A phase transition in human brain and behavior. *Physics Letters A* **169**, 134–44.

Kentridge, R. W. 1993. Cognition, chaos and non-deterministic symbolic computation: the Chinese room problem solved? *Think* **2**, 44–7.

Kentridge, R. W. 1994. Critical dynamics of neural networks with spatially localised connections. See Oaksford & Brown (1994), 181–214.

Kentridge, R. W. 1995a. Linking biologically realistic neural network models and psychology using grammatical inference. *Grammatical inference: theory, applications and alternatives*, S. M. Lucas (ed.), in press.

Kentridge, R. W. 1995b. Symbols, neurons and soap-bubbles and the neural computation underlying cognition. *Minds and Machines*, in press.

Li, W., N. H. Packard, C. G. Langton 1990. Transition phenomena in cellular automata rule space. *Physica D* **45**, 77–94.

Mitchell, M., P. T. Hraber, J. P. Crutchfield 1993. Revisiting the edge of chaos: evolving cellular automata to perform computations. *Complex Systems* **7**, 89–130.

Nicolis, G. & I. Prigogine 1989. *Exploring complexity*. San Francisco: W. H. Freeman.

Oaksford, M. & G. Brown (eds) 1994. *Neurodynamics and psychology*. **London: Academic Press.**

Packard, N. H. 1988. Adaptation at the edge of chaos. In *Dynamic patterns in complex systems*, J. A. S. Kelso, A. J. Mandell, A. F. Schlesinger (eds), 293–301. Singapore: World Scientific.

Pearlmutter, B. 1988. *Learning state-space trajectories in recurrent neural networks: a preliminary report*. Computer Science Department, Carnegie-Mellon University, Technical Report AIP-54.

Shastri, L. & V. Ajjanagadde 1989. *A connectionist system for rule based reasoning with multi-place predicates and variables*. Department of Computer and Information Science, School of Engineering and Information Science, University of Pennsylvania, Technical Report MIS-CIS-89-06 LINC LAB 141.

Shastri L. & V. Ajjanagadde 1993. From simple associations to systematic reasoning: a connectionist representation of rules, variables and dynamic bindings. *Behavioral and Brain Sciences* **16**, 417–94.

Uzunov, D.I. 1993. *Theory of critical phenomena: mean field, fluctuations and renormalization*. Singapore: World Scientific.

von der Malsburg, C. & E. Bienenstock 1986. Statistical coding and short-term plasticity: a scheme for knowledge representation in the brain. In *Disordered systems and biological organisation,* E. Bienenstock, F. Fogelman, G. Weisbuch, G. (eds), 247–72. Berlin: Springer-Verlag.

Williams, R.J. & D. Zipser 1989. A learning algorithm for continually running fully recurrent neural networks. *Neural Computation* **1**, 270–80.

Winograd, T. 1983. *Language as a cognitive process*. vol. 1. *Syntax*. Reading, Mass.: Addison-Wesley.

Wuensche, A. 1993. The ghost in the machine: basins of attraction in random boolean networks. In *Artificial life III*, C. G. Langton (ed.), 465–501. Redwood City, Calif.: Addison-Wesley.

Young, M. P., K. Tanaka, S. Yamane 1992. On oscillating neuronal responses in the visual cortex of the monkey. *Journal of Neurophysiology* **67**, 1464–74.

CHAPTER 11

Neural networks: the new statistical models of mind

Nick Chater

Introduction

Neural network, connectionist or parallel distributed processing models of cognition have rapidly become dominant in many areas of cognitive science (e.g. McClelland & Rumelhart 1986, Rumelhart & McClelland 1986a, Gluck & Bower 1988, Seidenberg & McClelland 1989, Elman 1990, Hinton & Shallice 1991). Yet the scope and power of neural network models, and their relation to other approaches to modelling cognition, have been controversial (Fodor & Pylyshyn 1988, Pinker & Prince 1988, Fodor & McLaughlin 1990). At one extreme, there is a hope, frequently expressed by cognitive scientists informally but rarely put down in print, that neural network models will sweep away other approaches to modelling cognition, and in particular the symbolic models that have until recently dominated cognitive psychology and artificial intelligence. At the other extreme is the view that cognitive or psychological explanation is necessarily pitched at a symbolic level, and that neural networks are hence irrelevant to such explanation (Fodor & Pylyshyn 1988, but see Chater & Oaksford 1990). Advocates of this view argue that neural networks are simply a rediscovery of old-style statistical methods, with well known limitations, in reaction to which the symbolic model of mind (Fodor 1975, Newell & Simon 1976) was originally developed.

In this chapter, I review theoretical work which shows that there is a close relationship between various kinds of neural network and statistical models. This work has been developed within the technical literature on neural networks, but has not received wide attention within the cognitive modelling community. Within this literature, neural networks are viewed as statistical models, although they are models of a novel and powerful kind. The connection with the familiar territory of statistics helps to clarify the status and power of neural network models in cognitive science. It should not, I will argue, be taken to suggest that neural network models are simply reinventions of failed models of the past. Rather, I suggest they should be seen as a new development within a rich and varied history of statistical models of cognition. Furthermore, the connection with statistics helps clarify the relationship between neural network and symbolic models of cognition, and makes it clear that they have separate concerns, rather than standing in competition. I shall be concerned almost exclusively with *inferential* statistics as opposed to purely *descriptive* statistics (i.e. not statistics as mere collection of numbers, or as tools for conveniently displaying data).

The structure of this chapter is as follows. I begin by outlining the scope of statistics in very broad terms, stressing the generality of statistical methods. I then turn to the relationship between statistical methods and neural networks, concentrating on neural network learning methods, and dealing with supervised and unsupervised methods in turn. Finally, I draw conclusions for the place of neural network models in the history of psychology and their relationship with other modelling approaches, in particular the symbolic approach.

What is statistical inference?

The elements of probability theory and statistics (I shall sometimes use "statistics" to refer to both of these, but distinguish the two when context requires it) are familiar to researchers in cognitive psychology and the cognitive sciences generally. However, statistics are frequently encountered in their role as tools for data analysis, rather than in their broader context as method for inference. It is in this latter context that statistical methods can plausibly be viewed as models of cognition (and we shall consider some aspects of the psychological tradition of statistical modelling, in relation to neural network models below). Moreover, because of the dominance of a limited "data analysis" view of statistics in certain areas of the cognitive sciences, the claim that neural networks might be just statistical models is sometimes viewed with incredulity. Hence, we begin by sketching the broader view of statistics as very general mathematical methods for uncertain inference, within which statistical methods as used in data analysis in the cognitive sciences form only a small part.

Statistical inference is founded upon the mathematical theory of probability, and the distinct statistical traditions differ on how this theory is understood. The interpretation of probability theory has been controversial since its very beginnings. Nonetheless, the most usual early interpretation of probability theory was as a tool for formalizing rational thought concerning uncertain situations, such as gambling, insurance and the evaluation of court-room testimony (Gigerenzer et al. 1989). Indeed, the very choice of the word "probability", which referred to the degree to which a statement was supported by the evidence at hand, embodied this interpretation – that is, "probability" originally signified "rational degree of belief". Jakob Bernoulli explicitly endorsed this interpretation when he entitled his definitive book *Ars conjectandi,* or the *Art of conjecture* (Bernoulli

1713). This "subjectivist" conception ran through the eighteenth and into the nineteenth centuries (Daston 1988), frequently without clear distinctions being drawn between probability theory as a model of actual thought (or more usually, the thought of "rational", rather than common, people (Hacking 1990)) or as a set of normative canons prescribing how uncertain reasoning should occur. In a sense, then, early probability theory itself was viewed as a model of mind.

As the distinction between normative and descriptive models of thought became more firmly established, probability theory was primarily seen as having normative force, as characterizing rationality; whether or not people actually followed such normative dictates was seen as a secondary question. A wide variety of arguments that purport to show that individual degrees of beliefs should obey the laws of probability calculus have been developed, based on betting quotients and "Dutch book" arguments (Ramsey 1931, de Finetti 1937, Skyrms 1977), theories of preferences (Savage 1954), scoring rules (Lindley 1982) and derivation from minimal axioms (Good 1950, Cox 1961, Lucas 1970). Although each argument can be challenged individually, the fact that so many different lines of argument converge on the very same laws of probability has been taken as powerful evidence for the view that degrees of belief can be interpreted as probabilities (e.g. see Howson & Urbach (1989) and Earman (1992) for discussions). The suggestion that probability theory can be viewed as a normative theory of uncertain reasoning sets the bounds of probability theory much wider than the confines in which it is frequently encountered in introductory textbooks. According to this view, probability theory is not just concerned with reasoning about coins, dice and accident rates, but is a calculus for rational thought.

Many inferential problems concern the relationship between models or hypotheses, and observation or data. Some of these problems are concerned with inferring the probability of various kinds of observation, given that the structure of the underlying model is known. So, for example, the model might be a fair coin, and the question of interest might be the probability that 50 heads or more will be obtained in 200 throws. Statistical inference, by contrast, applies in the opposite direction, using observed data to infer the structure of the underlying model. For example, given the observation of 50 heads in 200 throws, assessing whether the coin is unbiased, what its likely bias might be, and with what confidence the bias can be estimated, all involve statistical inference, since observed data are used to infer aspects of the underlying model.

The problem of inductive or statistical inference is very general, and arises, in different guises, in a variety of domains. In epistemology and the philosophy of science, the problem is that of choosing the hypothesis or theory which is best supported by a given body of empirical observations: this is the problem of *induction*. A particular approach to statistics, the Bayesian approach, is by far the most well developed formal account of inductive reasoning (e.g. see Horwich 1982, Howson & Urbach 1989, Earman 1992). In the context of psychology, cognitive science and artificial intelligence, machine learning, pattern recognition and the study of neural networks, statistical inference corresponds to the problem of *learning* underlying structure from experience. It is with this broad sense of the scope of statistics in view that the claim that the mind is an intuitive statistician (Gigerenzer & Murray 1987), or that cognitive processes can be viewed as statistical processes, can be understood. The claim is not merely that the mind performs t tests or ANOVAs (although this has been proposed (Kelley 1967)). It is that the dictates of statistical theory concerning inductive inference are descriptive, not just prescriptive, regarding certain aspects of thought.

The project of characterizing statistics is complicated by the variety of different statistical schools, many of whose differences stem, as noted above, from different interpretations of the probability calculus. So far, we have considered the subjectivist interpretation, according to which probabilities are primarily interpreted as concerning rational updating of degrees of belief. This viewpoint sees no fundamental distinction between inference from beliefs about hypotheses to beliefs about data (the standard probabilistic case), and statistical inference in the reverse direction. Bayes (1764) showed that inference in the two directions can be related by a simple corollary of the axioms of probability:

$$P(H_j|D) = \frac{p(D|H_j)P(H_j)}{\sum\limits_{i=1}^{n} P(D|H_i)P(H_i)} \tag{11.1}$$

This result is the foundation of Bayesian statistics, which allows the probability of a model or hypothesis H_j given data D to be estimated, given the probability of the data given each possible model or hypothesis H_i, and the prior probability of each H_i. By the application of Bayes's theorem, the normal laws of probability can be used to infer how probable each of a range of hypotheses is, given a data set, simply by mechanical calculation. Notice that the denominator is the same whatever hypothesis is under consideration, and acts as a normalization factor which ensures that the probabilities $P(H_i|D)$ sum to 1. It is often treated as a constant, and Bayes's theorem is then expressed, as above, by stating that $P(H_i|D)$ is proportional to $P(H_i|D)P(H_i)$.

According to a subjectivist interpretation, the prior probability $P(H_j)$ can be interpreted as an initial degree of belief in the hypothesis H_j. But for alternative views of probability, such as the frequentist interpretation (according to which probabilities are the limits of relative frequencies of repeated events (e.g. Fisher 1922, von Mises 1939)) and objectivist interpretation (according to which probabilities are objective properties of the world (Mellor 1971)), it is difficult to see how any sense can be made of such probability statements. For this reason, among others, various alternatives to Bayesian statistics have since been derived. The principal alternative schools are those of Fisher (1956, 1970) and Neyman and Pearson (e.g. Neyman 1950), and most standard statistical tests within the behavioural sciences (e.g. the t test, the ANOVA, χ^2 test) were developed by these schools (though the standard discussion of such tests in introductory statistical textbooks frequently blends incompatible elements of these approaches together – see Gigerenzer et al. (1989)). We shall focus on Bayesian statistical methods henceforth, since it is these, and related methods, that most closely relate to neural network models. Furthermore, the subjectivist, Bayesian approach relates probability and statistics most directly to problems of belief updating, and hence has the most natural relation to cognitive processing.

At this level of generality, it should be clear that there is no limitation on the nature or complexity of the models (hypotheses, theories) that can be assessed using Bayesian statistics, aside from the fact that they must be well enough specified that the probability of each data outcome can be calculated given that the model holds. That is, hypotheses or theories must constitute probabilistic models. (In practice, of course, many hypotheses are not well enough specified for this to be possible, and additional assumptions must be made in order to fill out the hypothesis or theory into a full probabilistic model, but we shall not be concerned with this issue here.)

Probabilistic models include deterministic models, which specify their data with probability 1, and models which are defined in terms of symbolic structures (e.g. sets of grammar rules), and learning for such models can proceed according to standard Bayesian procedures. Bayesian methods can also be adapted to assess parametrized classes of model (e.g. straight lines versus quadratic models in curve fitting (e.g. Young 1977)).

While there is in principle no limitation on model complexity, performing the appropriate calculations may be extremely difficult, involving severe mathematical and computational problems. Hence, in practice, researchers have been forced to concentrate on relatively simple underlying models. For example, in the domain of language, statistical research has focused on very simple stochastic models such as hidden Markov models (e.g. Huang et al. (1990) – note that the parameters of hidden Markov models are generally trained by maximum likelihood estimation, a Fisherian, rather than a Bayesian, method; however, it may be viewed as a special case of the Bayesian approach in which priors are uniform), and this has been true even when considering areas of natural language where such models have been shown to be inadequate (Chomsky 1957). It is this practical limitation that has led to the claim that "probabilistic" or "statistical" models of language or some other aspect of cognition are not able to capture its true structure. Taken at face value, the claim makes no sense, since any adequately specified model can, in principle, be used in statistical inference, and hence there is really no such class as the class of statistical models. What is meant by the claim is, presumably, that the simple stochastic models considered in current statistical studies are not adequate.

This means that if neural networks turn out to be closely related to statistical methods this does not necessarily mean that they are inadequate to model particular cognitive phenomena – for they are a new kind of statistical model, and must be considered on their own terms.

Neural networks and statistics

Neal (1993: 475) succinctly sums up the connection between neural networks and probability theory and statistics viewing "neural networks as probabilistic models, and learning as statistical inference". Since many neural networks used in cognitive modelling are feedforward networks trained with back-propagation, I shall concentrate on this case, considering the two halves of the view in turn. I then briefly consider unsupervised learning networks.

Supervised learning

Neural network architectures as probabilistic models

First let us consider how a neural network can be viewed as a probabilistic model (I follow Neal's (1993) development). Consider a neural network (Fig. 11.1) which takes vectors of real valued numbers, x, as input and produces real valued vectors, y, as output. Assuming that the network is deterministic, i.e. that the same input always results in the same output, the network architecture defines a function f, where $y = f(x, w)$, where $w = (w_1, \ldots, w_m)$ denotes the vector of m weights in the network. Let us now suppose that the target output, z, is just y with the addition of Gaussian noise, of fixed standard deviation σ (other noise functions can, of course, be considered, but this is the simplest). Once the input is specified, the probability of the target outputs is specified by

$$P(z|x,\sigma) \propto \exp(-|z - f(x,w)|^2 / 2\sigma^2) \qquad (11.2)$$

Thus, a neural network with a particular architecture and a given set of weights defines a probabilistic model: the output probabilities are fully specified given the input probabilities, in accordance with Equation 11.2. The neural network architecture (defined purely by the pattern of connections between nodes) thus defines a family of probabilistic models, parametrized by the weights w associated with the connections.

This formulation, while appropriate for modelling feedforward networks to be trained by back-propagation, is not, of course, a helpful way to analyze all supervised networks. For example, if the network has a stochastic dynamics, then the output of the network may itself be a probability distribution, rather than a particular deterministic state. For example, in the Boltzmann machine (Hinton & Sejnowski 1986) the goal is to produce an output probability distribution which models that observed during learning. In deterministic versions of stochastic dynamics, such as the deterministic Boltzmann machines (Peterson & Anderson 1987), real-valued output units are considered to denote the probabilities of each discrete binary output, rather than denoting a real-valued number.

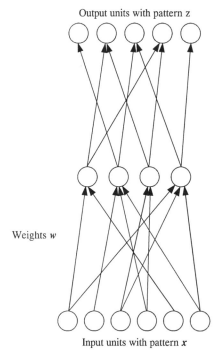

Output units with pattern z

Weights w

Input units with pattern x

Figure 11.1 A supervised learning system. The inputs x_i are transformed into outputs y_i, which are compared against targets z_i. One natural goal of such reconstruction is to minimize $|z_i - y_i|^2$. In statistical terms, minimizing least squares can be viewed as assuming that there is Gaussian noise on the output. As noted in the text, there is sometimes an additional error term, which punishes networks with large weights w.

Neural network learning as statistical inference

Before discussing the statistical interpretation, let us briefly summarize the back-propagation approach to training neural networks. Learning begins with a fixed network architecture and a specification of the target output, z_i, which is to be associated with each example input, x_i. Back-propagation adjusts the weights in the light of these data. Typically, weights are adjusted so that some error function is minimized, the most common error function being the squared difference between the actual network output and the target output summed over units and patterns:

$$E(w) = \sum_p |z_p - f(x,w)|^2 / 2\sigma^2 \tag{11.3}$$

In practice, it is sometimes found to be useful to allow weights to decay in proportion to their size, to discourage the network from developing extremely large weights, which sometimes lead to poor generalization. Hence the function to be minimized is modified to

$$E(w) = \left(\sum_p |z_p - f(x,w)|^2 / 2\sigma^2 \right) + \lambda |w|^2 \tag{11.4}$$

where λ is a constant which sets the amount of "weight decay" used.

If the weights are adjusted in sufficiently small steps in the direction according to $-dE/dw_i$, the overall error E decreases. Eventually, the weights will reach
a minimum, at which no local change decreases E. Unfortunately, there is no
guarantee that this will be a global minimum, and so the network may not necessarily achieve the lowest possible E value. This problem of "local minima" is,
however, a very general one, and typically applies in complex minimization
problems which are solved iteratively (e.g. it arises in training hidden Markov
models (Huang et al. 1990)).

The back-propagation learning algorithm is simply an efficient computational
scheme for calculating the $-dE/dw_i$ values, by passing an "error" signal from the
output units, where error is explicitly assigned, back through the rest of the network. It has the further advantage of being completely local – i.e. simple processes over the network units themselves serve to update the weights, and no
external controller is required. From an abstract point of view, all that matters is
that E is locally minimized somehow, and we shall not need to consider the
details of back-propagation below.

A number of authors have shown how this learning algorithm can be viewed
as statistical inference (e.g. Golden 1988, Buntine & Weigend 1991, Mackay
1992a, Neal 1992, Wolpert 1993). As noted above, we can consider the network
and weight values to define a probabilistic model from which the data are considered to be generated, and aim to choose the weights which correspond to the
most probable model, given the data $(x_1, z_1), \ldots, (x_n, z_n)$.

From Bayes's theorem it can be shown that

$$P\big(w \big| (x_1, z_1), \ldots, (x_n, z_n) \sigma\big) \propto P(w) P(z_1, \ldots, z_n | x_1, \ldots, x_n, \sigma, w) \qquad (11.5)$$

The probability of the data, given w and the assumption of Gaussian noise of
standard deviation σ, is

$$P(z_1, \ldots, z_n | x_1, \ldots, x_n, \sigma, \omega) \propto \exp\left(-\sum_p |z_p - f(x, w)|_2 / 2\sigma^2 \right) \qquad (11.6)$$

To calculate Equation 11.5 we must also specify some prior probability on w.
(We assume here that the variance σ is known. Buntine & Weigend (1991) show
that relatively minor modifications can deal with cases in which σ is unknown.)
If we are interested in favouring small weights, then a natural prior is to assume
that weight vectors are distributed in a Gaussian distribution, with standard
deviation ω, around 0. That is,

$$P(w) \propto \exp\left(-|w|^2 / 2\omega^2\right) \qquad (11.7)$$

Substituting Equations 11.6 and 11.7 into Equation 11.5 gives

$$P\left(w|(x_1,z_1),...,(x_n,z_n),\sigma\right) \propto \exp\left(-|w|^2/2\omega^2 - \sum_p |z_p - f(x,w)|^2/2\sigma^2\right) \qquad (11.8)$$

To maximize Equation 11.8 we minimize

$$E(w) = \sum_p |z_p - f(x,w)|^2/2\sigma^2 + |w|^2/2\omega^2 \qquad (11.9)$$

Thus we have a standard error function for back-propagation $E(w)$, as given in Equation 11.4, with $\lambda = 1/2\omega^2$. The parameter ω depends on the standard deviation of the Gaussian distribution of the priors. The smaller the standard deviation, the greater the bias towards networks with small weights.

Thus, we have a clear statistical interpretation of back-propagation learning. The strength of the weight decay term can now be understood in terms of how closely the prior distribution of weights is bunched around 0, i.e. it is determined by the value of ω. Interestingly, if the prior distribution is ignored, then the second term need not be considered, and we derive Equation 11.3. Thus, back-propagation without weight decay corresponds to computing the maximum likelihood weights, i.e. the weights according to which the data are most likely (Golden 1988).

A statistical interpretation of neural network performance is not just a mathematical curiosity. It makes sense of neural network learning, clarifies the assumptions underlying neural network performance, and provides insights into how neural network methods can be further developed. Thus, the use of least squares as a measure of error in statistical regression carries over as an appropriate measure of network error (the difference between the network's actual output and the specified output). In back-propagation networks, and variants, the weights are adjusted to perform gradient descent in this error. The statistical assumption underlying least squares is that the output value is subject to Gaussian noise; when this assumption is strongly violated, both statistical and neural network methods should ideally use an alternative error measure. For example, if the network outputs are known to be binary, an alternative measure, cross-entropy, is generally recommended as a more statistically appropriate error measure, and this has been widely used in neural network models. Here, then, statistics not only justifies standard methods, but suggests how they should be amended when necessary (see Hinton (1989) for discussion). Furthermore, a large range of new technical developments derive from the statistical interpretation (e.g. Mackay 1992a,b, Neal 1993, Wolpert 1993).

The statistical interpretation that we have considered amounts to viewing neural networks as a method for nonlinear regression, which is simply an extension of standard linear regression, which is a familiar data analysis tool in the behavioural sciences. Within conventional statistics, perhaps the most closely related approach to back-propagation is projection pursuit regression (Friedman & Stuetzle 1981), which has recently been related to neural network learning

(Intrator 1993). Standard linear regression aims to fit a straight line to a set of data points, so that least squares error is minimized, and this is justified as the maximum likelihood model, assuming Gaussian noise, just as in the network case described above. The analogue of weight decay in linear regression is systematically to favour lines with small regression components, and is known as ridge regression. The Bayesian analysis sketched above for neural networks with weight decay directly parallels a Bayesian rationale for ridge regression. Furthermore, linear regression is exactly modelled by a simplification of the standard back-propagation network – using no hidden units, and making the output units linear. Back-propagation affords a very considerable generalization over linear regression, since multilayered feedforward networks can learn to compute a very large class of nonlinear functions. Indeed, Hornik et al. (1989) have shown that any well behaved function can be approximated arbitrarily well by a neural network with sufficiently many hidden units.

Feedforward neural networks trained by back-propagation need not be viewed as a form of regression. With binary outputs, they can be viewed as classifiers, analogous to discriminant analysis. Indeed, with a single linear threshold unit (what Minsky & Papert (1969) termed a simple perceptron) they perform linear discriminant analysis between input points classified as 0 and input points classified as 1.

It is clear, then, that supervised networks, which are by far the most common network used in cognitive modelling, fit squarely in the tradition of conventional statistics, and are generalizations of familiar methods such as regression and discriminant analysis. We now turn to consider the statistical basis of unsupervised learning.

Unsupervised learning

Unsupervised learning methods involve finding structure in input data, with no specified "correct" output. The goal of the network is to extract interesting structure of some particular kind from the input. Unsupervised models have been much less used in modelling psychological data, although they have been viewed as a valuable source of hypotheses about aspects of human cognition (Kohonen 1984, Rumelhart & Zipser 1986, Ritter & Kohonen 1989, Finch & Chater 1992, Finch & Chater 1994; see also Ch. 12). An exception is Grossberg, who attempts to account for a large range of psychological data using rather elaborate unsupervised networks (e.g. Grossberg 1982). This work stands outside the mainstream of neural network research, and is beyond the scope of this chapter.

We shall briefly trace two connections between unsupervised learning and statistics. The first connection is simply that unsupervised learning methods frequently carry out identical or similar calculations to those of conventional statistical methods. For example, a one-layer feedforward network with lateral connections (Oja 1989) can learn to find principal components; competitive learning methods (such as that of Rumelhart & Zipser (1986)) can be viewed as computing slight variants of k-means cluster analysis (e.g. Krishnaiah & Kanal 1982).

The second connection has a deeper theoretical basis. Whereas supervised learning involves learning mapping between given input and target patterns, much unsupervised learning can be viewed as learning a mapping between input patterns and themselves – i.e. input and output are identical. In neural network terminology this is the "encoder" task. Since solving the encoder task is a special case of supervised learning, the statistical interpretation introduced above applies, and hence an appropriate error function is proportional to $|x_i - z_i|^2$, where z_i is the network output, and x_i is the pattern to be reconstructed.

The encoder task is trivial if a large enough network is available (in particular, when there are as many hidden units as input/output units) – the network can simply learn to perform the identity map. When the network is small, however, this is not possible, and to learn the task successfully the network must *compress* the input data into internal codes c_i, while losing as little information as possible. In order to compress data successfully, it is necessary to find structure within that data. To take a simple example, DeMers & Cottrell (1993) use standard back-propagation with a feedforward network to demonstrate that it is possible to compress input data which lie on a three-dimensional helix through a single hidden unit – thus, three-dimensional input data can be compressed onto a single dimension. In order to do this, the network must implicitly uncover the helical structure of the data, so that it can be represented by a single parameter. To take another example, Baldi & Hornik (1988) have shown that if a back-propagation network has just one hidden layer, then the units on that layer will extract the principal components of the input data (strictly, each of the n hidden units will find components which together span the subspace defined by the first n principal components, rather than finding exact principal components). It is because the goal of compression and reconstruction requires knowledge of the structure of the input that maximizing compression is an interesting goal of unsupervised learning. Indeed, there is also a direct theoretical connection between theoretical analysis of compression, in the minimum description length framework and Bayesian statistics (Rissanen 1983, Rissanen 1989), although I shall not consider this here.

In order to build a bridge between supervised and unsupervised learning, I have so far considered unsupervised learning methods which use standard feedforward networks trained by back-propagation. Of course, most unsupervised networks do not have this form; indeed, much interest in unsupervised learning concerns attempting to learn interesting structure without resorting to back-propagation and related methods. Nonetheless, the theoretical analysis sketched above can be used to derive many popular unsupervised learning algorithms.

Most unsupervised learning algorithms do not explicitly reconstruct the original input on a set of output units. Indeed, unsupervised networks generally consist only of input units, and what I shall call "feature" units which are

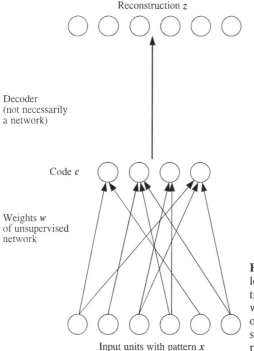

Reconstruction z

Decoder
(not necessarily
a network)

Code c

Weights w
of unsupervised
network

Input units with pattern x

Figure 11.2   An unsupervised learning system. The inputs x_i are transformed into code patterns c_i, which can be used to reconstruct the original input, giving z_i. One statistically natural goal of such reconstruction is to minimize $|z_i\text{-}x_i|^2$.

intended to display the structure implicit in the input. The above analysis can be applied by assuming a simple, fixed decoding mechanism, which maps feature unit patterns back onto the original input space (Fig. 11.2). Given this fixed decoding method, it is possible to calculate sum squared error as usual. Unsupervised algorithms can be viewed as adjusting their weights so that this implicit reconstruction can be as successful as possible – i.e. so that input data are compressed as well as possible.

For example, in competitive learning (Rumelhart & Zipser 1986) only a single feature unit is allowed to be active at any time, the unit whose weight vector is closest to the input pattern. The decoding mechanism for this network is simply to take the weight vector associated with the winning unit as indicating the input pattern. In order to minimize reconstruction error, it can be shown that the weights should move according to the standard competitive learning algorithm. Luttrell (1989, 1990, 1994) has extended this result to show that self-organizing maps similar (though not identical) to those of Kohonen (1984) can also be viewed as minimizing reconstruction error. Furthermore, networks which perform principal component analysis (Oja 1989) (without using full back-propagation) can also be understood in the same terms.

Discussion

I have outlined the close relationship between neural network models and statistics, and I now turn to considering the significance of this relationship for psychological theory. First, I shall attempt to put current neural network models in context in the history of psychology, arguing that they should be seen as descendants of previous statistical models of mind. Secondly, with the precursors of neural networks in mind, I shall draw out implications for the debate between neural network and symbolic approaches to cognition.

Relationship of neural networks to statistical models of mind

Neural networks are often portrayed as an entirely new and revolutionary approach to the mind (e.g. Clark 1989, Bechtel & Abrahamsen 1991); but by their critics they are frequently written off as associationism rediscovered (Fodor 1987, Fodor & Pylyshyn 1988). Neural networks do have close ties with a range of previous theories in psychology, including those based on associationist principles, although they are somewhat more complex in mathematical and computational terms. But they also have strong ties with a much broader tradition of modelling mental processes as involving statistical inference, and we shall briefly sketch some of these connections here (see Gigerenzer & Murray (1987) and Gigerenzer (1991) for further discussion of the tradition of statistical models of mind).

Perhaps the most well known statistical models have been outlined in the study of perception. The assumption that the mind makes psychophysical judgements and discriminations by using statistical techniques (based on Neyman–Pearson statistics) revolutionized psychophysics (Tanner & Swets 1954, Tanner 1965). The idea of the new "signal detection theory" was that the mind used statistical methods to take account of noise in perceptual stimuli. Earlier Brunswick (1943) had put forward a more general, but less mathematically sophisticated, doctrine of probabilistic functionalism, which held that mental statistical operations were necessary to integrate uncertain environmental cues. The methods of signal detection theory have since been applied to a broad range of cognitive processes, ranging from memory (Wickelgren & Norman 1966, Murdock 1982, Anderson & Milson 1989) to discriminating random from non-random patterns (Lopes 1981).

The study of similarity and categorization has also been influenced by statistical ideas. One statistically natural approach has been to model the environment as consisting of a number of distinct categories, which stochastically generate category examples. Given a particular category example, which must be classified, Bayes's theorem can be used to calculate the probability that it was generated by each of the possible categories. This approach to categorization gives rise to "likelihood" or "feature probability" models of categorization (Fried & Holyoak 1984, Anderson 1991). Thus, human categorization is viewed as involving the use of Bayesian statistics. Nosofsky (1990) has shown that the scope of this approach is actually rather wide, since it is mathematically extremely closely related to "exemplar" theories of categorization (Medin & Schaffer 1978, Nosofsky 1984, Estes 1986, Nosofsky 1986). Learning to categorize can itself be viewed as a (more difficult) problem of Bayesian inference, in which a particular number of types of generator category must be inferred. Recently, a psychological model of how this problem can be solved using Bayesian statistics has also been put forward (Anderson 1991). This kind of categorization model is formally closely related to mixture modelling approaches in neural networks (e.g. Jordan & Jacobs 1993).

Finally, statistical models have also been widely used in theorizing about human causal reasoning. For example, Kelley (1967) suggested that causal attribution was effected by conducting an intuitive ANOVA, and this approach has inspired a vast theoretical and experimental literature (e.g. see Cheng & Novick 1990).

The above discussion gives some idea of the breadth of the tradition of modelling mental processes in statistical terms, which stretches far beyond the confines of associationism. Hence, to suggest that neural networks lie within the tradition of statistical models of mind does not imply that they are simply a new form of associationism. Nonetheless, there are close connections between certain kinds of associative principle and particular neural network architectures. The best known relationship is, perhaps, that the Rescorla–Wagner law of classical conditioning is mathematically equivalent to the update rule for a single-layer neural network, one of the simplest neural network architectures (Gluck & Bower 1988). More sophisticated neural network-based models have also been used to attempt to provide new models of conditioning (Sutton & Barto 1981, Gluck et al. 1992).

Given the statistical interpretation of neural networks that we sketched above, neural network models can be viewed as lying firmly within this historical tradition of statistical models of cognition. But they do add something new. As I have argued, they add technical innovations so that the range of phenomena that can be modelled is much larger. Furthermore, neural networks are statistical algorithms implemented by a highly parallel processing architecture, which uses very simple processing units. Many, but by no means all, standard statistical methods can be efficiently implemented in this way; by implementing a statistical algorithm as a neural network we are automatically subject to an important constraint which appears to be a minimal condition for biological plausibility (Chater & Oaksford 1990).

Neural network models and symbolic theories of cognition

We are now in a position to reconsider the debate between symbolic and neural network approaches to cognition. As we noted above, advocates of neural networks sometimes argue that neural networks will entirely displace symbolic models; and defenders of symbolic approaches to cognition have countered that neural networks are simply irrelevant to psychological explanation, which should be couched exclusively in symbolic terms (Fodor & Pylyshyn 1988).

According to the arguments presented here, this debate should really be cast in broader terms, as a debate between statistical and symbolic approaches to mind. But, once cast in these terms, the debate appears spurious, since the two approaches are concerned with orthogonal issues. The advocate of statistical methods pursues the possibility that aspects of cognition can be understood in terms of the apparatus of probability, statistics, information theory and decision theory. The advocate of symbolic methods pursues the claim that aspects of cognition involve the formal manipulation of structured symbolic representations (Fodor 1975, Newell & Simon 1976, Pylyshyn 1984). These are independent and entirely compatible claims about the nature of mind; they do not stand in competition. As we noted above, statistics tackles the problem of induction; but it does not place constraints on what is induced – it could be the grammar of a language, or an everyday or scientific hypothesis, all of which might be internally represented in symbolic form. If the debate between statistical and symbolic ideas seems ill conceived, the debate between neural networks (a special case of statistics) and symbolic ideas seems equally ill conceived.

I suspect that there has been a tendency to adopt a more radical view because of the difficulties which have been encountered in pursuing a symbolic approach to mind. Symbolic methods dominated the computational study of mind from the beginning of the cognitive revolution, with high expectations in certain quarters that the problems of human cognition might rapidly be unravelled. This optimism was based on the hope that early successes in formal domains, such as mathematical or logical reasoning, or game playing, should readily scale up to model common-sense thought. In practice, this symbolic program has run into serious obstacles, in capturing the densely interconnected and defeasible character of human knowledge and in devising mechanisms to reason with such knowledge (see Oaksford & Chater (1991, 1993) for extended treatments; see Dreyfus & Dreyfus (1986), Fodor (1983) and McDermott (1987) for related positions). The problems with modelling everyday thought have led to an increasing focus on apparently specialized cognitive processes, such as syntactic processing, early vision and motor control. But even here there have been considerable difficulties in attacking real-world problems – symbolic parsers cannot cope with natural text, and vision systems are very brittle when faced with real images.

From the point of view of psychology, this disarray does not offer an appealing menu of computational methods which can be recruited as the basis of potential cognitive models. It is therefore tempting to believe that neural networks offer a radical alternative paradigm, within which these difficulties either do not arise, or can readily be resolved. In fact, however, most neural network models are simply unable even to represent the problems that symbolic approaches were formulated to deal with, let alone solve them. There are, for example, no neural network models which parse real text, or analyze real visual scenes – even to begin to tackle such problems appears to presuppose the ability to represent complex structured information, for which symbolic representation is the only candidate. Instead of taking up the problems of the old symbolic paradigm and showing how they can be solved with neural networks, in practice, connectionist cognitive science has simply shifted focus onto different problems, which appear to be amenable to neural network analysis. Interest in neural network cognitive modelling has, for example, focused on highly specific domains such as reading (Seidenberg & McClelland 1989, Bullinaria 1993, Plaut & McClelland 1993), learning the past tense of verbs (Rumelhart & McClelland 1986b, Plunkett & Marchman 1991), finding structure in simple sequential material, and modelling aspects of speech perception and word recognition (e.g. McClelland & Elman 1986, Waibel et al. 1987, Abu-Bakar & Chater 1993, Cairns et al. 1994).

The rise of neural network models has not, in reality, been a revolution against old approaches, but simply a shift of emphasis away from one set of problems, which appear intractable, to another set of problems that can, perhaps, more readily be tackled. In particular, the problems that have been eschewed are just those in which structured representations are required. This picture is strengthened by the fact that those neural network models which have dealt with problems previously tackled by symbolic methods have done so not by overthrowing the symbolic approach, but by implementing symbolic structures and processes in terms of neural networks. So, for example, neural networks have been used to implement semantic networks (Hinton 1981, Shastri 1985, Smolensky 1987), production systems (Touretzky & Hinton 1985), schemata (Rumelhart et al. 1986) and specialist knowledge representation formalisms such as μ-klone (Derthick 1987).

I have argued for an ecumenical position: neural networks and statistics give rise to important tools for studying learning and uncertain reasoning; symbolic methods allow us to model the representation and processing of complex information. Both or neither of these approaches to the mind may ultimately prove fruitful; but there is no incompatibility between these approaches, and, for now, it seems appropriate to pursue both. From our current perspective, it is difficult to see how cognitive theory will be possible without making sense of both approaches, and showing how they can be integrated: the richness of symbolic representations and processes appears to be indispensable in processing language, in vision or in modelling everyday thought, and the statistical inductive methods which show how the information stored in such representations can be adjusted in the light of experience, appears to be equally indispensable. Both symbolic methods and neural networks are distressingly weak when viewed in the context of the extraordinary complexity of the real problems, in perception, language and common-sense thought, that people routinely and effortlessly solve. Cognitive science is, I would argue, currently better advised to develop and pursue both theoretical approaches, rather than to attempt to struggle along with either alone.

References

Abu-Bakar, M. & N. Chater 1993. Processing time-warped sequences using recurrent neural networks: Modelling rate-dependent factors in speech perception. *15th Annual Conference of the Cognitive Science Society, Proceedings*, 191–7. Hillsdale, New Jersey: Lawrence Erlbaum.

Anderson, J. R. 1991. The adaptive nature of human categorization. *Psychological Review* **98**, 409–29.

Anderson, J. R. & R. Milson 1989. Human memory: an adaptive perspective. *Psychological Review* **96**, 703–19.

Baldi, P. & K. Hornik 1988. Neural networks and principal component analysis: Learning from examples without local minima. *Neural Networks* **2**, 53–8.

Bayes, T. 1764. An essay towards solving a problem in the doctrine of chances. *Royal Society of London, Philosophical Transactions* **53**, 370–418. Reprinted in *Biometrika* **45**, 296–315, 1958.

Bechtel, W. & A. Abrahamsen 1991. *Connectionism and the mind: an introduction to parallel distributed processing in networks*. Oxford: Oxford University Press.

Bernoulli, J. 1713. *Ars conjectandi*. Basel.

Brunswick, E. 1943. Organismic achievement and environmental probability. *Psychological Review* **50**, 255–72.

Bullinaria, J. A. 1993. Connectionist modelling of reading aloud. *2nd Workshop on the Cognitive Science of Natural Language Processing, Proceedings*. Dublin, 4–11.

Buntine, W. L. & A. S. Weigend 1991. Bayesian back-propagation. *Complex Systems* **5**, 603–43.

Cairns, P., R. Shillcock, N. Chater, J. Levy 1994. Lexical segmentation: the role of sequential statistics in supervised and un-supervised models. In *16th Annual Conference of the Cognitive Science Society*, A. Ram & K. Eiselt (eds), 136–41. Hillsdale, New Jersey: Lawrence Erlbaum.

Chater, N. & M. R. Oaksford 1990. Autonomy, implementation and cognitive architecture: a reply to Fodor and Pylyshyn. *Cognition* **34**, 93–107.

Cheng, P. W. & L. R. Novick 1990. A probabilistic contrast model of causal induction. *Journal of Personality and Social Psychology* **58**, 545–67.

Chomsky, N. 1957. *Syntactic structures*. The Hague: Mouton.

Clark, A. 1989. *Microcognition: philosophy, cognitive science and parallel distributed processing*. Cambridge, Mass.: Bradford Books/MIT Press.

Cox, R. T. 1961. *The algebra of probable inference*. Baltimore: The Johns Hopkins University Press.

Daston, L. 1988. *Classical probability in the enlightenment*. Princeton, New Jersey: Princeton University Press.

de Finetti, B. 1937. Foresight: its logical laws, its subjective sources. Translated in H. E. Kyburg & H. E. Smokler 1964 (eds), *Studies in subjective probability*. Chichester: John Wiley.

DeMers, D. & G. Cottrell 1993. Non-linear dimensionality reduction. See Hanson et al. (1993), 3–10.

Derthick, M. 1987. *A connectionist architecture for representing and reasoning about structured knowledge*. Department of Computer Science, Carnegie-Mellon University, Technical Report CMU-BOLTZ-29.

Dreyfus, H. L. & S. E. Dreyfus 1986. *Mind over machine: the power of human intuition and expertise in the era of the computer*. New York: The Free Press.

223

Earman, J. 1992. *Bayes or bust? A critical examination of Bayesian confirmation theory*. Cambridge, Mass.: Bradford Books/MIT Press.

Elman, J. L. 1990. Finding structure in time. *Cognitive Science* **14**, 179–211.

Estes, W. K. 1986. Array models for category learning. *Cognitive Psychology* **18**, 500–549.

Finch, S. & Chater, N. 1992. Learning syntactic categories using a neural network. *14th Annual Conference of the Cognitive Science Society, Proceedings*, 820–25. Hillsdale, New Jersey: Lawrence Erlbaum.

Finch, S. & Chater, N. 1994. Learning syntactic categories: a statistical approach. In *Neurodynamics and psychology*, G. D. A. Brown & M. Oaksford (eds), 294–321. London: Academic Press.

Fisher, R. A. 1922. On the mathematical foundations of theoretical statistics. *Royal Society of London, Philosophical Transactions A* **222**, 309–68.

Fisher, R. A. 1956. *Statistical methods and statistical inference*. Edinburgh: Oliver and Boyd.

Fisher, R. A. 1970. *Statistical methods for research workers*, 14th edn. Edinburgh: Oliver and Boyd.

Fodor, J. A. 1975. *The language of thought*. New York: Thomas Crowell.

Fodor, J. A. 1983. *The modularity of mind*. Cambridge, Mass: MIT Press.

Fodor, J. A. 1987. *Psychosemantics*. Cambridge, Mass.: Bradford Books/MIT Press.

Fodor, J. A. & B. P. McLaughlin 1990. Connectionism and the problem of systematicity: why Smolensky's solution doesn't work. *Cognition* **35**, 183–204.

Fodor, J. A. & Z. W. Pylyshyn 1988. Connectionism and cognitive architecture: a critical analysis. *Cognition* **28**, 3–71.

Fried, L. S. & K. J. Holyoak 1984. Induction of category distributions: a framework for classification learning. *Journal of Experimental Psychology: Learning, Memory and Cognition* **10**, 234–57.

Friedman, J. H. & W. Stuetzle 1981. Projection pursuit regression. *Journal of the American Statistical Association* **76**, 817–23.

Gigerenzer, G. 1991. From tools to theories: a heuristic of discovery in cognitive psychology. *Psychological Review* **98**, 254–67.

Gigerenzer, G. & D. J. Murray 1987. *Cognition as intuitive statistics*. Hillsdale, New Jersey: Lawrence Erlbaum.

Gigerenzer, G., Z. Swijtink, T. Porter, L. Daston, J. Beatty, L. Krüger 1989. *The empire of chance: how probability changed science and everyday life*. Cambridge: Cambridge University Press.

Gluck, M. A. & G. H. Bower 1988. From conditioning to category learning: an adaptive network model. *Journal of Experimental Psychology: General* **117**, 227–47.

Gluck, M. A., P. T. Glauthier, R. S. Sutton 1992. Adaptation of cue-specific learning rates in network models of human category learning. In *14th Annual Conference of the Cognitive Science Society, Proceedings*, 540–45. Hillsdale, New Jersey: Lawrence Erlbaum.

Golden, R. M. 1988. A unified framework for connectionist system. *Biological Cybernetics* **59**, 109–20.

Good, I. J. 1950. *Probability and the weighting of evidence*. London: Griffin.

Grossberg, S. 1982. *Studies of mind and brain: neural principles of learning, perception, development, cognition and motor control*. Boston: Reidell Press.

Hacking, I. 1990. *The taming of chance*. Cambridge: Cambridge University Press.

Hanson, S. J., J. D. Cowan, C. Lee Giles (eds) 1993. *Advances in neural information processing systems 5*. San Mateo, Calif.: Morgan Kaufman.

Hinton, G. E. 1981. Implementing semantic networks in parallel hardware. In *Parallel models of associative memory*, G. E. Hinton & J. A. Anderson (eds), 161–87. Hillsdale, New Jersey: Lawrence Erlbaum.

Hinton, G. E. 1989. Connectionist learning procedures. *Artificial Intelligence* **40**, 185–234.

Hinton, G. E. & T. J. Sejnowski 1986. Learning and relearning in Boltzmann machines. In *Parallel distributed processing: explorations in the microstructure of cognition*, vol. 1. *Foundations*, D. Rumelhart & J. L. McClelland (eds), 282–317. Cambridge, Mass.: MIT Press.

Hinton, G. E. & T. Shallice 1991. Lesioning an attractor network: investigations of acquired dyslexia. *Psychological Review* **98**, 74–95.

Hornik, K., M. Stinchcombe, H. White 1989. Multilayer feedforward networks are universal approximators. *Neural Networks* **2**, 359–66.

Horwich, P. 1982. *Probability and evidence*. Cambridge: Cambridge University Press.

Howson, C. & P. Urbach 1989. *Scientific reasoning: the Bayesian approach*. La Salle: Open Court.

Huang, X. D., Y. Ariki, M. A. Jack 1990. *Hidden Markov models for speech recognition*. Edinburgh: Edinburgh University Press.

Intrator, N. 1993. On the use of projection pursuit constraints for training neural networks. See Hanson et al. (1993), 15–20.

Jordan, M. I. & R. A. Jacobs 1993. Hierarchies of adaptive experts. See Hanson et al. (1993), 985–92.

Kelley, H. H. 1967. Attribution theory in social psychology. In *Nebraska symposium on motivation*, vol. 1, D. Levine (ed.), 192–238. Lincoln: University of Nebraska Press.

Kohonen, T. 1984. *Self-organization and associative memory*. Berlin: Springer-Verlag.

Krishnaiah, P. R. & L. N. Kanal (eds) 1982. Classification, pattern recognition and reduction of dimensionality. *Handbook of statistics*, vol. 2. Amsterdam: North-Holland.

Lindley, D. V. 1982. Scoring rules and the inevitability of probability. *International Statistical Review* **50**, 1–26.

Lopes, L. L. 1981. Decision making in the short run. *Journal of Experimental Psychology: Human Learning and Memory* **8**, 626–36.

Lucas, J. R. 1970. *The concept of probability*. Oxford: Oxford University Press.

Luttrell, S. P. 1989. Self-organisation: a derivation from first principles of a class of learning algorithms. *3rd IEEE International Joint Conference on Neural Networks, Proceedings*. Washington, DC, vol. 2, 495–8.

Luttrell, S. P. 1990. Derivation of a class of training algorithms. *IEEE Transactions on Neural Networks* **1**, 229–32.

Luttrell, S. P. 1994. A Bayesian analysis of self-organising maps. *Neural Computation* **6**, 767–94.

Mackay, D. J. C. 1992a. A practical Bayesian framework for backpropagation networks. *Neural Computation* **4**, 448–72.

Mackay, D. J. C. 1992b. The evidence framework applied to classification networks. *Neural Computation* **4**, 698–714.

McClelland, J. L. & J. L. Elman 1986. Interactive processes in speech perception: the TRACE model. See McClelland & Rumelhart(1986), 58–121.

McClelland, J. L. & D. E. Rumelhart (eds) 1986. *Parallel distributed processing: explorations in the microstructures of cognition*, vol. 2. *Psychological and biological models*. Cambridge, Mass.: MIT Press.

McDermott, D. 1987. A critique of pure reason. *Computational Intelligence* **3**, 151–60.

Medin, D. L. & M. M. Schaffer 1978. Context theory of classification learning. *Psychological Review* **85**, 207–38.

Mellor, D. H. 1971. *The matter of chance*. Cambridge: Cambridge University Press.

Minsky, M. & S. Papert 1969. *Perceptrons: an introduction to computational geometry*. Cambridge, Mass.: MIT Press.

Murdock Jr, B. B. 1982. A theory for the storage and retrieval of item and associative information. *Psychological Review* **89**, 609–26.

Neal, R. M. 1992. *Bayesian training of backpropagation networks by the hybrid Monte Carlo method*. Department of Computer Science, University of Toronto, Technical Report CRG-TR-92-1.

Neal, R. M. 1993. Bayesian learning via stochastic dynamics. See Hanson et al. (1993), 475–82.

Newell, A. & H. A. Simon 1976. Computer science as empirical enquiry. *Communications of the ACM* **19**, 113–26. Reprinted in M. Boden (ed.), *The philosophy of artificial intelligence*. Oxford: Oxford University Press, 1990.

Neyman, J. 1950. *Probability and statistics*. New York: Holt.

Nosofsky, R. M. 1984. Choice, similarity and the context theory of classification. *Journal of Experimental Psychology: Learning, Memory and Cognition* **10**, 104–14.

Nosofsky, R. M. 1986. Attention, similarity and the identification–categorization relationship. *Journal of Experimental Psychology: General* **115**, 39–57.

Nosofsky, R. M. 1990. Relations between exemplar-similarity and likelihood models of classification. *Journal of Mathematical Psychology* **34**, 393–418.

Oaksford, M. R. & N. Chater 1991. Against logicist cognitive science. *Mind and Language* **6**, 1–38.

Oaksford, M. R. & N. Chater 1993. Reasoning theories and bounded rationality. In *Rationality*, K. Manktelow & D. Over (eds), 31–60. London: Routledge.

Oja, E. 1989. Neural networks, principal components and subspaces. *International Journal of Neural Systems* **1**, 61–8.

Peterson, C. & J. R. Anderson 1987. A mean field learning algorithm for neural networks. *Complex Systems* **1**, 995–1019.

Pinker, S. & A. Prince 1988. On language and connectionism: analysis of a parallel distributed model of language acquisition. *Cognition* **28**, 73–193.

Plaut, D. C. & J. L. McClelland 1993. Generalization with componential attractors: Word and non-word reading in an attractor network. In *15th Annual Conference of the Cognitive Science Society, Proceedings*, 824–29. Hillsdale, New Jersey: Lawrence Erlbaum.

Plunkett, K. & V. Marchman 1991. U-shaped learning and frequency effects in a multilayered perceptron: implications for child language acquisition. *Cognition* **38**, 43–102.

Pylyshyn, Z. W. 1984. *Computation and cognition: toward a foundation for cognitive science*. Cambridge, Mass: Bradford Books/MIT Press.

Ramsey, F. P. 1931. *The foundations of mathematics and other logical essays*. London: Routledge and Kegan Paul.

Rissanen, J. 1983. A universal prior for integers and estimation by minimal description length. *Annals of Statistics* **11**, 416–31.

Rissanen, J. 1989. *Stochastic complexity in statistics inquiry*. Singapore: World Scientific.

Ritter, H. & T. Kohonen 1989. Self-organizing semantical maps. *Biological Cybernetics* **61**, 241–54.

Rumelhart, D. E. & J. L. McClelland (eds) 1986a. *Parallel distributed processing: explorations in the microstructures of cognition*, vol. 1. *Foundations*. Cambridge, Mass.: MIT Press.

Rumelhart, D. E. & J. L. McClelland 1986b. On learning the past tenses of English verbs. See McClelland & Rumelhart (1986), 216–71.

Rumelhart, D. E., P. Smolensky, J. L. McClelland, G. E. Hinton 1986. Schemata and sequential thought processes in PDP models. See McClelland & Rumelhart (1986), 7–57.

Rumelhart, D. E. & D. Zipser 1986. Feature discovery by competitive learning. In *Parallel distributed processing*, vol. 1. *Foundations*. D. E. Rumelhart & J. L. McClelland (eds), 151–93. Cambridge, Mass: MIT Press.

Savage, L. J. 1954. *The foundations of statistics*. New York: John Wiley.

Seidenberg, M. S. & McClelland, J. L. 1989. A distributed, developmental model of word recognition and naming. *Psychological Review* **96**, 523–68.

Shastri, L. 1985. Evidential reasoning in semantic networks: a formal theory and its parallel implementation. Department of Computer Science, University of Rochester, Report TR166.

Skyrms, B. 1977. *Choice and chance*. Belmont: Wadsworth.

Smolensky, P. 1987. *On variable binding and the representation of symbolic structures in connectionist systems*. Department of Computer Science, University of Colorado at Boulder, Technical Report CU-CS-355-87.

Sutton, R. S. & A. G. Barto 1981. Towards a modern theory of adaptive networks: expectation and prediction. *Psychological Review* **88**, 135–70.

Tanner Jr, W. P. 1965. *Statistical decision processes in detection and recognition*. Sensory Intelligence Laboratory, Department of Psychology, University of Michigan, Technical Report.

Tanner Jr, W. P. & J. A. Swets 1954. A decision-making theory of visual detection. *Psychological Review* **61**, 401–9.

Touretzky, D. S. & G. E. Hinton 1985. Symbols among the neurons: details of a connectionist inference architecture. *9th International Joint Conference on Artificial Intelligence, Proceedings*, 238–43.

von Mises, R. 1939. *Probability, statistics and truth*. London: Allen Unwin.

Waibel, A., T. Hanazawa, G. E. Hinton, K. Shikano, K. Lang 1987. *Phoneme recognition using time-delay neural networks*. ATR Interpreting Telephony Research Laboratories, Japan, Technical Report TR-1-0006.

Wickelgren, W. A. & D. A. Norman 1966. Strength models and serial position in short-term recognition memory. *Journal of Mathematical Psychology* **3**, 316–47.

Wolpert, D. H. 1993. On the use of evidence in neural networks. See Hanson et al. (1993), 539–46.

Young, A. S. 1977. A Bayesian approach to prediction using polynomials. *Biometrika* **64**, 309–17.

CHAPTER 12

Acquiring syntactic information from distributional statistics

Steve Finch, Nick Chater, Martin Redington

Introduction

Acquiring syntax appears to face the language learner with a "bootstrapping" problem. Acquiring syntactic rules presupposes that the syntactic categories in terms of which those rules are formulated have already been acquired; but syntactic categories only have meaning in virtue of the syntactic rules in which they figure. Learning syntactic categories and syntactic rules appear to be mutually interdependent. Consequently, the learner appears to be faced with what seems an impossible task: searching the entire space of category/rule combinations simultaneously.

Even if, as many theorists assume, the learner is equipped with a rich innate knowledge of grammatical rules and abstract grammatical categories, the problem of mapping lexical items onto such categories still remains. Indeed, only once the learner has learned to categorize the speech stream in terms of relatively complex categories will it be possible to bring any innate grammatical information to bear on the learning process. For example, the learner cannot assess word order constraints in the target language, or whether or not some linguistic feature such as "pro-drop" is allowed, until the linguistic input is represented in terms of distinct words each labelled with (at least an approximation to) its syntactic category. Therefore it seems that whether or not learners possess an innate store of grammatical information, the initial stages of language acquisition must be driven by linguistic input.

In considering the potential contribution of sources of information in the linguistic input, two questions may be asked: Can the putative source of information contribute in principle, and how could the relevant information be obtained? And is there empirical evidence to suggest that infants utilize the source of information? In this chapter, we are concerned almost exclusively with the first question, and hence our focus will be on the possible utility of different kinds of information, rather than the empirical case for the importance of each in child language acquisition.

Possible sources of syntactic information

There are three main sources of information in linguistic input which have been proposed as potentially useful in learning syntax, and which, in particular, may be useful in learning syntactic categories. These are based on distributional analysis of linguistic input, on relating the linguistic input to the situation or communicative context in which it occurs, and on the analysis of prosody.

Distributional or correlational bootstrapping

Various authors (Kiss 1973, Maratsos 1979, Maratsos & Chalkley 1981, Maratsos 1988, Finch & Chater 1992) have suggested that words of the same category tend to have a large number of syntactic regularities in common. For example, Maratsos suggests that word roots which take the suffix "-ed" typically take the suffix "-s" and are verbs. Words which take the suffix "-s", but not the suffix "-ed" are typically count-nouns. Consequently, if we take a large number of predicates such as *takes the suffix "-s"*, *takes the suffix "-ed"*, *takes the suffix "-ing"*, *appears immediately after "the"*, and so on, there will be strong correlations evident. These correlations can be used, through some statistical analysis, to find proto-word classes which can later be refined to word classes more consonant with a mature language theory. Various other approaches, based on measuring local statistics of large corpora of language, have also been proposed (Brill et al. 1990, Marcus 1991, Finch & Chater 1992, Finch & Chater 1993, Schütze 1993), and we shall consider these further below.

Simple distributional methods are sometimes associated with a general empiricist *tabula rasa* approach to language learning, which has been widely criticized (e.g. Chomsky 1959). However, this is not germane in the present context, since distributional methods are not proposed as a general solution to the problem of language learning, but rather as a possible source of information about syntactic structure. Furthermore, it may be that there are innate constraints on the possible distributional analyses which the learner can apply, and it is possible, though not necessary, that these constraints might be specific to the task of acquiring language. So distributional methods may, in some sense, embody prior knowledge.

Semantic bootstrapping

Grimshaw (1981) and Pinker (1984, 1987, 1989) hold that the mechanism for the initial classification of words makes use of a correlation between syntactic information and prior semantic categories provided by evolution or learning. This account presupposes concepts such as *possession*, *action*, *objecthood* and so on, in explaining the early acquisition of syntactic categories. They can also assume that complex conceptual representations already exist of external events, and dependencies between these representations and the sound stream can be exploited to infer low-level syntactic structure. Thus, since there is a strong correlation between, for example, being an object and being referred to by a noun, semantic categories, which might be expected to be innately present in descriptions of the world, need only be correlated with the speech sound stream in order to infer rough approximations to a mature syntactic classification. Also, the concept of *noun phrase* might be semantically bootstrapped by defining it to be "that which refers to an object", together with some innate assumptions about the relationship between language and the extant mental representations. These rough approximations can then be further subjected to various forms of semantic and distributional analysis in order to refine them to be consonant with a more mature linguistic theory.

Another, somewhat different approach which also stresses the importance of extralinguistic context is what Curtiss (1987) terms the "social interaction" model (Snow 1972, Bruner 1975, Nelson 1977, Snow 1988). This approach stresses the child's communicative intent and the importance of the development of appropriate communicative relationships with care-givers. The pragmatic purpose to which language can be put by the learner, or by care-givers, is thought to crucially affect the course of acquisition. Thus, nouns can be thought of as words which can be used to denote agents and patients, and verbs as words which can be used to denote actions, etc. (e.g. Schlesinger 1971, Braine 1976, Schlesinger 1988).

Prosodic bootstrapping

Morgan & Newport (1981) propose that learners exploit the mutual predictability between the syntactic phrasing of a sentence, and its prosody. Consequently, if the child takes note of how something is said, he or she has information about the "hidden" syntactic phrasing of the sentence that the child needs to find for a mature theory of language. Thus the syntactic structure of language is not so well "hidden" after all, and an approximation to it may be found by listening to how a sentence is spoken.

Assessing the potential contributions of information sources

In order to quantitively investigate the amount of information that can be gleaned by the language learner from each of these sources, it is useful to study the linguistic (and, for semantic approaches, extralinguistic) input actually received by the language learner. Looking at the structure of this input is important because some cues may seem to be very informative, but in fact occur very rarely, while other cues, which need not be useful on theoretical grounds, may in practice co-occur reliably with important aspects of syntactic structure.

It is difficult to assess the potential contribution of semantic factors in a quantitative fashion, since it is both extremely labour-intensive to record the extralinguistic context associated with even a small amount of linguistic input, and, furthermore, it is difficult to know what description of that context is likely to be relevant to the general cognitive apparatus of the language learner.

Prosodic information, since it is internal to the speech stream, may be more easily recorded, but is still labour-intensive to notate. There are currently no large (millions of words) corpora of conversation with detailed prosodic markings. In future, however, if such corpora are developed, it may be possible to give a quantitative assessment of the amount of information that prosody could potentially give the language learner.

Distributional methods can often be readily applied to language internal information, since word level corpora exist. In particular, unlike semantic and prosodic approaches, distributional analysis can be conducted over texts, represented purely as sequences of distinct words, and these are (at least for English) in almost unlimited supply. Also, reasonably large corpora of transcribed speech, such as the London–Lund corpus (Svartvik & Quirk 1980) and the CHILDES database (MacWhinney & Snow 1985) are also available. These are at least large enough to provide some validation of the performance of distributional methods which are primarily developed using text corpora.

In this chapter, we shall describe and illustrate the performance of a simple distributional technique. This has been applied to learning syntactic categories of lexical items and two- and three-word phrases (Finch & Chater 1992, Finch & Chater 1993), and here we report an extension of this approach to more complex phrasal structure.

While we focus on the distributional approach, it seems entirely likely that all of the types of information source (including semantic and prosodic sources) may be (perhaps highly) informative about syntactic structure and that, if so, the child may draw on them. It is simply that, for the reasons outlined above, such questions are very hard to investigate. Thus, we restrict the discussion here to quantitatively considering distributional methods on purely methodological grounds. For the same reasons, it is difficult to assess the potential importance of interactions between these sources and distributional information. Although we believe that such interactions may be very important, here we shall consider the potential role of distributional information in isolation. Notwithstanding, it is quite possible that the interaction of information sources is so important that any individual source is relatively weak when considered alone.

Objections to the feasibility of distributional methods

A number of arguments have been put forward which appear to undermine the feasibility of distributional methods (in isolation[1]) in category acquisition, and in syntax acquisition in general. Pinker (1984, 1987, 1989) makes the most cogent and influential case against distributional methods. He argues that these criticisms will apply to all distributional methods, illustrating his arguments by considering the work of Maratsos & Chalkley (1981). Pinker suggests that distributional methods have two fundamental problems. His *learnability* argument aims to establish that distributional methods are inadequate in principle, and his *efficiency* argument aims to show that they are unworkable in practice:

(a) *Learnability*. Since distributional methods work solely by examining observed utterances, they do not have access to negative evidence, and hence inevitably are unable to rule out overgeneral models of the language. "The child cannot use . . . absence as evidence, since so far as he or she is concerned the very next sentence could have [a positive example], and absence until then could have arisen from sampling error, or a paucity of opportunities for the adult to utter such sentences" (Pinker 1984: 48).

(b) *Efficiency*. This has two aspects. First, Pinker claims that there are too many possible distributional relations that are potentially relevant, and that exploring all these possibilities is combinatorially intractable. Second, he argues that distributional methods are liable to lead to inappropriate generalizations: "The child could hear the sentences *John eats meat*, *John eats slowly* and *the meat is good* and then conclude that *the slowly is good* is a possible English sentence" (Pinker 1984: 49). More generally, Pinker argues that since pertinent linguistic generalizations are not couched in terms of simple distributional properties such as preceding word, first word in sentence, and so on, inappropriate generalizations are inevitable.

We shall argue that neither of these arguments applies to distributional methods to solve the bootstrapping problem for natural language. To show this we present a range of simulation results which show that considerable amounts of information about both syntactic categories, and the categories of *phrases*, can be derived using a distributional analysis of a large, noisy, unlabelled corpus of English.

The learnability argument

The learnability argument is that negative evidence is essential to rule out overgeneral models of the language if a purely distributional approach is taken. If valid, this argument would seem to have extremely disturbing consequences for the feasibility of induction in many domains, not just syntax. In particular, the whole of empirical science is built exclusively on "positive evidence". There is, after all, no oracle which tells the physicist, chemist or biologist what does *not* happen; all that the scientist can do is observe what *does* happen (which is not the same every time a phenomenon is observed), and attempt to account for the data as well as possible. Thus, according to Pinker's account of distributional models, the language learner and the scientist are in just the same predicament. For both, it is never possible to definitively conclude that a phenomenon can be ruled out by distributional methods alone – the fact that it has not so far occurred may indeed have arisen from sampling error, or the like. The manifest possibility of scientific enquiry suggests that the learnability argument cannot be valid, either in general, or in the case of language learning.

Specifically, the problem with the learnability argument is that it does not take account of the fundamentally statistical character of inductive inference (whether these statistics are computed explicitly, or judged intuitively by the learner). Inductive inference involves choosing a model on the basis of a finite amount of data; it is not possible to find a model which is known to be correct, because there is always the possibility of later falsification, but it is possible to choose the model which is most probable, given the available data (using Bayesian statistical methods), to choose the model which makes the data most likely (using maximum likelihood methods), or to use some other criterion. Overgeneral models, which Pinker assumes cannot be ruled out without negative evidence, are rejected as highly improbable, since they predict the possibility of (classes of) data which are never observed (for a detailed discussion of inductive inference within a Bayesian model comparison framework see, for example, Earman (1992)). Pinker correctly describes methods which use the non-occurrence of tokens in a corpus as negative evidence as being dependent on the learning mechanism used, and therefore hard to evaluate, but does not go on to conclude that since the child certainly does have a learning mechanism, that it might well make use of non-occurrence as negative evidence.

For example, to return to Pinker's "slowly" example above, the use of distributional analysis might indeed derive the acceptability of "the slowly is good" from "John eats meat", "John eats slowly" and "the meat is good". However, empirically (in terms of the analysis that we shall describe below), the sequence "DET ADVERB-1 IS" is about 70 times less likely to appear than one would expect from chance if language was a random stream with lexical items appearing in proportion to how they actually appear. Here, "ADVERB-1" is the class of adverbs which includes "slowly", and "IS" is a class which includes "is, was, are, were, has, have". The non-appearance of this sequence is indicative of a syntactic constraint. Consequently, the non-occurrence of a sequence in a corpus can falsify (or make much less likely) a trivial hypothetical grammar.

Distributional methods can be shown to work

Although there may be no reason why distributional methods should not work in principle, Pinker's argument that they would be impracticable has yet to be addressed. The best way to answer this point is to provide a counter-example, where significant syntactic structure is demonstrably uncovered by linguistically naive distributional methods.

Recall that Pinker's main efficiency criticism is that relevant distributional statistics (e.g. subject/object relationships, head modifier relationships, etc.) are difficult to find, and that relationships which are easy to find (e.g. word adjacency) do not embody linguistic structure in a meaningful way, and consequently cannot be used to discover it. While we accept that highly linguistically relevant relationships are hard to find initially, we dispute the claim that simple relationships such as word adjacency cannot be exploited to find structure. Moreover, we shall show that the simple word-adjacency relationships can be used to infer much more linguistically perspicuous relationships encapsulating phrasal linguistic units of just the type which Pinker claims are most useful in discovering structure in natural language.

Finch & Chater (1992, 1993) proposed a tentative solution to the bootstrapping problem using distributional methods similar to that proposed by Kiss (1973). Kiss used a "most frequent first" approach, where the most frequent words appearing in a large corpus were clustered according to the similarity of statistical measurements of the lexical contexts in which they featured. This is in line with the view that it is not initially necessary to provide a theory which accounts for the acquisition of all of natural language in order to solve the bootstrapping problem, but rather just a significant part of it. The relations "last word", "next word", "last word but one" and "next word but one" were used as the basis of this classification. Although the methods used were not those proposed by Maratsos & Chalkley (1981), the spirit of the enterprise is similar – find some relationships which are highly correlated with syntactic structure, and use these to infer a syntactic classification for words. It was found that for the most frequent 2000 words, a highly linguistically perspicuous classification was uncovered, which featured all of the main word classes.

Pinker argues that one of the main problems with the efficiency of distributional bootstrapping is that there are potentially a very large number of distributional relationships which can be used to uncover linguistic structure.

> Perhaps, then, one can constrain the child to test for correlations only among linguistically relevant properties. There are two problems with this move. First of all, most linguistically relevant properties are abstract [e.g. syntactic categories, grammatical relations] ([this argument] owes its force to the fact that the contrapositive (roughly) is true: the properties that the child can detect in the input – such as serial positions and adjacency and co-occurrence relations among words – are in general linguistically irrelevant). (Pinker 1984: 49–50)

It may be true that the learner cannot formulate distributional generalizations in terms of linguistic abstractions, at least in the early stages of acquisition when presumably linguistic input cannot be parsed in appropriate linguistic terms.

Even if the relevant linguistic abstractions were available, then the results of distributional analysis would still only approximately specify the syntactic categories of individual lexical items (for instance, distributional tests in linguistics are useful heuristics for, rather than litmus tests of, category membership) (Radford 1988). For an information source to be useful, however, it does not have to be unequivocal. What is required is only that it is reliably statistically correlated with relevant linguistic regularities in real speech. Many perceptible relationships in the linguistic input, indeed the very examples that Pinker cites, have been shown to satisfy this requirement (regarding adjacency and co-occurrence relations, see Finch & Chater (1992), Finch (1993) and Schütze (1993); regarding serial position in sentences, see, for example, Hughes (1992)). Below we report our own recent work applying distributional methods to learning the syntactic categories of phrases, rather than just individual lexical items.

Finding phrasal categories

The rest of this chapter addresses the problem of uncovering syntactic structure at a higher level than just word classes. According to the standard view, the relevant level of linguistic analysis is a phrase-based one, where phrases are structured into trees, and are assigned labels, such as *noun phrase*, *prepositional phrase*, *sentence* and the like. We consider the degree to which it is possible to infer classes for *sequences* of words, as has previously been shown for word classes.

First, an initial classification of words is derived, and this classification is exploited to derive a classification of short (one-, two- and three-word) phrases. Then this classification is used to derive a syntactic classification of longer phrases.

Finch & Chater (1992) showed how a distributional analysis could roughly find syntactic categories.[2] They compiled a contingency table of 2000 common words against the contexts in which they appeared in a 40 million word corpus of USENET newsgroup articles. The context was simply defined to be the preceding two and following two words. To keep the computations tractable, attention was restricted to context words which were among the 150 most common words observed in the corpus. The context we used can therefore be thought of as four vectors of 150 dimensions, each dimension corresponding to one of the 150 most common words. The value of the vector is then given by the number of times the focal word appeared in the relevant relation (i.e. preceding, following, last but one, next but one). A definition of similarity between observed distributions of contexts was given (the Spearman rank correlation coefficient), and a cluster analysis performed to produce a hierarchical ontology of the words.

By stopping the hierarchical cluster analysis after only a certain number of links have been made, it is possible to find many classifications of words (i.e. partitions of the 2000-item word set). We stopped the classification when 500 categories remained (i.e. 75% of links were made), and chose the 100 most common of these as a classification of the "frequent part" of natural language. Our choice of these values is *post hoc* – not all values give such good results. In particular, if a very small number of categories is allowed, the distinctions between them have no obvious linguistic meaning. Nonetheless, linguistically meaningful results are obtained over a wide range of parameter values. Nearly all of these categories corresponded to linguistically coherent categories or sub-classes of categories. For instance, of the 100 categories, the two most common were (see Finch & Chater (1992, 1993) for more detailed results):

(a) **C1** the my your their his our its a an any some several another every these those such each no many most certain

(b) **C2** of in on at for with from by into through against about between without under within during via upon towards toward across among beyond regarding

The corpus can now be mapped from a sequence of lexical items to a sequence of what we call *C-level* categories. For example, every occurrence of "the" would be replaced by "C1". Sequences of length 1, 2 and 3 of these *C-level* categories were searched for in a large corpus, and the 3000 most common such sequences were chosen for distributional analysis. This time the context was defined to be the four surrounding *categories* rather than the four surrounding words. Again a cluster analysis was performed, and again this was terminated when 75% of the links had been made, resulting in a classification of *short sequences* (X1, X2, . . . , X150). Several of these *X-level* short sequences had interesting linguistic interpretations. For instance, one, which contained about 80 short sequences, seemed to correspond to short noun phrases, in that in new text they corresponded to word sequences such as "it", "the" "apparent size", "each article", "the mother", "the real data", "a scientific theory". Note that all the examples given here were randomly sampled from a corpus of USENET articles which were not included in the corpus used for categorization. There is a preference towards longer examples, mainly to avoid repetition. Another category corresponded to parts of the verb "to be", including as exemplars "has been", "will have been", "is", "are", "might be" and so on. Other linguistically perspicuous classes include prepositional phrases, *n*-bar phrases and parts of the verb "to have". There are many linguistically imperspicuous categories, however, but many of these correspond to apparently coherent classes, even though most linguists would not use them. For instance, one class includes "the top of", "the name of", "the person with" and so on. Another one, which was picked at random, includes "use the", "use at the", "break into these", "add an". This is not a perspicuous category, but if a noun phrase lacking a determiner is added, it becomes a simple verb phrase. This observation suggests the utility of a further stage of analysis, in which sequences of *these* categories are clustered together to find still higher-level structure.

Sequences of these X-level *short sequences* of length 1 and 2 were searched for in a corpus of 40 million words taken from USENET newsgroups (stripped of headers, footers and repetition). The 3000 most common of these sequences were chosen for analysis, and this time the context was the set of the surrounding four X-level categories. Since there are many ways to parse a stream of words into X-level categories, each focal sequence can have many different contexts associated with it (as opposed to one for the procedure above). For example, the sequence "the big black dog" can be parsed in many ways. In particular, if each constituent of the parse is to be a short sequence (of length 1, 2 or 3), this phrase can be represented by labelled bracketings of short sequences in seven ways: (X81: The) (X32: big) (X32: black) (X36: dog); (X81: The big) (X32: black) (X36: dog); (X81: the) (X32: big black) (X36: dog); (X81: the) (X32: big) (X36: black dog); (X81: The big black) (X36: dog); (X81: the) (X36: big black dog); (X81: The big) (X36: black dog). Each labelled bracketing, or *parse*, corresponds to a sequence of X-level categories: in this case, the sequences are X81 X32 X32 X36; X81 X32 X36; X81 X32 X36; X81 X32 X36; X81 X36; X81 X36; X81 X36.

If "the big black dog" was the left context of an item of interest, then although the immediately preceding category is always X36, the last but one category is either X81 (noun premodifier with determiner) or X32 (noun premodifier without determiner).

The method was applied to a corpus of 10 million words, and again a cluster analysis was performed and terminated when 75% of the links had been made, leaving 135 classes. Five of the eight most frequent classes correspond to coherent linguistic entities. The others are coherent, but end in determiners, and so are not classical constituents (although they would be categories in a categorial grammar). Table 12.1 shows some examples of word sequences found to be in these five phrasal classes. We also give a small selection of random sequences from the corpus to show that the elicited categories really do uncover significant structure relative to random selection. We label with "*" those sequences which could not be analyzed as their description by a linguist, and by "?" those which are strange. In parentheses we give the percentage of un-starred examples.

As can be seen from Table 12.1, the classification is not entirely accurate, but remember that our goal is not to find a correct classification of language immediately, but rather to find significant amounts of structure which can later be refined by other methods which might make use of semantic and prosodic information. In many of the classes, over 80% of members could have the same syntactic category (recall that our aim is not to "parse" sentences, but rather to find what structure might be posited as a plausible arc by, for example, a chart-parser). Thus, it seems that distributional methods can provide significant information concerning phrasal level syntactic structure, even when used in isolation.

The important point is that some non-trivial structure of language has been learned, and that this has been done by applying non-language-specific distributional techniques to raw language data. For information, in a corpus of 500 000 valid words (i.e. words in the most frequent 2000 words in the corpus), 1 730 000 constituents were discovered. Of these, 1 400 000 were coherent

Table 12.1 This table shows some token sequences of four of the classes found by empirically clustering word sequences according to the similarity of their contexts of occurrence.

Simple sentences: what is a context, that's a different story, you will also receive a copy, we could hold some events, you must continue, we have the chance, some groups have no names, you start out, * you have any problems, the project should work, the old version is still available, ? I think it, I will have the car, you are standing, there's always the chance, I kept them, it would be appropriate, I think there's a piece, there is a french culture office, ? I would argue, * it is called, the bar could be seen, it's ok, the conference is over, I was talking to a friend. (92%)

Verb phrases: give away, pick them up, buy some audio tapes, suggest a company, have a new book and manual, get away from it, ask them to change the entry, change the entry, * think the world, can't remember what day, got nothing, disagree, do something, look around for people, go, even understand the questions, really want an argument, be appropriate, need to move out, get the information, try to send them, know of a place, live, read about them, don't have my copy, get to the question, get back on this, ? tell this, make them, go around, change the subject, know it, call the previous owner, give the name of their version, * believe that their version, can't see anything, have to have messages. (97%)

Noun phrases: ? the situation theory and its applications, the natural language group, some sort of code, * this since it, a new reference to the database, their hands, ? the bar with their parents, that day, the logical structure of natural languages, * me for a game, * it on line, a case, some of my stuff, a change of date, what parts, the rights to them, the number one, the end of the world, ? the money on a government, * the name of product, something similar, a fairly normal life, ? a dog to the club, * the point where it, what number, * some areas this, that way, the attention, * that names, * that names and references, many of the good responses, ? several friends in this, several friends in this area, any of the above equipment, ? another in your opinion. (77%)

Infinitival complements: to accept this attitude, * to allow laser printer, to be about her, to be at an end, to buy more, to call them, to change the name, to come up, to find the problem, to get a piece, to get me back, to get over it, to have a baby, to hear more, to keep it, to leave an engine, to mention me, to mention the groups in question, to pay the high prices, to play them, to read in the shell window, ? to replace the include, to run their own bbs, to start, to start a discussion, to take it, to take over the world, to take this out, to use the drive, to use the old mode, to wait for the music, to write the software, to have brought it. (97%)

Prepositional phrases: of this network, for the family, in this, ? between the state, in the story, in your question, of the story, back to the list, of this article, with a person, in our state, in an order, in our culture, with the image, of the country, on the net, on this, to the parents, of india, at the door, in certain parts, in the same area, * in article, for the us, ? in the general, with the local party, for the community, in the original post, on the individual, on engineering, of the free world, in those countries, to the other states, by the indian army, on men, * by one of her, by an individual, of the world, on a host, ? in the history, down in the country, of ancient times, of small discussion, from him, on his relationship, of news, via this, in india, * without star. (94%)

Random sequences: whether such political rubbish should temporarily as visitors, issue on, the involvement of subhash, this message, looks like because of, that is dharma from, to the, so i checked that too, my situation was, why those who keep their, information about i feel you, problem is that, if others, who promote and protect the, shouldn't you reserve judgement, it is a, islam was categories, where at least 80% of their member tokens were considered possible to have the same simple category in categorial grammar. Of the word tokens, 90% were in some category of length at least 2, and 80% of these were coherent categories. The categories listed above covered 35% of the corpus (i.e. 35% of all word tokens appeared in at least one of the above categories). This figure increases to 65% if another two highly coherent categories (noun groups and verb phrases) are included.

Conclusions

Distributional methods have been shown to be able to uncover significant linguistic structure at several levels in natural language. In particular, we have demonstrated the relative ease of distributionally bootstrapping abstract linguistic entities including approximations to all word classes, relatively simple noun phrases, verb phrases, prepositional phrases and sentences. Although much "fine-grain" structure in natural language, such as verb subcategorization frames, has not been demonstrated, it is plausible that more sophisticated distributional methods will be capable of finding more subtle regularities. Indeed, some verb subcategorization information has been acquired, since although "disagree" is classified as a verb phrase, other single verbs such as "do" or "buy" classified as simple verb phrases only if followed by a candidate object.

This work suggests a number of interesting avenues for future research. These methods can be applied to corpora which more accurately reflect the input received by the child. The CHILDES database (MacWhinney & Snow 1985) provides over 2 million words of transcribed care-giver speech, which, while large enough to find an initial classification of words (see Redington et al. 1993), is too small to apply the techniques described here in full.

Another interesting question concerns the applicability of these methods to languages other than English. Whilst many of the grammatical regularities indicated in English by word order are, in other languages, more reliably indicated by various morphological regularities (e.g. case marking and so on), there is no reason why the general method described here should be restricted to exploiting word sequence regularities. Other sources of distributional regularity, such as inflectional ending, morphological structure and so on, might be exploited to derive syntactic information. However, it should also be noted that even for languages which do not have mandatory word order constraints, word order is still probably highly informative of syntactic category, so the methods used here may work well even with these languages. Both of these research avenues are currently impeded by the paucity of very large machine readable corpora.

The success of distributional methods in discovering syntactic categories at the lexical and phrasal level raises the question of the scope of such methods in other areas of language acquisition, including the acquisition of grammatical rules. The general problem of language acquisition appears so difficult, and to be solved so effortlessly by the child, that we suspect that many sources of information, possibly including an innate universal grammar (Chomsky 1980), may be involved.

Notes

1. Pinker allows that distributional analysis may have some role in language acquisition, when supplemented by other, more important sources of information – in particular, semantic information. While, as noted above, we suspect that it is highly plausible that information is integrated in this way in child language acquisition, we argue here that distributional information can be a surprisingly valuable source of information even when considered in isolation.
2. Note that this method assigns a single syntactic category to each lexical item. Since many lexical items are syntactically ambiguous, the challenge of capturing all possible readings remains. This is an important topic for further research, worked on, for example, by Kupiec (1993).

Acknowledgements

Portions of this chapter are based on Finch & Chater (1994).

References

Braine, M. D. S. 1976. Children's first word combinations. *Monographs of the Society for Research in Child Development* **41**, 25–67.

Brill, E., D. Magerman, M. Marcus, B. Santorini 1990. Deducing linguistic structure from the statistics of large corpora. DARPA *Speech and Natural Language Workshop*. Hidden Valley, Pa.: Morgan Kaufmann.

Bruner, J. 1975. The ontogenesis of speech acts. *Journal of Child Language* **2**, 1–19.

Chomsky, N. 1959. A review of B. F. Skinner's verbal behavior. *Language* **35**, 26–58.

Chomsky, N. 1980. *Rules and representations*. Boston, Mass.: MIT press.

Earman, J. 1992. *Bayes or bust*. Cambridge, Mass.: Bradford Books/MIT Press.

Finch, S. 1993. *Finding structure in language*. PhD thesis, Centre for Cognitive Science, University of Edinburgh.

Finch, S. P. & N. Chater 1992. Bootstrapping syntactic categories. *14th Annual Conference of the Cognitive Science Society, Proceedings*, 820–25. Hillsdale, New Jersey: Lawrence Erlbaum.

Finch, S. P. & N. Chater 1993. Learning syntactic categories: a statistical approach. In *Neurodynamics and psychology*, M. Oaksford & G. D. A Brown (eds), 295–322. London: Academic Press.

Finch, S. P. & N. Chater 1994. Distributional bootstrapping: from word class to proto-sentence. *16th Annual Meeting of the Cognitive Science Society, Proceedings*, 301–6. Hillsdale, New Jersey: Lawrence Erlbaum.

Grimshaw, J. 1981. Form, function, and the language acquisition device. In *The logical problem of language acquisition*, C. L. Baker & J. McCarthy (eds). Cambridge, Mass.: MIT Press.

Hughes, J. 1992. The statistical inference of parts of speech. Unpublished paper, Department of Computer Science, University of Lancaster.

Kiss, G. R. 1973. Grammatical word classes: a learning process and its simulation. *Psychology of Learning and Motivation* **7**, 1–41.

Kupiec, J. 1993. An algorithm for finding noun phrase correspondences in bilingual corpora. *31st Annual Meeting of the Association of Computational Linguists, Proceedings*, 17–22.

MacWhinney, B. & C. Snow 1985. The child language data exchange system. *Journal of Child Language* **12**, 271–95.

Maratsos, M. 1979. How to get from words to sentences. In *Perspectives in psycholinguistics*, D. Aaronson & R. Rieber (eds). Hillsdale, New Jersey: Lawrence Erlbaum.

Maratsos, M. 1988. The acquisition of formal word classes. In *Categories and processes in language acquisition*, Y. Levy, I. M. Schlesinger, M. D. S. Braine (eds), 31–44. Hillsdale, New Jersey: Lawrence Erlbaum.

Maratsos, M. & M. Chalkley 1981. The internal language of children's syntax. In *Children's language*, vol. 2, K. E. Nelson (ed.). New York: Gardner Press.

Marcus, M. 1991. The automatic acquisition of linguistic structure from large corpora. In *1991 Spring Symposium on the Machine Learning of Natural Language and Ontology, Proceedings*, D. Powers (ed.). Stanford, Calif.

Morgan, J. & E. Newport 1981. The role of constituent structure in the induction of an artificial language. *Journal of Verbal Learning and Verbal Behaviour* **20**, 67–85.

Nelson, K. 1977. Facilitating children's syntax acquisition. *Developmental Psychology* **13**, 101–7.

Pinker, S. 1984. *Language learnability and language development*. Cambridge, Mass.: Harvard University Press.

Pinker, S. 1987. The bootstrapping problem in language acquisition. In *Mechanisms of language acquisition*, B. MacWhinney (ed.). Hillsdale, New Jersey: Lawrence Erlbaum.

Pinker, S. 1989. *Learnability and cognition*. Cambridge, Mass.: MIT Press.

Radford, A. 1988. *Transformational grammar*, 2nd edn. Cambridge: Cambridge University Press.

Redington, F. M., N. Chater, S. Finch 1993. Distributional information and the acquisition of linguistic categories: a statistical approach. *15th Meeting of the Cognitive Science Society, Proceedings*, 48–53. Hillsdale, New Jersey: Lawrence Erlbaum

Schlesinger, I. M. 1971. Production of utterances and language acquisition. In *The ontogenesis of grammar*, D. I. Slobin (ed.). New York: Academic Press.

Schlesinger, I. M. 1988. The origin of relational categories. In *Categories and processes in language acquisition*, Y. Levy, I. M. Schlesinger, M. D. S. Braine (eds). Hillsdale, New Jersey: Lawrence Erlbaum.

Schütze, H. 1993. Word space. In *Advances in neural information processing systems 5*, S. J. Hanson, J. D. Cowan, C. L. Giles (eds). San Mateo, Calif.: Morgan Kaufmann.

Snow, C. 1972. Mother's speech to children learning language. *Child Development* **43**, 549–65.

Snow, C. E. 1988. The last word: questions about the emergence of words. In *The emergent lexicon*, M. Smith and J. Locke (eds). New York: Academic Press.

Svartvik, J. & R. Quirk 1980. *A corpus of English conversation*. Lund: LiberLaromedel Lund.

SPEECH AND AUDITION

Paul Cairns

The variety of chapters in this section is indicative of the diversity of subdomains that exist under the aegis of speech and audition. Each of the chapters is concerned with a different facet of the human acoustic/language processor, but in ensemble they characterize the spread of issues which current psycholinguistic speech research addresses.

Within the domain of speech research a dichotomy has traditionally been made between production and perception, although arguably the domains are interlinked and may be driven by shared modality-independent representations. In this section the emphasis is on perception, with three chapters in this area (Ch. 13 by Leslie Smith, Ch. 14 by Mukhlis Abu-Bakar and Nick Chater and Ch. 15 by Paul Cairns, Richard Shillcock, Nick Chater and Joseph Levy). Production is represented by Trevor Harley and Siobhan MacAndrew's model (Ch. 16).

One of the fundamental concerns in speech is accounting for variation in the signal. The most typical examples of the problem are: variation that arises through extraneous noise; characteristics of individual speakers; and effects relating to speech rate. The chapters in this section address variation using different techniques: the central concern for Abu-Bakar and Chater is how it is possible for listeners to recognize syllables and speech sounds that are spoken at different rates, while Smith employs digital signal-processing techniques to normalize auditory signals; Cairns et al. use noisy data to train their network, which learns to compensate for the disruption.

In contrast to models of written language processing (see Section II on models of reading), in speech the temporal dimension of the signal, and concomitant processing, must be dealt with. Harley and MacAndrew choose to abstract away from the dynamic nature of the signal itself, but do investigate the time-course of processing in the production of word forms. Cairns et al. and Smith adopt models that are capable of handling temporal signals, while Abu-Bakar and Chater, as already stated, make temporal problems the focus of their work.

The need to process temporal information has stimulated interest in the application of recurrent neural networks to cognitive modelling. Three of the four models in this section use some sort of recurrent network. Both Cairns et al. and Abu-Bakar and Chater opt for the use of fully recurrent networks trained with *back-propagation through time*, while Harley and MacAndrew use the *interactive activation* architecture in which connection strengths are set by hand and no learning takes place. Smith eschews neural networks in his chapter, but calls on signal-processing techniques to model the function of neural components of the auditory processing system.

Linguists have traditionally been concerned with the definition of *levels* of structure in language, i.e. features, phonemes, syllables, words, phrases, etc. In terms of linguistic levels the four chapters in this section provide a broad coverage: Smith deals with raw speech waveforms which are processed in accord with what is known about the human auditory system; one level up, Abu-Bakar and Chater deal with idealized, unsegmented, representations of acoustic elements such as formant frequency and voice onset time; at a slightly higher level of abstraction Cairns et al. use segmental input in which each segment is a bundle of phonetic features. This input is used in order to bootstrap higher-order structure such as words; finally, Harley and MacAndrew are concerned with the production of words as atomic units.

The chapters by Cairns et al. and Smith share a similar topic: that of *segmentation*, although the nature and purpose of the segmentation is different in each case. Smith is concerned with showing how bursts of acoustic energy can be used to define perceptual units such as musical notes or stretches of speech. On the other hand, Cairns et al. investigate the use of slightly higher-level distributional cues in uncovering word and syllable boundaries. Both these chapters also share a computational philosophy in using *unsupervised* systems which discover structure in their input by virtue of interactions between the structure of the system and the data. Structure emerges without the need for the system to be trained with labelled input, the approach is thus *bottom-up*. In contrast, Abu-Bakar and Chater use a *supervised* technique in which the model is trained to perform the desired mapping.

I will now briefly summarize the chapters in this section.

In Chapter 13, Smith presents a data-driven technique for segmenting continuous speech waveforms so as to produce chunks of input that have a putative semantic correspondence, providing an auditory scene analysis. The model is applied to speech, as well as to the segmentation of notes in musical input. Some of the techniques used are based on processing known to occur in the cochlea. The model relies on traditional signal-processing techniques rather than neural networks.

In Chapter 14, Abu-Bakar and Chater detail a series of experiments that investigate the capabilities of recurrent neural networks for processing temporal data that vary in rate of presentation, in particular so-called *time-warped* sequences. Their experiments first test the networks using arbitrary localist patterns, before applying them to speech data, focusing on the differentiation of /pi/and /bi/ or /ba/ and /wa/ at varying rates of presentation.

In Chapter 15, Cairns et al. present a recurrent network that is trained on a large corpus of conversational speech, and that learns to encode the temporal dependencies between phonetic elements in the input. This network is then used as a tool in modelling two psycholinguistic phenomena: phoneme restoration; and the acquisition of lexical segmentation.

In Chapter 16, Harley and MacAndrew use an interactive activation network to model what they refer to as *lexicalization* – by which they mean the process of accessing the semantic representation of a word in the mental lexicon and outputting its stored phonological representation. They assess the extent to which the model accounts for diverse empirical data in lexicalization such as speech errors, imageability effects and frequency effects.

CHAPTER 13

Onset/offset filters for the segmentation of sound

Leslie S. Smith

Introduction

This work investigates temporal segmentation of single-source sound, that is, the segmentation of a single stream of sound (in the sense of Bregman (1990)) using the AIM human auditory processing model (Patterson & Holdsworth 1990) followed by neurobiologically inspired onset and offset filters. It extends the work reported in Smith (1994). Temporal segmentation is one aspect of an overall auditory scene understanding or synthetic music comprehending or speech interpretation system. The techniques used are entirely bottom-up: that is, they are driven by the data, without explicit higher-level knowledge of the sound, or of what is expected in the way of segmentation, and are based on models of the inner ear and cochlear nucleus. They are therefore applicable to any sound stream. The methods used are not neural models in that single neurons are not simulated, but are neural in the sense that the onset and offset filters discussed in a later section model the output of a population of neurons.

But is segmentation necessary? Given that the aim is to interpret or comprehend the sound, the alternatives would be to process the sound stream either all at once, or else in fixed-length sections. Many sound streams are too extended to permit the former approach, since one needs some results of the interpretation before the sound stream finishes: for example, interpreting continuous speech or a musical phrase. The latter approach would require the sections to be shorter than the shortest element of sound, and would present problems when a section contained parts of more than one sound element. We believe that low-level interpretation is carried out time segment by time segment, and that the time segment end-points are derivable from the gross structure of the sound stream.

The segmentation produced in this way corresponds to sudden increases and decreases in energy in the different frequency bands of the signal. This frequently corresponds to notes in music; however, it does not correspond to a segmentation of continuous speech into words since words frequently are run together. In speech, the segmentation produced tends to follow the opening and closing of the lips, or the starting and stopping of sound associated with consonants. Although one cannot build up words reliably from these segments (since words in continuous speech are not separately enunciated so that, for example, "ice cream" sounds exactly the same as "I scream") the same segments do tend to occur when the words are spoken by different speakers with the same accent. To that extent, this segmentation technique offers a data-driven processing technique for breaking down continuous speech into short pieces. Spoken language does consist of a sequence of words, but this is not reflected reliably in the speech signal: to recreate the word sequence, it will generally be necessary to segment the data (for example using the techniques described here), synthesize these segments, and resegment, using higher-level knowledge (for example, knowledge of the words themselves).

The techniques described in the following section are very general. Although inspired by the processing which appears to occur in the cochlea and cochlear nucleus, they are not neural models, in that their behaviour represents an abstraction of what the neural system appears to be doing rather than modelling the neurons themselves. Nonetheless, they retain the general-purpose nature of the neural original. The closest approach is in Wu et al. (1989), where a more complex but more faithfully biological approach was taken for onset channels. Although the techniques used are very simple and bottom-up, they could easily be incorporated into a more complex system which was partly top-down. The next section in this chapter applies the techniques to speech and a variety of musical sounds. The final sections discuss the results and how this work might be extended.

Methods

The methods used are outlined in Figure 13.1.

The sound signal was acquired using an AKG D109 microphone, in an ordinary office environment. This was digitized at 22 050 samples per second, 16 bits linear, using a Singular Solutions A/D64x. The resulting file was used as input to the basilar membrane movement module of the AIM human auditory processing model (Patterson and Holdsworth 1990). The parameters were set so as to give about 32 bands of output, from 100 Hz to about 10 kHz, with the audiogram switched off. This gives an output of about 32 channels of digitally filtered signal, with the centre frequency of each band being approximately logarithmically spaced between the start and end frequencies (Fig. 13.2). Each channel is relatively wide-band. The bands, and their widths are based on what is known of the human cochlea (Moore and Glasberg 1983, Patterson and Holdsworth 1990).

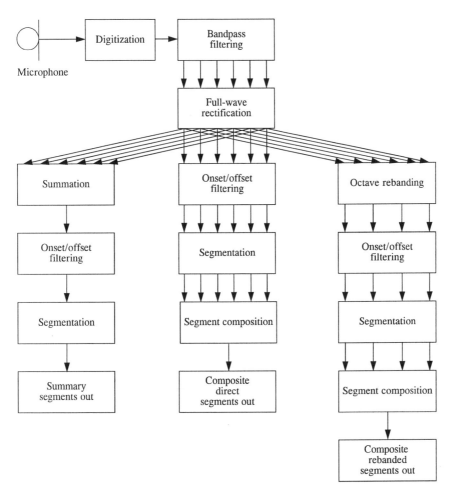

Figure 13.1 Outline of the processing of the sound. On the left is shown simple summary segmentation. In the centre is shown onset/offset filtering then segmentation of each band, followed by composition of segments. On the right is shown the rebanding of the rectified channel outputs prior to onset/offset filtering and segmentation of these new bands, followed by segment composition.

The output of each band was then full-wave rectified to give a measure of the energy in each band. This can be thought of as modelling in outline the effect of a population of inner-cochlear hair cells. In the first experiments, the outputs of all the bands were summed. This gives a measure of the total signal energy, modelling, again in outline, the total activity in the auditory nerve. This summary signal was used as input to the onset/offset filters, and the output of this filter used to segment the signal (described in the following sections). In later experiments (see below), two more sophisticated approaches were taken. In the first, each channel was segmented separately, and the results combined to

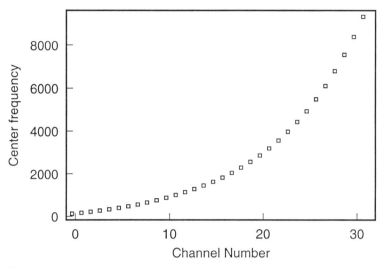

Figure 13.2 Plot of centre frequency of each channel. The parameters used in the AIM model were minimum frequency = 100 Hz, maximum frequency = 10 kHz, density = 1, minimum bandwidth = 24.7 Hz and quality = 9.265. The bandwidth is the minimum bandwidth + characteristic frequency/quality.

produce a composite direct segmentation. In the second, the channels were formed into a smaller number of bands, and these separately segmented, and the results combined to produce a composite banded segmentation. In all cases, it was possible to return to the original sound signal, and to hear what each segment actually sounded like. Listening to each segment was very useful in the development of the segmentation techniques, and in their assessment.

The onset/offset filter

The onset/offset filters used are an abstraction of cochlear nucleus cells responsive to onsets and offsets (Pickles 1988, Blackburn & Sachs 1989, Blackwood et al. 1990, Brown 1992). Onset has also been found to be an important psychophysical grouping criterion (reviewed in Brown 1992). Brown suggests that the biological onset and offset cells are organized into a two-dimensional map, with one dimension being frequency, and the other excitatory (for onset) or inhibitory (for offset) delay. In this chapter, we shall deal initially with a much simpler system, one in which the onset/offset filter is applied to the whole signal. This is really taking simplification too far, and later we shall look at applying the onset/offset filters channel by channel.

The filters used are based on the ideas applied to visual processing in Marr & Hildreth (1980). Two different filters have been experimented with, namely a difference of exponentials filter (DoE) and a half difference of Gaussians (HDoG). Both filters produce their results by taking the difference between an average over recent time, and an average over a longer sample of recent time. This

250

approach is more noise immune than one which uses differences directly.

Each average is computed by convolving the filter with the summary signal $s(x)$ described above:

$$A_z(t,k) = \int_0^t fz(t+x,k)s(x)\,dx \qquad (13.1)$$

where $f(x, k)$ is the convolving function. z is either E or G, depending on whether a DOE or an HDOG filter is being used. The k parameter of f defines the particular moving average being used. For the DOE filter,

$$f_E(x,k) = k\exp(-kx) \qquad (13.2)$$

and for the HDOG filter,

$$f_G(x,k) = \sqrt{k}\exp(-kx^2) \qquad (13.3)$$

These particular functions have been chosen so that their integral is a constant independent of k. Both functions have their maxima at $x = 0$, and tail away rapidly towards 0 as x increases. Thus, $A_z(t, k)$ is a moving average of $s(x)$, most strongly influenced by the value of $s(x)$ near t.

We define the onset/offset operator as the difference between a pair of averages. Such an operator could be defined in terms of three parameters, t, $k1$ and $k2$; however, noting that the positive average will always be a shorter-term average than the negative average, and because we want to consider families of such onset/offset operators, we define the operator as

$$O(t,k,r) = A_z(t,k) - A_z(t,k/r) = \int_0^t \left[f_z(t-x,k) - f_z(t-x,k/r) \right] s(x)\,dx \qquad (13.4)$$

where $r > 1$. Thus, k defines the short-term average, and k/r defines the longer-term average. This filter has the appropriate property of giving a 0 output for constant input. The convolving functions are illustrated in Figure 13.3. Although

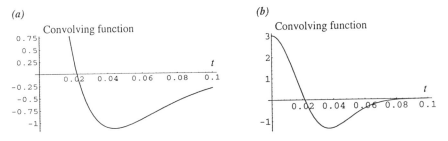

Figure 13.3 (a) DOE convolving function ($f_E(t,k) - f_E(t,k/r)$), with $k = 50$ and $r = 1.2$. The value at $t = 0$ is 8.33. (b) HDOG convolving function ($f_G(t,k) - f_G(t,k/r)$) with $k = 1200$ and $r = 1.2$.

they have the same intercept, the DoE filter is much sharper at the start, and has a longer tail. These factors conspire to make the DoE output rather less smooth than the HDoG one, and to make the HDoG filter more sensitive to offsets (in the sense that it produces a noticeable minimum).

The output of the filter is not intended to represent the output of any specific cell; rather, it is a mixture of the output of a population of onset and offset cells. It has a high positive value when a large number of onset cells (and very few offset cells) would fire, and it has a highly negative value when a large number of offset cells (and very few onset cells) would fire.

Segmentation using the onset/offset filter output

The output from the filter rises at the start of a sound, and falls when the sound is decreasing. Precisely what each filter is sensitive to depends on the filter type and parameters. The very sharp peak of the DoE filter at $t = 0$ makes its output less smooth than an HDoG with the same intercept, and its longer tail makes it less sensitive to offsets. This can be seen in Figure 13.4c,e. Clear minima marking the initial decay of each note can be seen in Figure 13.4e (HDoG) but not in Figure 14.4c (DoE).

Exactly how these outputs should be used to perform segmentation of a sound is not immediately clear. One could consider a sound to start when the output crosses 0; however, this makes the system very sensitive to any extraneous noise, and since, as is clear from Figure 13.4c–e, the rise in output is very steep for a sound of reasonable volume, we prefer to identify the start of a sound as being when the output exceeds some threshold.

Identifying the end of a sound is more difficult. Although some sounds have a clear finish (such as a note on a flute or a saxophone), many do not, and one often considers the end of the sound as being the start of a new sound, certainly when the notes are played quickly. Examples are notes played on a glockenspiel, or single plucked notes on a guitar. Listening to a single note excerpted from the middle of a sequence of notes from such an instrument, one is aware of the previous note continuing. (Whether this is due to habituation, or is some form of high-level learned effect is unclear.) If one listens to the whole sequence, one does not normally hear each note continue under the new note. This suggests that using the start of the new note in a sequence as the end of the previous note is reasonable. Often, the exact placing of the end of a sound is rather arbitrary, particularly for short sounds in an echoing environment, such as a hand clap in a bare room.

Using the output of the onset/offset filter, the immediately obvious choices for ending a segment are where the filter output recrosses 0, or where it has a negative minimum. The former marks the point at which the filter declares that the two averages are the same, which could be because the sound is constant, or because the degree of onset matches the degree of offset (e.g. when a sound which had a recent onset is now decreasing in intensity), and the latter marks the maximum rate of offset. The former is clearly too early: the sound is still present, and the latter may well also be too early, since the maximum rate of offset of a sound may occur while the sound is still present. This is particularly true for sounds which have a rapid attack and decay at their start, followed by a slower sustained period, and a slow final decay. Many plucked and percussive sounds have this form (see Fig. 13.4a,b). Speech tends to have both a rapid onset and offset, reflecting the opening and closing of the vocal tract.

Segmenting using the summary onset/offset filter output

The summary signal was further processed to improve the segmentation it provided. Since the perceived end of a segment can appear to depend on what follows it, we considered pairs of adjacent segments produced using the negative minimum of the summary onset/offset filter to mark the segment end. These were generally not contiguous. To make them contiguous, the program must decide whether to extend the end of first segment forwards, extend the beginning of the second segment backwards, or to insert another "quiet" segment between the two segments. If there is a large gap between the two segments, it is reasonable to insert another segment to mark this gap. We need only decide on how large a gap we should consider to be large enough: we will write G_quiet for this gap length, in milliseconds. After some experimentation, the following was found to be a reasonable compromise. First, look forward from the minimum marking the initial estimate of the end of a segment to the crossing of a value which marks the start of the next segment, seeking out additional minima. Choose the last minimum which is within a certain predefined ratio (K_minmin) of the largest minimum and extend the first segment to this point. If the time between this point and the start of the next segment is less than G_quiet, then extend the next segment backwards to this point. Otherwise, insert an additional "quiet" segment from this new end of the first segment to the start of the second segment. This technique was used with G_quiet = 50 ms and K_minmin} = 0.4.

Segmentation by combining channel segmentations

The summary onset/offset signal can be considered as a measure of a population encoding for a population of onset and offset cells whose input is the complete signal. The technique described in the previous section is equivalent to considering that there is one onset cell (which fires when a threshold is crossed) and some offset cells in the system (whose output is processed to find an appropriate offset), followed by some intelligent post-processing.

Two alternatives to this were investigated. In the first, the rectified output of each channel was used as input to an onset/offset filter, thus emulating the existence of a range of onset/offset cells each with input from some part of the auditory nerve. In the second, the rectified outputs of the channels were combined firstly by adding up all the channels at the low end of the frequency spectrum into one channel and adding up all the channels at the high end of the spectrum into another channel, and then rebanding the intermediate channels into a smaller number of channels using octave relationships. The idea was to model an alternative form of grouping of the signals on the auditory nerve, where the signals are grouped due to the co-occurrence of pulses caused by the nerve fibres being stimulated by low-frequency signals with considerable harmonic content. That such signals exist is not in doubt: that such co-occurrence causes this form of grouping of fibres is not supported by any neurobiological evidence, although one might expect it to happen, following von der Malsburg & Singer (1988).

To be more precise, all the channels between the minimum frequency band (F_minimum) and the low critical frequency (F_LoCritical) were combined into one band. All the channels between the high critical frequency (F_HiCritical) and the maximum frequency band (F_maximum) were combined into one band. Both of these were normalized by dividing by the number of frequency bands combined.

The number of the other bands (N_octavebands) was variable, but was normally small. The fundamental resonant frequencies of the channels associated with each band was calculated as follows. For the lowest band, the fundamental frequency (F_O) was set to F_LoCritical. For each band above this, the fundamental was set by

$$F_i = F_i - 1 * \exp\left(\log(2) / N_{\text{octavebands}}\right) \qquad (13.5)$$

for $i = 1$ to (N_octavebands – 1). This results in a sequence of N_octavebands frequencies between F_LoCritical and rather less than 2 * F_LoCritical, with logarithmic spacing.

The aim of the banding technique was to group together channels containing harmonics of each F_i. This was achieved by calculating which harmonic was nearest to the characteristic frequency of each channel. The distance between the characteristic frequency and the harmonic frequency was computed by finding what fraction the difference between the two frequencies was of the filter bandwidth at that harmonic, computed as in the legend of Figure 13.2 (Patterson & Holdsworth 1990). The original intention was to weight the contribution of each channel by a factor inversely related to this weight. However, we found that simply including a band if the distance was less than or equal to 1, and ignoring it if the distance was greater than 1, was simpler and effective.

In both cases, the technique for forming the composite segmentation was the same. First, each channel was segmented simply, that is, from the crossing of the onset threshold, past the maximum of the onset, to the next minimum. These segmentations were combined by using the start of the first segment on any channel as the start of the composite segment, then merging this with any segments which started before the end of this segment on other channels. The estimate of the end of the composite segment was taken to be the end of the last of these segments. This process was repeated until no more segments could be merged. The start of the next segment was taken to be the start of the first segment on any channel after the end of the composite segment, and the whole process repeated until a complete set of segments had been found. In contrast to the technique used in the previous section, the segments are not generally contiguous.

Experimental results

We report here the results of applying these techniques to a short piece of speech and to some single-note musical instruments. The instruments used are a single note at a time plucked guitar, and a flute and a saxophone, played both tongued and slurred.

The short piece of speech segmented was a male voice saying "Department of Computing Science". The results of this segmentation are shown in Table 13.1. Quiet segments have been omitted from the first column. Clearly, this results in some parts of the sound being omitted; however, listening to these segments shows that they consist of the end of the previous segment, followed by (frequently) a plosive, followed (sometimes) by the start of the next segment. They are very difficult to render into a phonetic translation: they sound like a flap or a click followed by the plosive, and often not like speech at all. The segments shown do sound like parts of the phrase. In the second set of columns, it is clear that the segmentation technique fails to break up the sound properly. From inspection of the signal, this is the result of the energy in the different bands starting and stopping at different times. This is not a problem when the signals are all summed, as in the first set of columns, or when the signals are rebanded, as in the last set of columns. In general, when successful, the speech has been broken at consonant boundaries, corresponding to interruptions in voicing.

Table 13.1 The segmentation of the utterance "Department of Computing Science". Times are in 0.5 ms units. The signal was divided into 30 bands from 100 Hz to 8 kHz. The HDoG filter used $k = 1000$ and $r = 1.2$ throughout. The banded signal used F_LoCritical=200Hz, F_HiCritical=5kHz and N_octavebands=5.

Summary HDoG			Composite direct			Composite rebanded		
Segment		Phon.	Segment		Phon.	Segment		Phon.
Start	End	trans.	Start	End	trans.	Start	End	trans.
721	934	/di/	628	994	/di/	717	933	/di/
1091	1358	/aː/	1061	1466	/paː/	1087	1362	/aː/
1552	1358	/ən/	1495	1825	/mən/	1561	1731	/ən/
2023	2207	/ɔv/	1851	1967	/t/	1866	1954	/t/
2399	2501	/kɔm/	2007	2309	/ɔv/	2020	2209	/ɔv/
2801	2861	/pju/	2385	2750	/kɔm/	2393	2579	/kɔm/
2861	3027	/uː/	2753	4153	/pjuːtɪŋsajɛn/	2773	3025	/pju/
3158	3561	/tɪŋs/	4292	4489	/s/	3105	3292	/tɪŋ/
3562	4134	/ajɛn/	4502	4611	/s/	4310	3621	/s/
						3652	3814	/ajɛ/
						4324	4465	/s/

255

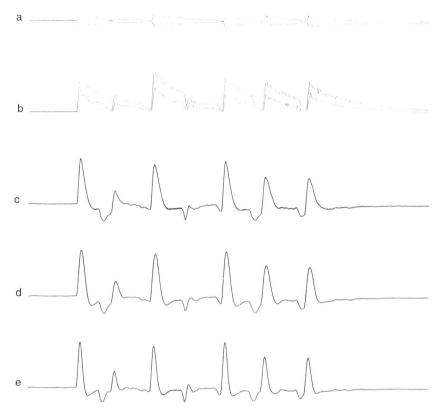

Figure 13.4   The plucked guitar sound. (a) The outline of the original signal. (b) The signal after rectification and summation. The results of various onset/offset filters: (c) a DOE filter, with $r = 50$ and $k = 1.2$; (d) and (e) an HDOG filter, with $k = 1.2$ and $r = 600$ and $r = 1200$, respectively.

Figure 13.4 shows the effect of applying the processing techniques to a plucked guitar. It is clear that the notes are plucked with varying intensities, though this is less obvious when listening. The transition from note to note is quite clear from the rectified summary signal: all the notes except note 3 result in a strong onset signal at the note's start, allowing the summary segmentation algorithm to pick out the notes. The onset of note 3 tends to get lost in the offset of note 2, unless a faster filter is used, as in Figure 13.4e.

Table 13.2 shows the segmentations that result from the filters used in Figure 13.4c–e. The first one uses a DOE filter with an intercept of 21 ms, and the second uses an HDOG with an intercept of 30 ms. These have very similar outputs, partly because of the extended tail of the DOE. The third filter uses an HDOG with an intercept of 21 ms. The short tail of the HDOG makes this a faster filter than the DOE with the same intercept, and as a result it picks out the onset of note 3, which otherwise gets lost in the offset of note 2. The direct composite segmen-

Table 13.2 The result of segmenting the plucked guitar sound. (a) shows the sound segments as found by ear. (b) shows segmentation produced for one DOE filter, and for two different HDOG filters. Columns 3 and 4 show the segmentation start and end times using only the the simple segmenting technique, and columns 5 and 6 show the use of the more sophisticated technique discussed the text.

(a)	Note	Start	End
	0	600	1020
	1	1060	1560
	2	1580	2020
	3	2020	2480
	4	2500	3000
	5	3040	3580
	6	3580	5180

(b)	Filter	Segment	Start	End	Start	End
	Summary	0	631	870	0*	631
	DOE	1	1108	1259	631	989
	$k = 50$	2	1602	1838	989*	1108
	$r = 1.2$	3	2538	2766	1108	1585
		4	3062	3271	1585	2520
		5	3617	3876	2520	2971
		6			2971	3563
		7			3563	4259
	Summary	0	637	845	0*	637
	HDOG	1	1116	1270	637	1003
	$k = 600$	2	1610	1817	1003*	1116
	$r = 1.2$	3	2548	2742	1116	1587
		4	3070	3270	1587	2525
		5	3627	3827	2525	2982
		6			2982	3583
		7			3583	4185
	Summary	0	633	792	0*	633
	HDOG	1	1103	1226	633	982
	$k = 1200$	2	1602	1778	982*	1103
	$r = 1.2$	3	2093	2169	1103	1585
		4	2539	2688	1585	2054
		5	3058	3238	2054	2520
		6	3614	3763	2520	2961
		7			2961	3561
		8			3561	4172
	Composite	0	553	1857		
	direct	1	2043	3042		
	HDOG	2	3048	3773		
	$k = 1000$	3	3780	3876		
	$r = 1.2$	4	3960	4229		
		5	4263	4469		
		6	4485	4556		
		7	4579	4762		
		8	4780	4851		
		9	4879	5035		
		10	5038	5119		
	Composite	0	635	864		
	rebanded	1	1089	1340		
	HDOG	2	1456	1818		
	$k = 600$	3	2057	2297		
	$r = 1.2$	4	2512	2765		
		5	2928	3052		
		6	3060	3280		
		7	3594	3901		
		8	4229	4325		
		9	4900	5005		

* Segments which are created by the addition of a "quiet" segment between two "note" segments. Times are in 0.5 ms units.

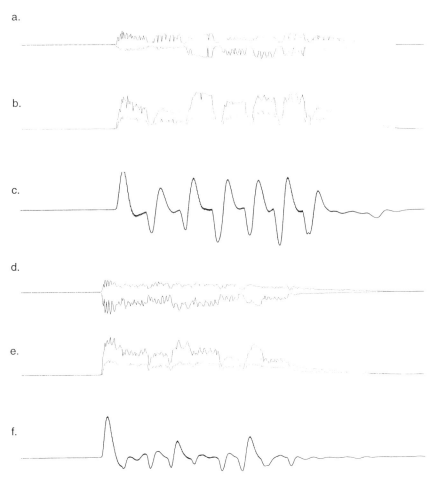

Figure 13.5 The saxophone sound. (a) Original tongued saxophone sound signal. (b) Summary rectified signal from (a). (c) Summary onset/offset signal output from an HDOG filter with $k = 600$ and $r = 1.2$. (d) Original slurred saxophone sound signal. (e) Summary rectified signal from (d). (f) Onset/offset signal output from an HDOG filter with $k = 600$ and $r = 1.2$.

tation techniques used on the guitar shows clear evidence of undersegmentation, putting together notes 0–2, and 3–4, then oversegmenting note 6. The rebanded composite technique fares much better, and does not undersegment at all, although it still oversegments note 6.

Figure 13.5 shows the waveforms for tongued and slurred saxophone sounds. The ends of the tongued sounds are clearly visible, and this is reflected in the segmentation produced in Table 13.3. However, the direct composite segmentation technique again undersegments, gathering nearly all the notes into one segment. The slurred saxophone sound is more difficult to segment, both visually and by the techniques here. As can be seen in Table 13.4, none of the

Table 13.3 Segmentation of the tongued saxophone sound by ear and by an HDoG filter, a composite filter applied to all bands, and a composite filter applied to the rebanded signal. Parameters are 30 bands, from 100 Hz to 8 kHz, with rebanding parameters F_LoCritical = 200 Hz, F_HiCritical = 5 kHz, and N_octavebands=5, and onset/offset parameters $r = 600$ and $k = 1.2$. Times in 0.5 ms units.

By ear	Summary, G600/1.2	Quiet	Composite direct	Composite rebanded
1280–1720	1315–1803		1305–3891	1313–1595
	1803–1865	✓		
1800–2180	1865–2258			1860–2169
2200–3120	2258–2693			2294–2556
	2693–2771	✓		
2700–3120	2771–3036			2767–2980
	3036–3130			
	3130–3187	✓		
3120–3460	3187–3542			3012–3430
	3542–3601	✓		
3540–3920	3601–3951			3533–3821
	3951–4017	✓		
3920–5040	4017–4849		3904–4387	4008–4221
			4750–4863	
			4952–5112	

Table 13.4 Segmentation of the slurred saxophone sound by ear and by an HDoG filter, a composite filter applied to all bands, and a composite filter applied to the rebanded signal. Parameters are 30 bands, from 100 Hz to 8 kHz, with rebanding parameters F_LoCritical = 200 Hz, F_HiCritical = 5 kHz and N_octavebands = 5, and onset/offset parameters $r = 600$ and $k = 1.2$. Times in 0.5 ms units.

By ear	Summary, G600/1.2	Quiet	Composite direct	Composite rebanded
1120–1420	1145–1236		1136–1455	1143–1648
1420–1800	1458–1711	✓	1464–2437	1706–1864
	1711–1842			
1800–2080	1842–1898	✓		1883–2676
	1898–2130			
2080–2420	2130–2453			
2420–2780	2453–2509	✓	2447–3418	2711–2862
	2509–2839			
2780–3010	2839–2907	✓		2891–3490
	2907–3126			
3020–3400	3126–3449			
3400–3720	3449–3627	✓	3558–3821	3601–3821
	3627–3823			
3720–4880			3826–4227	3884–4037
			4289–4575	
			4831–4950	

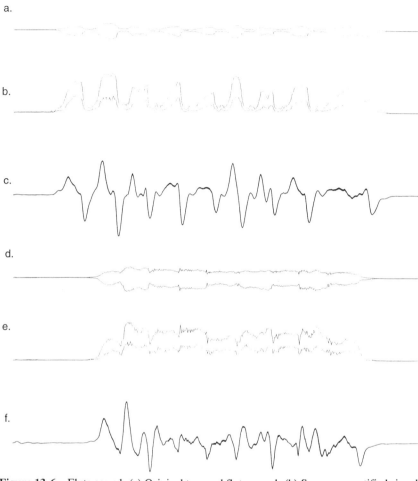

Figure 13.6 Flute sound. (a) Original tongued flute sound. (b) Summary rectified signal from (a). (c) Onset/offset signal output from an HDoG filter with $k = 1000$ and $r = 1.2$. (d) Original slurred flute sound signal. (e) Summary rectified signal from (d). (f) Onset/offset signal output from an HDoG filter with $k = 1000$ and $r = 1.2$.

techniques perform well: the summary segmentation and the rebanded composite techniques are better than the direct composite technique. The reasons will be discussed in the next section.

Figure 13.6 shows the waveforms for tongued and slurred flute sounds. As with the saxophone, the tongued segments are quite visible, unlike those from the slurred notes. The system performs well on the tongued notes (Table 13.5). There is a tendency towards oversegmentation with the summary techniques, even with $k = 600$. Again the direct composite technique tends to undersegment, and again the rebanded composite technique does much better. On the slurred notes (Table 13.6), the summary segmentation technique oversegments. The direct composite technique undersegments, as usual, and the rebanded composite

Table 13.5 Segmentation of the tongued flute sound by ear and by an HDoG filter with $k = 600$ and $r = 1.2$, an HDoG filter with $k = 1000$ and $r = 1.2$, a composite filter applied to each band ($k = 600$ and $r = 1.2$), and a composite filter applied to rebanded signal ($k = 600$ and $r = 1.2$). Parameters are: 30 bands, from 100 Hz to 10 kHz, with rebanding parameters F_LoCritical = 200 Hz, F_HiCritical = 5 kHz and N_octavebands = 5. Times are in 0.5 ms units.

By ear	Summary, G600/1.2	Quiet	Summary, G1000/1.2	Quiet	Composite direct	Composite rebanded
600–1180	675–1148		662–1129		645–1170	669–1155
	1148–1283	✓	1129–1259	✓		
1220–1700	1283–1544		1259–1526		1185–1328	1229–1329
	1544–1671		1526–1652		1330–1674	1349–1533
	1671–1809	✓	1052–1796	✓		1535–1672
1740–2080	1809–2029		1796–1949		1714–1808	1784–2154
	2029–2157		1949–2044	✓	1821–2688	
	2157–2324	✓	2049–2138			
			2138–2218	✓		
2240–2620	2324–2668		2218–2282			2315–2668
	2668–2845	✓	2282–2646			
			2646–2758	✓		
2760–3140	2845–3181		2758–2805		2701–3617	
	3181–3329	✓	2805–3165			2748–2835
			3165–3318	✓		2837–3219
3260–3590	3329–3601		3318–3581			3315–3602
	3601–3813	✓	3581–3694	✓		
3720–4100	3813–3696		3694–3767		3647–4127	3699–3788
	3696–4105		3767–3955			3799–3971
	4105–4226	✓	3955–4085			3994–4109
			4085–4185	✓		
4220–4580	4226–4639		4185–4621		4138–4663	4208–4640
	4639–4748	✓	4621–4717	✓		
4760–5700	4748–4829		4717–4817		4685–4832	720–5297
	4829–5294		4817–5282		4868–5625	5396–5631
	5294–5441	✓	5282–5383	✓		
	5441–5632		5383–5399			

technique, again, performs best. The summary technique tends to be guided by the variations in the envelope, which do not correspond to the note changes, but the rebanded composite technique can overcome this problem.

Discussion

The techniques used here can be applied to any sound. We do not pretend that this is how sound is really segmented in the human ear: although there is a basis in the biology for this approach, there are many more onset and offset cells and

Table 13.6 Segmentation of the slurred flute by ear and by an HDoG filter with $k = 1000$ and $r = 1.2$, a composite filter applied to each band ($k = 1000$ and $r = 1.2$), and a composite filter applied to the rebanded signal ($k = 1000$ and $r = 1.2$). Parameters are 30 bands, from 100 Hz to 10 kHz, with rebanding parameters F_LoCritical = 200 Hz, F_HiCritical = 5 kHz and N_octavebands = 5. Times are in 0.5 ms units.

By ear	Summary G1000/1.2	Quiet	Composite direct	Composite rebanded
1280–1700	1236–1734		1157–2017	1221–1750
1700–2180	1734–2019		2036–2726	1758–2021
	2019–2075	✓		2040–2220
	2075–2216			
2180–2640	2216–2271	✓		2244–2429
	2271–2428			2464–2877
	2428–2605			
2640–3110	2665–2767		2738–2978	2908–3332
	2767–2877		3000–3076	
	2897–3131			
3110–3570	3131–3357	✓	3088–4334	3352–3611
	3357–3600			
3540–4101	3600–3652	✓		3633–3846
	3052–3845			3940–4185
	3845–3940	✓		
	3940–4053			
4010–4580	4053–4182		4335–4987	4196–4531
	4182–4246	✓		
	4246–4593			
4580–5000	4593–4765			4612–4767
	4765–5046			4802–4967
5000–5680	5046–5223	✓	5010–5503	5033–5455
	5223–5589		5541–5642	

many other cells with different types of responses in the cochlear nucleus (Blackburn & Sachs 1989). This work represents a very simple application of an onset/offset-based technique to the sound segmentation problem.

The approach is purely data-driven. It is completely ignorant of any higher-level information that might help to drive such a segmentation (such as prosody in speech (Cutler 1990), or information about a particular instrument). This may appear to throw away far too much that might be helpful; however, since the approach is entirely data-driven, it can provide a first approximation at segmentation, one which can later be modified using additional information. A different data-driven system is described by Andre-Obrecht (1988), which uses changes in the statistical structure of the signal, as modelled by a parametric description of a sliding 8 ms window on the signal, and is applied only to speech segmentation. Their system is again entirely data-driven, but makes no attempt at all at biological plausibility. The system described by Wu et al. (1989) is also data-driven, and uses onsets and offsets, but is oriented to finding acoustic events in speech.

We do not attempt to describe the statistical structure of the signal, concentrating purely on signal onsets and offsets, and our aim is (initially, at least) segmentation.

The summary segmentation system breaks up the speech signal into pieces which do correspond to phonetically identifiable pieces. If one considers the quiet segments as well, the whole utterance is divided into non-overlapping segments; however, this is not an effective segmentation, as the quiet segments are simply the spaces between the identified segments, and listening to these shows them to be difficult to identify. The banded composite segmentation technique gives better results; additionally, we believe that the simpler direct multiple segmentation could be improved by using a better technique for combining the segments. The technique used tends to extend segments indefinitely, as is also visible in the results for segmenting the guitar, flute and saxophone.

The summary segmentation system deals well with simple plucked sounds, and with tongued blown sounds. It is also successful with simple hand claps and a glockenspiel (Smith 1993). This is at least partly because it identifies the start and end of bursts of energy, and these percussive sounds are made up from well defined bursts of energy, with an envelope which is characterized by a sudden onset, and a more gradual (though frequently still quite rapid) offset. Given that the filter intercept approximates to the duration of the energy in the pulse, this technique works well. For sounds with a different envelope, the system may not work as well. Internal changes in intensity inside a musical note can divide up the note (as in the second note in Table 13.5); however, if onset/offset filters with a range of intercepts were applied, then a filter whose intercept closely approximated the note length would not suffer from this problem. (However, since the note length is much larger than the intercepts used, and the filter's rise time is dependent on the intercept, the resulting start of the segment would be delayed.) Certainly, when listening to a musical note, it is possible to be aware simultaneously that a note changes in intensity or tone, without considering it as more than one note, suggesting the existence of multiple concurrent timescales of interpretation.

The slurred saxophone and flute experiments point out one of the major problems with the simple system described here: a sequence of notes without any spaces between them may not be segmented correctly. For the summary segmentation technique, this is because the whole spectrum is summed together. We had hoped that the composite techniques would fare better. However, the direct composite technique fails, we believe, for two reasons: first, the signals in the channels do not start and stop at the same time, and secondly because the segment composition technique used places the same weight on segments produced by different channels, even though they have very different signal strengths. These two factors conspire to make the direct composition technique undersegment. The banded composite technique fares better, perhaps because there is more commonality in start and stop times across a note and its harmonics, and perhaps because it simply uses fewer bands.

(a)

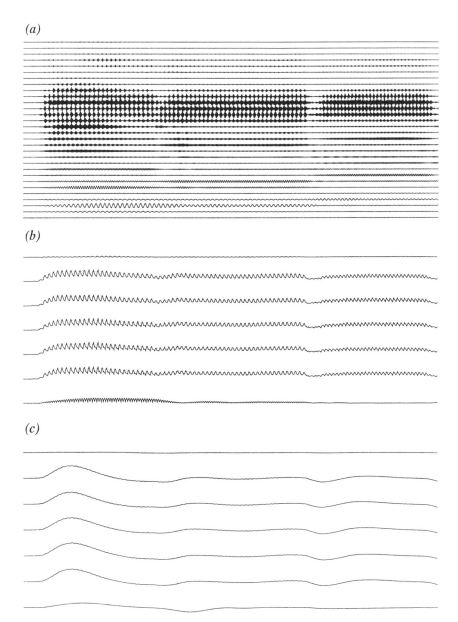

(b)

(c)

Figure 13.7 (a) The signal in each band for a 550 ms block of the slurred saxophone sound. (b) The banded signal for a 550 ms block of the slurred saxophone sound. The parameters used were F_LoCritical = 220 Hz, F_HiCritical = 8 kHz, N_octavebands = 5. (c) The output of the onset/offset filter (HDoG, $k = 600$, $r = 1.2$) applied to the banded signal.

However, for the slurred saxophone, the results from both the composite techniques are disappointing. Examination of the original signal suggests why: the energies in the different frequency bands start and stop at different times. Figure 13.7a shows this for the first three notes. Generally, the low-frequency elements finish after the higher-frequency ones. This is sufficient to mislead the direct composite segmentation. The rectified banded signal, shown in Figure 13.7b, covers the same time period, and again shows how the lowest-frequency band leads the others, particularly at the end of the first note. The similarity of the middle traces suggests that sharper and possibly more bands be used. The effect of the skew in energy starts and stops on the onset/offset signal, shown in Figure 13.7c, is clearly sufficient to mislead the segmentation system.

Conclusions and further work

A prototype system which can segment speech and some musical sounds played by single-note instruments has been demonstrated. Three different versions have been used: a summary version which sums the rectified signal prior to applying the onset/offset filter and segmenting, and two composite versions which perform segmentation on a number of bands, and then combine these. In the first (direct) composite version, the bands are used as they are, and in the second (rebanded) composite version the signal is rebanded into a smaller number of octave-related bands, as well as a low and a high band.

The results from the simplest summary segmentation are surprisingly good. This reflects a view of sound as made up of bursts of energy, with (quiet) spaces between. The composite direct technique is disappointing, although the composite rebanded technique fares better. For sounds for which the envelope is less informative, such as the slurred flute, the composite rebanded technique does better than the summary onset; however, for others, such as the slurred saxophone, it does not. Figure 13.7b suggests one reason for the shortcomings of the rebanded composite technique: the traces near the centre are virtually identical. This suggests that more and sharper filters should have been used, thus allowing the bands to be sharper, and so bring out the changing spectral structure. The primary problem with this is processing time and memory: as it is, the filtered 30-band sounds take up about 1 Mb per second!

The system described here extends the work in Smith (1993) by applying the onset/offset filter to different spectral areas. In this earlier work, it was suggested that a range of filters with varying intercepts should be applied. Using onset/offset filters with short intercepts could be useful in identifying the actual start and end of energy bursts, where the approximate position had been found by using onset/offset filters with a larger intercept. In addition, using onset/offset filters with very short intercepts (about 3 ms) one can locate the amplitude modulations of higher-frequency bands (Fig. 13.7a). These are characteristic of certain musical instruments, and of speech, and would be useful in the characterization of the segments for later interpretation. It is worth noting, however, that using very sharp filters can tend to remove these modulations, as they are resolved into sidebands of the central frequency.

The segment composition technique used needs further work. Currently it uses only the segment duration, so that segments from parts of the signal with low energy are treated the same as segments from parts of the signal with high energy. To follow the biology, there should be some form of population coding to the onsets and offsets, so that the strength of the onset, and the power in the channel are used in producing the composite segmentation. Other, less biologically justifiable, techniques could be applied to producing the segmentation from the onset/offset signals, such as the "weak string" of Blake and Zisserman (1987). However, from the point of view of understanding how humans segment sound, techniques based on cochlear nucleus cell behaviour are more reasonable.

Acknowledgements

I thank Lawrence Gerstley for supplying the sounds, and the other members of the Centre for Cognitive and Computational Neuroscience for useful discussions, and the University of Stirling for a 6 month sabbatical during which much of this work was done. I should also like to thank the attendees at the Second Neural Computation and Psychology Workshop for useful discussions.

References

Andre-Obrecht, R. 1988. A new statistical approach for the automatic segmentation of continuous speech signals. *IEEE Transactions on Acoustics, Speech and Signal Processing* **36**, 29–40.

Blackburn, C. C. & M. B. Sachs 1989. Classification of unit types in the anteroventral cochlear nucleus: PST histograms and regularity analysis. *Journal of Neurophysiology* **62**, 1303–29.

Blackwood, N., G. Meyer, W. Ainsworth 1990. A model of the processing of voiced plosives in the auditory nerve and cochlear nucleus. *Institute of Acoustics, Proceedings* **12**(10), 423–30.

Blake, A. & A. Zisserman 1987. *Visual reconstruction*. Cambridge, Mass.: MIT Press.

Bregman, A. S. 1990. *Auditory scene analysis*. Cambridge, Mass.: MIT Press.

Brown, G. 1992. *Computational auditory scene analysis*. Department of Computing Science, University of Sheffield, Report TR CS-92-22.

Cutler, A. 1990. Exploiting prosodic probabilities in speech segmentation. In *Cognitive models of speech processing*, G. T. M. Altmann (ed.), 105–21. Cambridge, Mass.: MIT Press.

Marr, D. & E. Hildreth 1980. Theory of edge detection. *Royal Society of London B, Proceedings* **207**, 187–217.

Moore, B. C. J. & B. R. Glasberg 1983. Suggested formulae for calculating auditory-filter bandwidths and excitation patterns. *Journal of the Acoustic Society of America, Journal* **74**, 750–53.

Patterson, R. & J. Holdsworth 1990. *Air applications model of human auditory processing, progress report 1*, annex 1. Cambridge: Medical Research Council Applied Psychology Unit.

Pickles, J. O. 1988. *An introduction to the physiology of hearing*, 2nd edn. London: Academic Press.

Smith, L. S. 1993. *Temporal localisation and segmentation of sounds using onsets and offsets*. University of Stirling, Stirling, CCCN Technical Report CCCN-16.

Smith, L. S. 1994. Sound segmentation using onsets and offsets. *Journal of New Music Research* **23**, 11–23.

von der Malsburg, C. & W. Singer 1988. Principles of cortical network organisation. In *Neurophysiology of neocortex*, P. Rakic & W. Singer (eds), 69–99. New York: John Wiley.

Wu, Z. L., J. L. Schwartz, P. Escudier 1989. A theoretical study of neural mechanisms specialized in the detection of articulatory-acoustic events. In *Eurospeech '89, Proceedings*, J. P. Tubach & J. J. Mariani (eds), 235–50, Paris.

CHAPTER 14

Time-warping tasks and recurrent neural networks

Mukhlis Abu-Bakar & Nick Chater

Introduction

Finding structure in real-world acoustic signals, whether from the perspective of engineering or psychology, is difficult because not only must an underlying sequence of elements be discerned, but, frequently, the duration of those elements may be "warped" in complex ways. Furthermore, in some contexts, such as speech recognition, this warping is especially problematic, since the very identity of sequence elements may be given by duration-based cues, which warping will distort. The ability to cope flexibly and successfully with time-warped material is crucial if neural networks, or any other computational technique, are to be applicable to a wide range of real-world temporal processing tasks.

Utterances are frequently time warped because speakers speed up and slow down when they talk rather than maintain a constant rate of speech. The variation in rate that occurs in conversational speech can be quite substantial (Miller et al. 1984). More often, time warping distorts the temporal structure of words: a segment of a word may be compressed, another stretched, while others remain durationally invariant to changes in the speech rate. The problem is more complex at the phonemic level. As articulation time is altered due to changes in the speech rate, certain acoustic properties that specify the identity of phonetic segments are modified, since they are themselves temporal in nature. For instance, a short duration of some property may specify one phonetic segment while a longer duration specifies another (Lisker & Abramson 1964). Thus, time warping potentially confounds temporal cues to phoneme identity.

Within engineering, there have been various approaches to solving time-warping problems (e.g. hidden Markov models (Huang et al. 1990), dynamic time warping (Lipmann et al. 1987) and dynamic rate adaptation (Nguyen & Cottrell 1993)). The present work attempts to apply connectionist tools to the problem of time warping. To the extent that connectionist methods are psychologically plausible, this work also gives an attractive approach to modelling aspects of human speech perception.

Using recurrent back-propagation

Recurrent neural networks have been widely used in modelling sequence processing (e.g. Elman 1990), including a wide range of problems drawn from speech processing (e.g. Norris 1990, Shillcock et al. 1991, Watrous et al. 1990). Recurrent networks are attractive for such problems since their behaviour depends on the entire sequence of inputs, rather than just the current input, although there are various ways in which feedforward networks can be modified in order to handle sequential material (see Chater (1989) for a review).

We use a standard recurrent neural network architecture, in which the units in the hidden layer are connected to all other hidden units by weights which operate with a delay of one time step. This kind of recurrent network is often thought of as involving an additional set of units, the "context" units, to which the hidden units at the previous time step are copied. According to this conception, the context units are treated simply as additional input units to the network. This kind of recurrent network can be trained in a variety of different ways, the most common being Elman's (1990) "copy-back" scheme, which uses a computationally cheap approximation to gradient descent in error to change the weights. We use recurrent back-propagation (Rumelhart et al. 1986), which computes gradient descent more exactly by "unfolding" the recurrent network into a sequence of serially connected feedforward networks, and then trains the resulting network using standard back-propagation. The only additional constraint on learning is that the weights in each "incarnation" of the recurrent network in the unfolded feedforward network are constrained to be the same, so that it is possible to fold the trained feedforward network back up into a recurrent network.

In general, the larger the number of unfoldings used, the more exactly the network computes true gradient descent, although the benefits of additional unfoldings begin to tail off after some point, because very deep feedforward networks are very slow to train. It is also important to note that the number of unfoldings used in training does not place a strict limit on the distance back in the sequence to which the network can learn to be responsive. Even if the network is trained as a feedforward network unfolded through n time steps, the "context" units in the final unfolding are likely to contain information about the inputs at earlier time steps, and the network may therefore learn to become sensitive to this information. Nonetheless, although under certain circumstances networks can learn to respond to information which is very much more temporally distant than the number of unfolded time steps, in practice, performance is generally rather poor for such distant items (see Chater & Conkey (1994) for discussion). Training used conjugate gradient descent, and was implemented on the Xerion simulator (van Camp & Plate 1993).

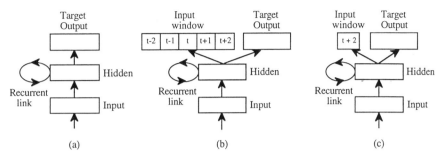

Figure 14.1 Folded versions of (a) a standard recurrent network (network A), (b) a network with five additional input windows at the output layer (network B), and (c) a network with a single additional input window at the output layer (network C).

Comparing architectures

Perhaps the main advantage of using recurrent networks, as opposed to other neural network methods, is their ability to treat temporal material, such as speech, as a sequence of events and take input one at a time. We trained the networks by feeding them with the sequences one input pattern at a time and keeping the target output pattern present throughout the presentation of each sequence. The production of the correct output as the sequence is presented indicates that the sequence is classified successfully. If performance is optimal, correct classification should occur after the "recognition point" of the category is reached, i.e. when enough of the sequence has been encountered that it can be classified unambiguously (Norris (1990) uses a model of this kind to capture cohort effects in word recognition).

To see how well networks can make such classifications with time-warped stimuli, we compared the basic recurrent network architecture (network A) with two minor variants (networks B and C). These latter networks contained additional input windows of different sizes at the output layer (Fig. 14.1). In one (network B), this window contained nodes representing inputs at the previous two and next two time slices and the current input (cf. Maskara & Noetzel 1992, Shillcock et al. 1992). For the other variant of the network (network C), the nodes in the window represented input at the $t+2$ time step only. In contrast to the target output, which remained constant, these additional outputs changed with the presentation of each input. The idea was to force the network to pay attention to the individual elements being presented in succession for a specified window and/or to prepare the net to accept inputs that arrive at a specified time in the future. The following set of experiments was designed to test how well these various networks classify sequences presented at a rate they are not familiar with.

Non-duration-based stimuli

For the experiments in this section, we used sequences which are unique in the sequential order of their constituent elements and whose respective identities remain unaffected by changes in the duration of these elements. Two training versions and one test version of 27 sequence types were built from all possible combinations of the numbers 1, 2 and 3, with each version representing different rates of input (fast, medium, slow). The set of stimuli presented at the intermediate rate was the test set. The three numbers were implemented as three-bit binary elements. Thus, 100 stood for 1, 010 for 2 and 001 for 3.

The basic network consisted of an input layer of three input nodes, a single hidden layer of either 30 or 36 nodes, and an output layer of 27 nodes. The two variants of the network contained an additional 15 and three output nodes, respectively. The networks were unfolded for 13 cycles during training. We ran every simulation twice with a different weight start for each attempt. Batch learning was employed.

Experiment 1: simple variation of input

We varied the duration of the constituent elements of the sequence types following the scheme used by Norris (1990). Table 14.1 illustrates the temporal composition of a model sequence type across the three rates. The period over which each constituent member of a sequence type appears (or the number of time it is successively presented) is captured by the relation $N_i = i$, where i is the rate, and N_i the number of successive presentations of each constituent member at the specified rate. Thus, in the "fast" series, each member of a sequence type remains constant for one time slice. In the "medium" series, this is extended to two units of time each, and in the "slow" series, each constituent member lasts for three units of time.

For all the 54 training stimuli, it is possible to determine by hand at which point in a stimulus the sequence type is recognizable. With a few exceptions, this does not normally require that the entire stimulus be processed. It would be interesting if the nets can capture this sequence structure by identifying the sequence type at the point in time when the stimulus item becomes unique. The crucial test, however, is how to generalize from this sequence structure to new stimuli presented at an intermediate rate.

All stimuli were preceded by an input pattern in which all three bits were set to 0. In the long version of the stimuli, where each constituent member of a

Table 14.1 A model sequence type at the three presentation rates (experiment 1).

Rate	t1	t2	t3	t4	t5	t6	t7	t8	t9
1: Fast	A	B	C						
2: Medum	A	A	B	B	C	C	C		
3: Slow	A	A	A	B	B	B	C	C	C

sequence type was repeated three times, they were followed by a final 0 input pattern. In the short version where each constituent member of a sequence type appeared only once, they were followed by seven input patterns which were set to 0. In the test stimulus where the constituent members appeared twice, they were followed by four of these 0 input patterns. As an illustration, sequence type 321 would be presented to the network in the slow and fast modes as follows:

Slow mode			Fast mode			
0	0	0	0	0	0	at time t1
0	0	1	0	0	1	at time t2
0	0	1	0	1	0	at time t3
0	0	1	1	0	0	at time t4
0	1	0	0	0	0	at time t5
0	1	0	0	0	0	at time t6
0	1	0	0	0	0	at time t7
1	0	0	0	0	0	at time t8
1	0	0	0	0	0	at time t9
1	0	0	0	0	0	at time t10
0	0	0	0	0	0	at time t11

Results

Training ceased when the sum-squared error of the training decreased by less than 0.0001. Total training time was usually between 1000 and 2000 iterations. To assess network response, we looked for the node with the highest activation at the point the sequence type can be identified. The activation of this node must be higher than that of other competing nodes by a criterion value of at least 0.25; otherwise there will be no winner, and the search for the winning node moves to the next time step. A stimulus is accepted as being correctly classified if the winning node corresponds to the target node and that its distance from the other nodes is maintained right up to the end of the stimulus and one time step after.

All three nets successfully fulfilled these criteria for all the training stimuli, whatever the hidden unit population. With the test stimuli, however, the nets

Table 14.2 Successful identification of test stimuli for each network type and hidden unit population (experiment 1).

Network type	Hidden units	Targets correct	On time
A	30	25	15
	36	27	21
B	30	23	15
	36	24	15
C	30	25	16
	36	26	18

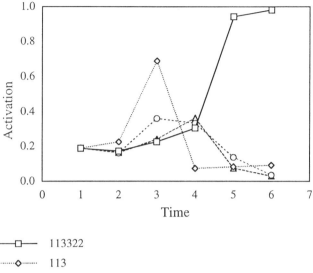

—□— 113322

········◇········ 113

----○---- 113311

----△---- 113333

Figure 14.2 Activity of test stimulus 113322 (solid line) with 132 as the underlying sequence type. Activations of three other competing sequence types are also shown (experiment 1).

achieved a slightly lower rate of success (Table 14.2). Consider the results from simulations with 36 hidden units since these show better performance. Only network A maintained 100% correct response for the test stimuli, whereas networks B and C had three and one wrong classifications, respectively. In terms of the number of timely responses, network A managed a high 21, while networks B and C only 15 and 18, respectively. Thus, network A performed more consistently with both sets of stimuli than networks B and C.

Figure 14.2 demonstrates the recognition process of the test (medium) version of sequence type 132 (appeared as 113322). Notice that the sequence of input up to the third time step is an exact copy of the fast version of sequence type 113. Since the net has seen this stimulus item during training, it responds appropriately by exciting the output node of this sequence type at its unique point (i.e. at the third time step). However, when another 3 comes along at the following time step, the net is forced to revoke its decision, for this creates a novel string 1133. This change is depicted by the downward shift in the activation value of the sequence type 113. Activations of other sequence types continue to remain low. Interestingly, when 2 is presented at the fifth time step, the net promptly and correctly activates sequence type 132 without waiting for the last input pattern to arrive. This is an optimal performance where the strategy is to accumulate just enough information to make the final decision.

Experiment 2: complex variation of input

For many complex temporal stimuli, including speech, changes in rate do not result in a simple compression and expansion of the signal as modelled in the previous section. Rather, the time warping is quite complex. One case in point concerns the absolute and relative durations of vowels in conversational speech. An increase in speech rate has been shown to reduce the duration of a long vowel (e.g. [i]) more than a short vowel (e.g. [I]), so that the durational difference between the two vowels is reduced at the faster rate of speech (Miller 1981). In this experiment, we modified the earlier stimuli to reflect such complexity.

Instead of varying the duration of all the constituent members of a sequence type equally via a single linear function, three different functions were used. Two versions of these stimuli were constructed. In the first version, X, the duration of the first constituent member remained constant across rates following the relation $N_i = 1$. For the second member, the duration was specified by the same linear function used previously, namely, $N_i = i$; and for the last member, a nonlinear function $N_i = 2^{(i-1)}$ was used. For all the functions, i is the rate and N_i the number of successive presentations of the constituent member at the specified rate. The resulting temporal configuration can be found in Table 14.3. A second version, Y, was created from the first version by switching the functions associated with the first and third constituent members. The motivation behind having two versions of the complex variation of sequence types was the intuition that a left-to-right processing model of this kind will exact a higher cognitive cost if the transition from one element to another occurs much later in the sequence than if it occurs earlier in time, since the system must learn to attend to temporally more distant information. We wanted to confirm this intuition. The nets were trained with the two versions separately. As in the previous experiment, the medium series in both versions served as the test set.

Table 14.3 Two versions (X and Y) of a model sequence type at the three presentation rates (experiment 2).

Version	Rate	t1	t2	t3	t4	t5	t6	t7	t8
X	1: fast	A	B	C					
	2: medium	A	B	B	C	C			
	3: slow	A	B	B	B	C	C	C	C
Y	1: fast	A	B	C					
	2: medium	A	A	B	B	C			
	3: slow	A	A	A	A	B	B	B	C

Results

All the nets classified both versions of the training stimuli correctly, irrespective of hidden unit population. Generalization to the test stimuli, however, was not uniform across stimulus versions and hidden unit population (Table 14.4). With the exception of network C, simulations with 36 hidden units for the other two

Table 14.4 Successful identification of each version of test stimuli for each network type and hidden unit population (experiment 2).

Network type	Hidden units	Targets correct	On time
Version X			
A	30	27	27
	36	27	27
B	30	27	27
	36	27	26
C	30	27	27
	36	27	27
Version Y			
A	30	22	22
	36	25	23
B	30	20	9
	36	22	13
C	30	26	21
	36	22	18

nets produced better results. And, as expected, version Y, as opposed to version X, proved more difficult for all three nets. Simulations with version X produced perfect scores on correct classification, and almost perfect scores on getting the classifications correct on time, but with version Y the recognition rate varied between 20 and 26 while that of timely responses between 9 and 23. Comparatively all round, network B performed less well than the other two networks.

Duration-based stimuli

In the preceding experiments, the duration of the constituent members of a sequence type made no difference to the identity of that sequence type that we altered by time warping. In this section, we consider a set of sequences whose identity depends on the duration of these very elements. This occurs in a variety of ways in natural speech, and is extensively discussed in the speech production and perception literature (see Miller (1981) for a review). One commonly cited example involves the voicing distinction between /bi/ and /pi/ as specified by the voice onset time (VOT). These syllables can be differentiated simply by the duration of this property: /b/ having typically shorter VOTs than /p/. More importantly, however, as speaking rate changes from fast to slow and the individual words become longer, the criterion VOT value that distinguishes /b/ and /p/ also moves towards longer values (Miller et al. 1986). Potentially, due to this variation, the mapping from acoustic signal to phonetic percept is difficult, but, interestingly, listeners adjust for these variations with apparent ease. Our goal is to work towards a first approximation of this "rate normalization" process.

Although primarily intended as an abstract test, our stylized stimuli were loosely patterned after the synthesized syllables used by Volaitis & Miller (1992). Two contrasting stimuli, /bi/ and /pi/, took the form

/bi/ → 211133333333333444444
/pi/ → 211111111111333444444

where the numbers represent the states of various acoustic properties; in this case, 2 may be taken to refer to the release burst, 1 to F1 cutback, 3 to transition, and 4 to steady state. The duration of a particular property is specified by the number of times the corresponding state is repeated. /pi/ is derived from /bi/ simply by lengthening the latter's VOT (counted from the onset of the burst till the onset of transition). This involves extending the F1 cutback by cutting into the transitions. Localist representations of 4-bit patterns were used for the states. The basic network consisted of four input units, five or ten hidden units, and two output units. The two variant networks, B and C, contained an additional 20 and four output nodes, respectively. In this and the next set of simulations, the nets were unfolded for 36 time cycles during training.

Experiment 3: non-overlapping stimuli

From the production data of Miller et al. (1986), it appears that within a place of articulation, there is some overlap in the distribution of VOT values for voiced and voiceless consonants across different speech rates. However, in this section, we assume no overlap of VOT values between categories across rates. Thus, recognition should be a straightforward task from the processing point of view: a VOT that lasts for a certain time range specifies one segment, and another if it extends beyond that range. Six /bi/–/pi/ pairs were constructed across six rates, as shown in Figure 14.3. One pair (23 time steps syllable duration) was set aside as test material.

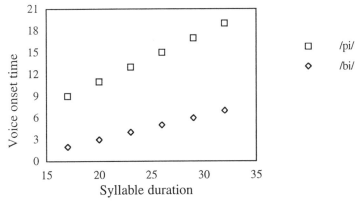

Figure 14.3 Distribution of /bi/ and /pi/ tokens for the non-overlapping case (experiment 3).

Results

All three nets were able to handle effectively both the training and test stimuli. As expected, the strategy employed by the nets operates in a straightforward fashion: information about VOT alone triggers the contrast between /b/ and /p/ for this group of non-overlapping stimuli. Syllable duration is thus irrelevant in the distinction and exerts no influence over the processing task.

Experiment 4: overlapping stimuli

We now consider the more realistic case in which VOT values overlap over a certain range, as in natural speech. Figure 14.4 shows the relationship between VOT and syllable duration for 14 /bi/ and /pi/ stimuli. Six of these stimuli are within the overlap range (/pi/-1 and /bi/-3, /pi/-2 and /bi/-5, and /pi/-3 and /bi/-7, where /pi/-*n* denotes /pi/ at rate *n*, with rate 1 being the fastest rate and 7 the slowest). Every /bi/–/pi/ pair in this range has an identical VOT but different syllable duration, as illustrated below. Sequence U is a /bi/ syllable presented at a slower speech rate (as specified by a longer syllable duration) than sequence V, a /pi/ syllable, but their VOT values are the same. To recover the intended voicing feature specified by the VOT value, the nets have to consider the entire stimulus.

U   /bi/   21113333344444444444
V   /pi/   21113333344444

Two /bi/–/pi/ pairs (rates 2 and 5) were set aside as test material. Of these, /bi/-5 and /pi/-2 have identical VOT values but different syllable duration.

Results

All three networks were successful in learning to classify the training stimuli including those within the overlap range. However, only networks B and C were able to generalize to all the test stimuli appropriately. The stimuli in the overlap

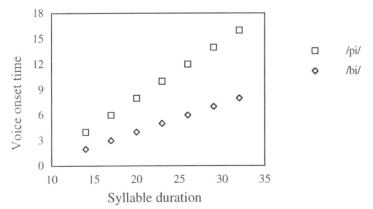

Figure 14.4   Distribution of /bi/ and /pi/ tokens for the overlapping case (experiment 4).

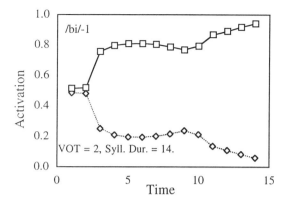

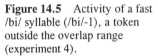

Figure 14.5 Activity of a fast /bi/ syllable (/bi/-1), a token outside the overlap range (experiment 4).

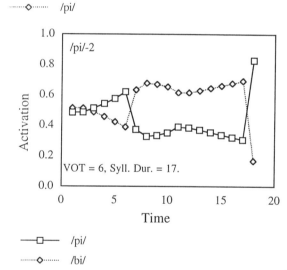

Figure 14.6 Activity of a test /pi/ syllable (/pi/-2), a token from the overlap range (experiment 4).

range proved difficult for network A: it classified both /bi/ and /pi/ as /bi/. Nevertheless, the fact that the other nets can make appropriate generalizations with this kind of stimuli was encouraging.

The networks' on-line processing reveals that the process of identifying voiced and voiceless tokens that lie outside the overlap region is straightforward, as it was with the previous set of stimuli, with performance nearly optimal at the offset of the VOT (Figure 14.5). The processing of the tokens in the overlap region, however, proceeded differently, and in two stages. Figure 14.6 shows the on-line processing of a voiceless token from the overlap range (/pi/-2). In the first stage, VOT is calculated. The nets show an initial preference for /pi/ by gradually increasing the activation of /pi/ through the entire length of the VOT. Upon reaching the end of the VOT, however, the activations for the voice and voiceless tokens switch direction, triggered by the possibility that the given VOT duration is uncharacteristically short for a /pi/ stimulus thus favouring /bi/ instead. This is the second stage where vowel duration is considered. At the end of the vowel, the activations again switch direction, this time cued by the possibility that the given vowel duration against the earlier information about VOT can only match a /pi/ rather than a /bi/ stimulus. Thus, syllable duration, in this case, is critical for the identification of the voiceless tokens.

Discussion

What these experiments reveal about the basic computer science of recurrent networks is encouraging. The type of problem dealt with here is tractable using relatively simple networks. Without additional input windows at the output layer, the networks work well in accommodating the shorter non-duration-based time-warped sequences as well as the longer duration-based sequences whose constituent elements do not overlap in time. With additional windows of input units trained to remember and predict inputs, the networks handle well duration-based stimuli (overlap and non-overlap) but not non-duration-based stimuli. However, with a single input window at $t + 2$, the recurrent network can accommodate the full range of test cases we have considered with an appreciable degree of accuracy. We might speculate that having to predict the future forces the network to encode the structure of the input material more carefully, and that this, as a side-effect, results in a representation which is useful for the classification task. However, having too many time steps to predict and remember might have forced the network to devote too much attention to these tasks, and therefore to neglect the classification task.

In all four experiments the nets needed only to interpolate from training data in order to perform well on the test data. It is therefore not clear if the nets can cope with input which is more extreme than that in their training sets. The ability to *extrapolate* from training data is seemingly crucial with respect to real speech data, for there is no guarantee that the training set will contain stimuli covering the wide range of speaking rates. A follow-up experiment was thus carried out with network C, and the findings suggest that extrapolation is an achievable task for this network. In the experiment, we used stimuli from experiment 4, but instead of training the net with stimulus pairs at rates 1, 3, 4, 6 and 7, and testing it on pairs at rates 2 and 5, we trained the net on stimulus pairs at rates 2, 3, 4, 5 and 6 and tested it on pairs at the extreme rates (1 and 7). A particularly interesting case is the test item /pi/-1 which has the same VOT value as training item /bi/ -3. Despite being trained on the longer /bi/-3 stimulus, the net successfully classified /pi/-1 on the basis of its shorter overall duration. Thus, the net was able to extrapolate to data at rates not in the training set.

The strategies arrived at by the nets are interesting as psychological models, even if it can be argued that recurrent networks, trained by algorithms like back-propagation through time, are not particularly psychologically plausible. The present findings are significant in that they offer a plausible account of the cor-respondence between the way in which a contextual variable alters VOT values and the way in which the variable is used to restructure phonetic categories in perception. We have shown a mechanism whose strategy is to pick up early in the stimuli any information that is relevant to the contrast being judged, and to alter any initial decision if later information proves important for the distinction. Specifically, where no overlap is present, and the range of VOT is distinct between /bi/ and /pi/ across different speech rates, syllable duration is an unnec-essary aid to phoneme distinction. But where there is overlap in the VOT distribu-tion, as one would find in real speech, the mechanism discriminates between stimuli on the basis of whether they are within or outside the overlap region of the VOT continuum; syllable duration is critical only when processing tokens from the overlap range. This raises some questions about the nature of the human speech-processing system. First, in the face of changing speech rates, is the system sensitive to the structural distribution of temporal properties such as VOT that provide cues to phonetic contrasts? In particular, does the system treat differently tokens that belong to the overlap region and those that do not? Sec-ondly, assuming that the system can make a voicing decision partway through the syllable, is the initial decision made on the VOT and then changed once the syllable has been processed, or is the decision postponed until processing of the entire syllable is completed? These questions require empirical study which is beyond the scope of this chapter.

Extension and application to speech perception

In this section, we describe simulations that apply the network's strategy (using network C) on another set of contrast, namely /b/ and /w/. These consonants are distinguished by the abruptness of their onsets or changes in transition duration (hereafter referred to as TD). In the studies of the production of these conso-nants, the onset for /b/, as in the syllable /ba/, was reported to be more abrupt than that for /w/ (Dalston 1975). Perceptually, the standard contrast effect has also been reported for these phonetic categories: as syllable duration increases, the /b/–/w/ boundary moves towards transitions of longer duration (Miller & Liberman 1979). This boundary, however, shifts in the opposite direction when the increase in syllable duration is effected by adding a final transition corre-sponding to a third phonetic segment, as in /bad/. The simulations here demon-strate how the network responds to the combined effect of syllable duration and syllable structure.

Stimuli

Eleven pairs of "speech-like" stimuli that resemble /ba/ and /wa/ syllables co-varying in syllable duration and TD values were constructed. For every pair

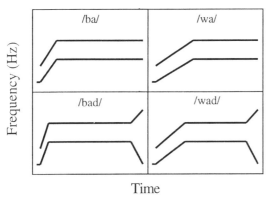

Figure 14.7 Schematic two-formant representations of /ba/, /bad/, /wa/ and /wad/ at a single rate. Each consists of a brief prevoicing (first formant only), a variable duration of formant transition appropriate for /b/ or /w/, a subsequent period of steady state formants, and, for the CVC stimuli, a final period of transition corresponding to /d/. Notice the variations in TD across category and syllable structure.

of /ba/ and /wa/ syllables of a certain syllable duration, we created a corresponding pair of /bad/ and /wad/ syllables of similar overall duration. The timing of the phonetic segments was constrained such that, for any given syllable duration, the TD value for syllables with a CVC (consonant–vowel–consonant) structure was always shorter than for those with a CV (consonant–vowel) structure, as shown in Figure 14.7. This production pattern was directly derived from the perceptual findings of Miller & Liberman (1979) with respect to the /b/–/w/ distinction when the rate and syllable structure were altered (cf. Summerfield 1981, Volaitis & Miller 1991). The "formant frequencies" which we modelled the stimuli from can be found in Abu-Bakar & Chater (1993a).

The relationship between TD and syllable duration for all stimuli is shown in Figure 14.8. Notice that CVC syllables are located along individual distributional curves separate from syllables of the CV type. However, curves which hold syllables with the same syllable–initial consonants are pulled closer together. Twelve tokens of varying syllable duration and structure were reserved as test items. Of these, /wa/-2, /wad/-4, /ba/-7 and /bad/-9 have identical TD values but different syllable duration.

In the simulations reported here, 58 unfoldings were used. The input layer which consisted of 31 units can be thought of as falling into two groups. One group represents the frequency of the first formant, and the other represents the frequency of the second formant. We use a simple localist-style coding to represent this frequency information. Each unit in a particular bank represents a particular frequency, and if a formant has frequency F, then all and only the units which represent frequency values F and less will be active. Sixty hidden units were used, which seems to be approximately the smallest number of units that can learn the task successfully. The target output window of the net has six units; one each for /ba/, /wa/, /bad/ and /wad/, and another for /b/ and /w/. The last two can be conceived of as phoneme detectors. They were included to allow for some independence between the identification of the syllable–initial phonetic segments and the classification of the syllables. This target output is in addition to another bank of units which is trained to continually predict the next but one pattern in the input sequence (recall Fig. 14.1c).

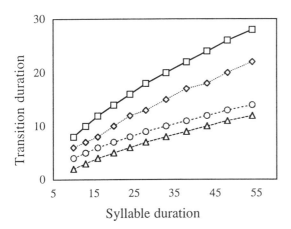

—□—	/wa/
········◇········	/wad/
----○----	/ba/
----△----	/bad/

Figure 14.8
/ba/–/bad/–/wa/–/wad/ stimuli
used in the final experiment.

Results

Training stopped after about 400 iterations, by which time the network had successfully classified all the training stimuli and correctly generalized to the test tokens. Figure 14.9(left) shows the on-line recognition of a /wa/-2 stimulus, one of the four test tokens with identical TD values. Since the TD is ambiguous between /b/ and /w/, the activation of the /w/ detector remains very low for much of the duration of the syllables. It shoots up only at the offset of the vowel segment, which also happens to be the end of the syllable. It is also at this point that the /wa/ syllable is distinguished from the other syllables (/wad/, /ba/ and /bad/) by way of an abrupt increase in the activation of the node representing the /wa/ syllable. In Figure 14.9(right), we have a /wad/-4 stimulus of the same TD value as /wa/-2. As with the latter, the activation of the /w/ detector is low initially and shoots up at the offset of the steady state section of the syllable. But here, the recognition point does not coincide with the end of the syllable, for there is still the final transition following the vowel. Thus, a parsimonious explanation for the syllable–initial distinction is one that takes into account the relationship between the TD and the adjacent vowel and not overall syllable duration. Indeed, we found that for all the stimuli in the overlap region, irrespective of whether they are CV or CVC syllables, the pattern is the same, i.e. the identity of the syllable–initial phoneme as well as the syllable is processed in relation to the syllable's CV component.

Figure 14.10 illustrates the processing of a group of fast /ba/ and /bad/ syllables (/ba/-2 and /bad/-2) whose TD values are outside the category overlap. As

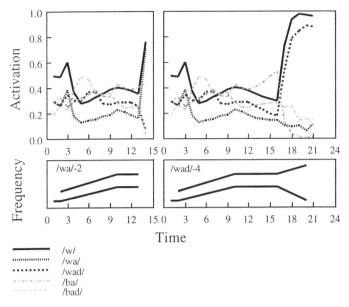

Figure 14.9 Activity of /w/ in /wa/-2 (top left) and /wad/-4 (top right), and the corresponding temporal representation of the stimuli at their respective rates (bottom panels). The activity of the /b/ detector is purposely omitted here while those of /ba/ and /bad/ are included.

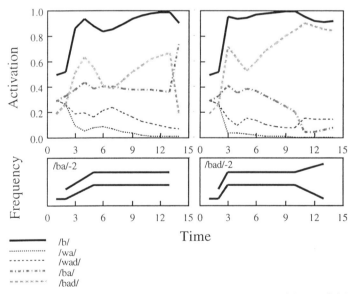

Figure 14.10 Activity of /b/ in /ba/-2 (top left) and /bad/-2 (top right), and the corresponding temporal representation of the stimuli at the given rate (bottom panels). The activity of the /w/ detector is purposely omitted here while those of /wa/ and /wad/ are included.

the figure shows, the identity of the syllable–initial /b/ is obvious right from the start since the TD values are exclusive to the /b/ category. In contrast to syllable–initial phoneme identification, however, syllable recognition occurs late. For the /bad/ syllables, activation increases gradually and reaches the maximum just before the onset of the final transition, whereas for the /ba/ syllables, activation remains low almost throughout the syllable but increases abruptly at the end of the vowel (which also coincides with the offset of the syllable). The recognition of other stimuli outside the category overlap follows the same description, i.e. syllable–initial phoneme recognition occurs as soon as the TD is determined, while syllable identification takes place only at the end of the vowel.

Discussion

There are two main results of the present simulations. First, the present findings are similar to those obtained from the preceding series of simulations in that rate information, specified by the overall stimulus duration, is important for the phonetic distinction between the syllable–initial segments, but only for items in the category overlap. Second, and more importantly, the network picked up the durational constraints we imposed on syllable structure. In the way that we have manipulated the CV and CVC stimuli, syllable structure, apart from giving information about overall syllable duration, also provides specific information about the duration of the phonetic segments that constitute the syllable. Our experiments show that this was crucial to the network in two instances: first, when making the relevant phonetic contrast for those items in the category overlap, and, secondly, when distinguishing between syllables of different structure but identical syllable–initial phoneme for all tokens along the TD continuum.

The work also speaks on the issue of whether information provided by syllable structure is used by listeners during perception. Miller & Liberman (1979) have shown that this was indeed the case for their subjects. Interestingly, however, their proposal that listeners calculate the number of phonetic segments to determine the articulation rate of a given syllable and use this to influence the phonetic categorization of an initial consonant does not fit with the network performance. The behaviour of our network suggests a more general strategy which involves making a durational contrast between the cue provided by the TD and the following adjacent segment, a vowel in this instance. Syllable structure, as noted earlier, reconfigures the vowel duration and TD with respect to the overall syllable duration, as is the case when a third segment is appended to the CV syllable (see Volaitis & Miller 1991). Possibly, this is geared to maintain perceptual intelligibility in much the same way that language communities intentionally regulate vowel length in order to enhance perceptually the closure duration cue for voicing distinctions (Kluender et al. 1988). The resulting configuration, particularly the relationship between a vowel and the initial TD, is learned by the network, and this information is used both to capture the relevant phonetic contrast and to generalize to new stimuli.

This explanation is consistent with the results obtained by Newman & Sawusch (1992). In examining the effects of adjacent and non-adjacent phonemes on the perception of the syllable–initial contrast between /sh/ and /ch/, cued by the duration of friction, in the /shwaes/–/chwaes/ series, they demonstrated that varying the /w/ duration produced the standard contrast effect, i.e. a longer /w/ made the initial segment seem shorter, or sound more like a /ch/, while a shorter /w/ made the initial segment seem longer, or sound more like a /sh/. Variation in the duration of the non-adjacent vowel, on the other hand, had no contrastive effect. In relation to the /bad/–/wad/ stimuli used in our computational experiments, the distant segment that does not contribute to the contrast effect is the final transition. The same explanation can also be used to interpret the data obtained by Volaitis & Miller (1991). In studying the role of syllable structure on the perception of VOT in /di/–/ti/ and /dis/–/tis/ syllables in the context of changing speech rate, they found that listeners adjusted for changes in VOT in relation to the syllable's CV duration, and not to its overall duration, which is consistent with our model.

Conclusion

Motivations for dealing with time-warped sequences have been discussed, and a successful recurrent network model has been described that can accommodate a range of rate-varying stimuli. Application of this model to a specific problem in phonetic perception has also been examined with encouraging results. To the extent that recurrent networks embody a learning account, one remaining issue raised by the model is whether the strategy of contrasting segmental durations is dependent on what is learned about the properties of actual speech, or is the strategy that is hardwired in the auditory system (Diehl & Walsh 1989). Recently, we completed another series of studies with these networks (Abu-Bakar & Chater 1994a,b) to investigate the viability of the model in simulating other phenomena in speech perception, namely shifts in category boundaries due to rate, experience and selective adaptation, and alteration to the internal structure of phonetic categories as a consequence of changes in speaking rate. The results from this work bring further implications for spoken language processing and models of perception and categorization of human speech (see Abu-Bakar & Chater 1994c).

Acknowledgements

Most of the material contained in this chapter has appeared elsewhere (Abu-Bakar & Chater 1993a,b). We gratefully acknowledge the helpful comments of an anonymous reviewer on an earlier draft of the manuscript. We thank the Department of Psychology, University of Wales, Bangor, and the Centre for

Cognitive Science, University of Edinburgh, for the extensive use of computer facilities. Correspondence should be addressed to the second author.

References

Abu-Bakar, M. & N. Chater 1993a. Studying the effects of speaking rate and syllable structure on phonetic perception using recurrent neural networks. *Irish Journal of Psychology* **14**, 426–441.

Abu-Bakar, M. & N. Chater 1993b. Processing time-warped sequences using recurrent neural networks: modelling rate-dependent factors in speech perception. *15th Annual Conference of the Cognitive Science Society, Proceedings*, 191–7. Hillsdale, New Jersey: Lawrence Erlbaum.

Abu-Bakar, M. & N. Chater 1994a. Distribution and frequency: modelling the effects of speaking rate on category boundaries using recurrent neural networks. *16th Annual Conference of the Cognitive Science Society, Proceedings*, 3–8. Hillsdale, New Jersey: Lawrence Erlbaum.

Abu-Bakar, M. & N. Chater 1994b. Phonetic prototypes: modelling the effects of speaking rate on the internal structure of a voiceless category using recurrent neural networks. *3rd International Conference on Spoken Language Processing, Proceedings*. Tokyo: The Acoustical Society of Japan.

Abu-Bakar, M. & N. Chater 1994c. A recurrent neural network model of rate effects in phonetic perception. Unpublished paper, Department of Linguistics, University of Wales, Bangor.

Chater, N. 1989. *Learning to respond to structure in time*. Research Initiative in Pattern Recognition, RSRE Malvern, Technical Report RIPRREP/1000/62/89.

Chater, N. & P. Conkey 1994. Sequence processing with recurrent neural networks. In *Neurodynamics and psychology*, M. Oaksford & G. D. A. Brown (eds), 269–94. London: Academic Press.

Dalston, R. M. 1975. Acoustic characteristics of English /w, r, l/ spoken correctly by young children and adults. *Acoustical Society of America, Journal* **57**, 462–9.

Diehl, R. L. & M. A. Walsh 1989. An auditory basis for the stimulus-length effect in the perception of stops and glides. *Acoustical Society of America, Journal* **85**, 2154–64.

Elman, J. L. 1990. Finding structure in time. *Cognitive Science* **14**, 179–211.

Huang, X. D., Y. Ariki, M. A. Jack 1990. *Hidden Markov models for speech recognition*. Edinburgh: Edinburgh University Press.

Kluender, K. R., R. L. Diehl, B. A. Wright 1988. Vowel-length differences before voiced and voiceless consonants: an auditory explanation. *Journal of Phonetics* **16**, 153–169.

Lippmann, R. P., E. A. Martin, D. P. Paul 1987. Multi-style training for robust isolated-word speech recognition. *IEEE International Conference on Acoustics, Speech, and Signal Processing, Proceedings*.

Lisker, L. & A. Abramson 1964. A cross-language study of voicing in initial stops: Acoustical measurements. *Word* **20**, 384–422.

Maskara, A. & A. Noetzel 1992. Forced simple recurrent neural networks and grammatical inference. *14th Annual Conference of the Cognitive Science Society, Proceedings*, 420–7. Hillsdale, New Jersey: Lawrence Erlbaum.

Miller, J. L. 1981. Effects of speaking rate on segmental distinctions. In *Perspectives on the study of speech*, P. D. Eimas & J. L. Miller (eds), 39–70. Hillsdale, New Jersey: Lawrence Erlbaum.

Miller, J. L. & A. M. Liberman 1979. Some effects of later-occurring information on the perception of stop consonant and semivowel. *Perception and Psychophysics* **25**, 457–465.

Miller, J. L., F. Grosjean, C. Lomanto 1984. Articulation rate and its variability in spontaneous speech: a re-analysis and some implications. *Phonetica* **41**, 215–55.

Miller, J. L., K. P. Green, A. Reeves 1986. Speaking rate and segments: a look at the relation between speech production and perception for the voicing contrast. *Phonetica* **43**, 106–15.

Newman, R. S. & J. R. Sawusch 1992. Assimilative and contrast effects of speaking rate on speech perception. *Acoustical Society of America, Journal* **92**(suppl. 2), SP11.

Nguyen, M. & G. W. Cottrell 1993. A technique for adapting to speech rate. *IEEE Workshop on Neural Networks for Signal Processing, Proceedings*.

Norris, D. 1990. A dynamic-net model of human speech recognition. In *Cognitive models of speech processing: psycholinguistic and computational perspectives*, G. T. M. Altmann (ed.), 87–104. Cambridge, Mass.: MIT Press.

Rumelhart, D. E., G. E. Hinton, R. J. Williams 1986. Learning internal representations by error propagation. In *Parallel distributed processing: explorations in the microstructures of cognition*, vol. 1. *Foundations*, D. Rumelhart & J. McClelland (eds), 318–62. Cambridge, Mass.: MIT Press.

Shillcock, R., J. Levy, N. Chater 1991. A connectionist model of word recognition in continuous speech. *13th Annual Conference of the Cognitive Science Society, Proceedings*, 340–3. Hillsdale, New Jersey: Lawrence Erlbaum.

Shillcock, R., G. Lindsey, J. Levy, N. Chater 1992. A phonologically motivated input representation for the modelling of auditory word perception in continuous speech. *14th Annual Conference of the Cognitive Science Society, Proceedings*, 408–13. Hillsdale, New Jersey: Lawrence Erlbaum.

Summerfield, A. Q. 1981. Articulatory rate and perceptual constancy in phonetic perception. *Journal of Experimental Psychology: Human Perception and Performance* **7**, 1074–95.

van Camp, D. & T. Plate 1993. *Xerion neural network simulator*. Department of Computer Science, University of Toronto.

Volaitis, L. E. & J. L. Miller 1991. Influence of a syllable's form on the perceived internal structure of voicing categories. *Acoustical Society of America, Journal* **89**(suppl. 2), SP10.

Volaitis, L. E. & J. L. Miller 1992. Phonetic prototypes: influence of place of articulation and speaking rate on the internal structure of voicing categories. *Acoustical Society of America, Journal* **92**, 723–735.

Watrous, R., B. Ladendorf, G. Kuhn 1990. Complete gradient optimisation of a recurrent neural network applied to /b/, /d/, /g/ discrimination. *Acoustical Society of America, Journal* **87**, 1301–9.

CHAPTER 15

Bottom-up connectionist modelling of speech

Paul Cairns, Richard Shillcock, Nick Chater, Joseph P. Levy

Introduction

Low-level phonological information plays an important role in spoken word recognition. It is certain that listeners are highly sensitive to simple sequential phonological patterns – thus any speaker of English can instantly say which of the following are possible words in their language: /snarp/, /mplaf/, /krad/, /sakf/. Furthermore, one intuitively rates legal examples such as /sfɪp/ as being less "normal" than items such as /stɪp/. Although it might be possible for such judgements to be mediated by rapid calculation of some lexical intersection, for reasons of computational efficiency it would be advantageous for this type of information to be represented in summary form in a component of the human language processor. We believe that this sublexical information impinges upon higher-level processes such as lexical access, and there is evidence to show that this is the case (e.g. Jakimik 1979, Foss & Gernsbacher 1983). In this chapter we describe a system that learns to encode this type of information. Our system is not a model of any particular psycholinguistic process in itself; rather, it can be used as a tool to represent and assess the possible influence of sublexical phonological information in specific cognitive processes.

In recent years, statistically based models have come to dominate computational psychological modelling, with neural networks playing an increasingly prominent role. However, it is frequently the case that such work is prone to a particular misunderstanding about the nature of statistical modelling: in order to stand as a valid model of human behaviour, and particularly *learning* behaviour, a statistical system must be derived from input that is representative of genuine natural language input – yet this fact seems to be forgotten in much connectionist modelling work. Thus, a model which employs a "toy" training set, of say a dozen lexical items, can only make defensible claims about human behaviour to the extent that its input is statistically representative of natural input –

improbable given such a small sample. Linked to this issue is the tendency to use training data that are idealized and noise-free. This unrealistic stance is presumably motivated by a belief that working with real data is difficult, and will tend to obscure the structure of interest to the particular modelling task at hand. One of the goals of this chapter is to show that real, noisy data can be used in psycholinguistic modelling, and that such data allow our models to claim increased ecological validity. To this end we present a large corpus of phonologically transcribed speech that is derived from the London–Lund corpus (LLC) of conversational English (Svartvik & Quirk 1980). This corpus serves as input to a neural network that learns the sequential phonological statistics (*phonotactics*) of the English speech stream.[1] The rich source of data that the corpus provides is representative of the input that humans have access to, and therefore we can assess the extent to which the data could be of use in particular psycholinguistic processes.

Use of the corpus facilitates the investigation of the feasibility of bottom-up modelling. There has been much debate as to whether (parts of) the human language processor are *interactive* or *modular* (e.g. Fodor 1983, Tanenhaus & Lucas 1987, Elman & McClelland 1988). The fundamental difference between interactive and modular accounts is that only the former allow top-down influence of higher-level information on lower-level processing. However, it is important to note that both accounts can make use of bottom-up information, and thus bottom-up modelling cannot directly refute the viability of either account. However, bottom-up modelling can be broadly supportive of the modular account if it shows that some particular psycholinguistic phenomenon can be given an encapsulated bottom-up explanation, since it is generally a goal of scientific investigation to invoke as parsimonious an explanation of the data as possible. Furthermore, use of a strictly bottom-up system allows us to propose learning models in which higher-level representations are *bootstrapped* by lower-level information with the minimum *a priori* domain knowledge. Another benefit of the full-scale approach that our use of the corpus brings is our ability to model existing data without the need to use simplifying re-encodings. Thus, we are able to feed original experimental stimuli directly into our network.

We choose to use a recurrent neural network for encoding the phonotactic information. Such devices are ideal for the multidimensional probability estimation which this domain demands. Unlike its symbolic equivalent the *hidden Markov model* (HMM – see Huang et al. 1990) in which segmental categories must first be reconstructed in the input before the probabilistic model can be applied, the network can, in theory, solve the categorization problem and the transitional modelling problem simultaneously. The network is simply trained to predict the next chunk in the input corpus. Learning is *unsupervised* at the computational level: the net comes to encode knowledge that reflects the statistical properties of its input without being told explicitly what information is relevant. In this sense, any regularities and structure that the network is sensitive to are *emergent* properties of the data. Thus, our models are bottom-up because the network is not trained to map between different linguistic levels; only sublexical information is ever presented.

The rest of this work is structured as follows: first, we describe our phonological input corpus and its derivation; secondly, we outline the neural network model that was developed; and finally, we apply our model to two very different psycholinguistic domains: phonemic restoration, and the acquisition of lexical segmentation.

A phonological retranscription of the LLC

The LLC is a large body of English conversation transcribed orthographically and available on-line. Because of its size (around 460 000 words) an automatic method was developed for its phonetic transcription (Fig. 15.1). First, the words are replaced by their phonemic citation forms using an on-line dictionary. Then these forms are input to a set of rewrite rules that introduce phonological alternations into the string (e.g. assimilation, vowel reduction). None of the rules uses word boundary information to specify its context of application – this fact is important for our segmentation model. The output from the rule set is a corpus of 1.5 million phonetic segments.

It is, of course, impossible to recreate the original speech data, but this method has two main advantages: first, we need a very large corpus of conversational speech if its statistics are to be representative – at present there is no sufficiently large corpus with a genuine phonological transcription; second, this method provides a higher-order approximation to genuine data, when compared with a corpus derived from a phonemic dictionary in combination with word frequency counts. For example, our data will be representative of the distribution of strings of closed-class words such as "if I can".

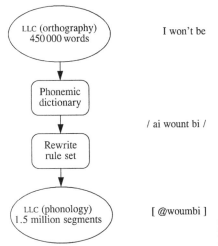

Figure 15.1 Retranscription of the LLC. Here the symbol "@" stands for "schwa".

This segmentally transcribed version of the corpus was then retranscribed in terms of nine binary features that represent the cognitive elements of government phonology (see Kaye et al. 1985, Kaye et al. 1990, Harris & Lindsey 1995). The nine elements are seen as cognitive categories which are universals and which are sufficient for the representation of speech. The sound patterns of the acoustic input are mapped onto these elements. Harris & Lindsey advance a realizational autonomy hypothesis, which allows each of the elements to be phonetically realized, independent of the other elements. Work by Williams & Brockhaus (1992) has shown how the government phonology elements can be automatically extracted from the speech stream, so we have reason to believe that coding in this way represents a step further towards ecological validity.

In summary, our retranscribed version of the LLC represents a large source of genuine, and noisy (in terms of false starts, hesitations and speaker change-overs), speech data that are essential for representative modelling of spoken word processing.

The model

Choice of architecture

In implementing a system that encodes phonotactics we decided to use a technique that is computationally *unsupervised*. A *supervised* system is trained to encode information by using input that is pre-classified or labelled to denote its relevant aspects. In contrast, an *unsupervised* system is one in which the representations that the system builds are not dictated by some external "teacher", but instead arise from the interaction between inherent statistical structure in the data, and the computational characteristics of the system.

The reason for using an unsupervised system is simply that it allows one to create learning models that are strictly bottom-up, in the sense that no higher-level information is employed either in operation, or, most importantly, during training. Such a model, by virtue of being purely bottom-up, can be used (a) to support modular processing theories, and (b) in modelling acquisition. To understand why strictly bottom-up mechanisms are important for modelling acquisition, consider training a network to classify nouns and verbs by presenting a large corpus of these items and somehow indicating to the network when an item is a noun or verb: this supervised approach would not be a good candidate for a model of how infants discover the noun/verb distinction, since presumably in real life nobody is signalling their grammatical function explicitly. So, using an unsupervised system allows the model to be strictly bottom-up, and hence a suitable candidate for modelling development in the case where one wishes to make minimal assumptions about prior knowledge.

There are many ways of constructing unsupervised statistical systems that will uncover structure in input (e.g Draper & Smith 1981, Kohonen 1988, Linsker

1988). However, here we have need of a technique that will uncover structure in time. One such method that has been used in other psycholinguistic modelling work is the *prediction task* (e.g. Elman 1990). The central idea is that if a system is capable of predicting the next element in a time series, then that system must have knowledge of the underlying statistical structure of the sequence. The prediction task method is unusual in that, to use Marr's notion of computational levels (Marr 1982), it uses a system that is supervised at the *implementational* level (e.g. a back-propagation network in which a "teacher" specifies the target: the next slice of input) to obtain a model that is unsupervised at the *computational* level. We call such a system *self-supervised* because the input and desired output are specified by a single stream of data. A self-supervised system builds representations and extracts information based on the salient emergent characteristics of the input corpus. In other words, the system will try to reduce the mismatch between its predictions and reality – this can only be done if it can learn something of the sequential structure of its input. Which aspects of structure are important must be decided by the system itself.

To tackle the prediction problem, one must use an architecture that can deal with temporal representations. A flexible architecture that meets this requirement is the *simple recurrent network* (SRN) (Elman 1990), which comprises a standard feedforward neural network, augmented with a set of *copy-back* or *state* units that permit a limited feedback within the system. In operation, the activation values of a layer (typically the hidden layer) are *copied-back* onto the context units, on a one-to-one basis. If the context units are then connected to the hidden layer with standard feedforward modifiable connections the network will have access to a "memory" of previous hidden unit states, and can respond to constraints which may be defined over any number and combination of previous inputs, over any time period. In practice, learning is generally much more successful when constraints are relatively local. SRNs have been used productively in modelling a range of aspects of language processing (e.g. Cleeremans et al. 1989, Elman 1990, St John & McClelland 1990, Elman 1991, Shillcock et al. 1991, Norris 1993).

SRNs can be trained using standard back-propagation, since their feedback connections are not modifiable. The back-propagation learning algorithm was developed for feedforward networks, where it can be shown to perform gradient descent in error space – that is, the weights are changed slightly in the direction which reduces error as much as possible. For SRNs, the *copy-back* method is computationally cheap, but computes only a rough approximation to true gradient descent.

An alternative method of encoding time in a neural network is to use *back-propagation through time* (BPTT; see Rumelhart et al. 1986). Using this technique, standard feedforward networks can be augmented with arbitrary modifiable feedback connections. To train these networks, the standard back-propagation algorithm can be simply extended. In essence, the network is "unfolded" in time: during training, activation is passed forwards through the network for a fixed number of iterations. Then, the error signal is passed back through the net for the same number of iterations with the error terms being summed for each weight. Finally, the weights are changed by an amount proportional to the iterated sum of required weight changes. One disadvantage of this method is that the number of unfolded time steps is fixed in advance, so one must be sure to use a number of time steps that is larger that the maximum periodicity of the relevant dependencies in the input. In our case, we know that phonotactic dependencies are relatively local, so this is not a problem. However, BPTT would be unsuitable for input with unbounded dependencies. Another disadvantage of BPTT is that it is computationally more expensive to train, since each node in the net must store its entire input and output histories on each time step until the weights are updated.

One important difference between the SRN and BPTT techniques that is not often appreciated is that the SRN must learn to encode time *directly* in its hidden unit representations, at least if the period of the dependency is more than one time step. In a BPTT net, on the other hand, time is explicitly present, as it is in the "moving window" technique of NETtalk (Sejnowski & Rosenberg 1987). This is because BPTT allows the error signal to be back-propagated through longer stretches of time than the SRN (see Chater & Conkey (1994) and Chater & Conkey (1992) for a detailed comparison). In practice, this means that temporal tasks will be learnt much more quickly by a BPTT net, although the hidden unit representations developed by an SRN may be sufficient motivation for selecting that technique (for instance in an unpublished experiment by the authors it was found that both BPTT and SRNs could learn to be delay lines of 1 place, but SRNs were very difficult to train when the delay was 2 or more, while BPTT showed no deterioration in learning ability).

Since we are interested in obtaining the best prediction that we can, we used the computationally more expensive method: BPTT. The network is *self-supervised* in the sense that the input and target output for network training are specified by a single stream of data. While here we have assumed the use of a neural network, Cairns et al. (1994) describe the use of more "traditional" statistical methods.

Network training

The network architecture is shown in Figure 15.2. There are nine input units corresponding to the nine government phonology elements, 60 hidden units and 27 ($= 9 \times 3$) output units which are conceptually divided into three blocks. The hidden units have feedback connections that loop back onto themselves. The net is unfolded for ten time steps, so this is the maximum span of a sequential dependency that it can become sensitive to (strictly speaking, the maximum span that can be used is eight time steps since update in BPTT is synchronous, and thus activation takes two time steps to pass through the two-layer net).

The network must echo the current slice of input, remember the previous, and

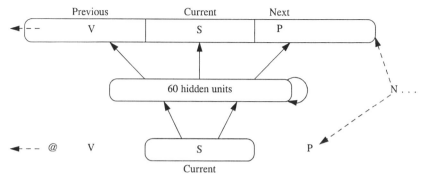

Figure 15.2   The network. The solid arrows between layers indicate complete connectivity with modifiable uni-directional links. The dotted arrows show how the input corpus arrives over time to specify the input and output target.

predict the next. These three tasks serve two functions: (a) using multiple tasks has been shown to improve learning and generalization in recurrent networks and other systems; (b) we are free to use any of the output subgroups in modelling. In this chapter we restrict ourselves to the prediction and autoassociation units. While the prediction part of the output task obviously requires knowledge of English phonotactic structure, the recall, and autoassociation, outputs do not, as things stand, require any use of domain-specific knowledge. Thus, if one trained the net with English, but tested with a stretch of Spanish, one would not expect the recall and autoassociation performance to differ (but prediction would presumably be impaired).

As a method for forcing the network to look at context when deciding on recall and autoassociation outputs, we added noise to the input stream by randomly flipping features from 0 to 1 (or vice versa) with a certain probability (p was selected so that on average 1.5 features were distorted in each segment). Given this noisy situation, when autoassociating a /d/ for example, the network cannot be sure that the input is not a corrupted /t/. However, by looking at the surrounding context, suppose /s_r/, then it may be possible to make a more informed decision. Of course, in reality the task is more difficult since the network's currency is features not phonemes. Another advantage of our use of this type of noise distortion is that it adds empirical validity to the task, since the real world is inevitably noisy.

The net is trained using BPTT, a steepest descent procedure, and a cross-entropy error measure (see Hinton (1989) – cross-entropy is a good measure to use when one wishes to interpret continuous-valued outputs as probabilities of binary decisions). Training comprises two passes through a training stretch of the corpus 1 million phonemes in length (with different noise on each pass), and thus 2 million phonemes in total.[2] It was found that better results were obtained if the learning rate was decayed as training proceeded (possibly as this facilitates a sort of annealing), so ε was reduced from 0.1 to 0.0004 as a function of number of examples trained.

295

Training was halted when the generalization error on a test stretch of the corpus was judged to have reached the asymptote on inspection of the error graph for the test corpus.

After training, the network is able to echo the current slice of input, and remember the previous, with reasonable accuracy. It correctly reproduces around 85% of current segment categories, and around 65% of previous segments (recall that the use of noise makes these tasks non-trivial). However, due to the nature of the task, prediction in categorial terms is poor – only about 10% off all segments are correctly predicted. However, the network has come to encode a wealth of information about the statistical temporal regularities of the input, as shall be seen in the following sections in which we outline its use in psycholinguistic modelling.

Simulating phonemic restoration

Elman & McClelland (1988) conducted an ingenious series of experiments which aimed to circumvent the methodological problems in testing whether the Ganong effect (Ganong 1980), which claims to show a word superiority effect in phonemic restoration, is genuinely caused by the top-down influence of lexical information – whether there really is *cognitive penetration* of perceptual mechanisms. Their logic was to consider whether abstract lexical level representations could be shown to interact with a perceptual process known to be operative at the phonological level: *compensation for coarticulation* (see Mann & Repp 1981). This would not occur in the modular account, which states that lexical level representations may directly affect behaviour, but do not influence lower phonological representations, and hence cannot affect any cognitive processes defined over such representations.

Elman & McClelland assumed that compensation for coarticulation is defined over phonological representations: when a speaker produces a /s/, for instance, this affects the production of the next segment, and listeners compensate for this by moving their criterion for the perception of the following segment. For example, given a series of synthetically produced tokens between /t/ and /k/, listeners move the category boundary towards the /t/ following a /s/ and towards the /k/ following a /ʃ/.

Following observations of the behaviour of the (interactive) TRACE model of word recognition (McClelland & Elman 1986), Elman & McClelland considered the case where this effect occurs across word boundaries, a word-final /s/ influencing a word-initial /t/ as in "Christmas tapes" for instance. If lexical level representations genuinely influence phoneme level representations, the compensation of the /t/ should still occur when the /s/ relies on lexically driven phoneme restoration for its identity. Phoneme restoration had previously been studied within single words (e.g. Warren 1970, Warren & Sherman 1974), making it difficult to distinguish the interpretation involving genuinely interactive lexically driven phoneme restoration, in which the phoneme level representations are really changed, from the interpretation in which the response is simply based on the lexical level. Exploitation of the compensation for coarticulation effect allowed Elman & McClelland to distinguish between these interpretations.

Subjects heard pairs of words such as "Christmas tapes" or "foolish capes", in which the last segment of "Christmas" or "foolish" was replaced by a synthetic segment judged in isolation to be midway between /s/ and /ʃ/. The first segment of "tapes"/"capes" was a synthetic segment drawn from a cline between /t/ and /k/. Subjects were required to report the identity of the second word. Their responses revealed that the restored identity of the ambiguous fricative at the end of the first word affected the identification of the ambiguous stop at the beginning of the second word in a way which paralleled the compensation effect when genuine fricatives were present. Elman & McClelland's interpretation of this effect was that the final fricative was being restored, causing a real change in sublexical representations, in so far as there were consequences for the perception of the first segment of the next word. There was no other way for the identity of the second word to be affected by the prior context.

Despite the fact that further empirical studies have suggested that this restoration-plus-compensation phenomenon might be less reliable than was first suggested – specifically, some sort of degradation of the speech signal is often necessary for the effect to be observed (McQueen 1994, McQueen et al. 1993) – the original demonstration is still of substantial theoretical interest.

Modelling of the phenomenon

There are two possible ways in which bottom-up models might be compatible with this experiment: first, it might be that compensation for coarticulation is itself not exclusively mediated by phonological representations, but directly influenced by lexical factors, as recent work (McQueen 1994) suggests. If it is the case that top-down information affects *compensation*, then this would leave open the question of whether *restoration* is amenable to top-down influence or not; secondly, the apparently lexical effect on phoneme restoration might actually be due to processes defined solely at the phonological level or below, rather than interaction with the lexical level. It is this second explanation which we will argue for, with evidence from the connectionist simulations described below. First we detail other explanations of the phenomenon that have been put forward.

The original explanation for the data came from TRACE, in which the interactive activation between lexical and phoneme nodes served to restore the word-final fricatives, and thus maintain compensation for coarticulation. An alternative explanation was that of Massaro (1992) who showed how his *fuzzy logical model of perception* (FLMP) could predict the compensation results with surprising accuracy. The core of the FLMP is the description of how the various perceptual and contextual information sources are integrated probabilistically to produce a categorization decision. In simulating Elman & McClelland's results, the FLMP integrates the information from nine variables: the seven points on the "tapes"/"capes" continuum; the identity of the final fricative; and the lexical context (i.e. "Christmas"/"Spanish"). Thus, the FLMP model uses lexical information, and as such is not bottom-up. Given this, it is hard to distinguish the empirical claims that the FLMP makes from those of TRACE. Although Massaro is insistent that his model is non-interactive, it is hard to see what he means by this, especially since he claims that "the FLMP explanation is in terms of perceptual processes, and is not simply a result of a post-perceptual decision mechanism".

Norris has presented two small-scale connectionist simulations intended to model the phenomenon, in principle, with a recurrent network (Norris 1992, Norris 1993). The network was trained to map a feature level input onto a phoneme and/or word level output. The simulations employed a lexicon of 12 CVC (consonant–vowel–consonant) words (/sɔs/, /ʃɔs/, etc.) constructed from a set of eight phonemes.

In the first simulation, feature level inputs are mapped both to a localist phonemic output and a localist lexical output. A parallel to the compensation for coarticulation effect was observed, within the limits of the lexicon used: the percentages of /t/ and /k/ responses to the first phoneme of the second word depended on the identity of the first word, as in Elman & McClelland's original experiment. Norris suggests that although the test phase of the network consists only of feedforward activation the apparent top-down effect may stem from the fact that lexical influences are pre-compiled into the network's lower weights (weights before the hidden units, which are common to lexical and phonemic processing; after the hidden units the weights become exclusive to the lexical and phonemic outputs, respectively). This pre-compiling of lexical influence is able to occur during training, when, of course, error is back-propagated down through the network as the model is exposed to pairs of words.

In this first simulation the lower level of the network duplicates lexical information, as Norris recognizes. Therefore, the major theoretical claim is not at all bottom-up, in the sense that *bottom-up* proscribes the use of higher-level information from the lexicon. In fact, this simulation would seem closest to the work of Massaro in that it shows how different lexical and sublexical information sources can be integrated, albeit in a pre-compiled rather than interactive manner.

However, in a second simulation (Norris 1993), the lexical part of the training was ignored and the network was only required to map feature level input to phonemic output. The restoration/compensation phenomenon still emerged, albeit more weakly. Norris suggests that the network owed its ability to simulate the phenomenon simply to its capacity to take the prior featural context into account via its recurrent connections. The network was using contextual information to determine the identity of phonemes in the feature level to phoneme level mapping. (The network was encouraged to use context by the use of noise in training: the particular feature level descriptions were made unreliable.) However, there are problems with the model: the small number of lexical items used for training, together with the clear delimitation of the initial word boundary, means that the end of the first word is completely predictable, making restoration and compensation inevitable. There is no demonstration that the network has arrived at a general solution, and indeed this would be remarkable given the very small training set. It is much more likely that the net has simply rote learned the sequences, and that this simulation does not demonstrate anything more interesting than the ability of a recurrent network to learn sequences.

It is not clear that Norris's explanations can be sustained with input which reflects a full-sized lexicon and phonemic inventory, together with the uncertainty of word boundaries in normal speech processing. However, we are in a position to assess the influence of context with our full-scale and representative model, which we detail in the next section.

Modelling phoneme restoration with phonotactics

We assessed the hypothesis that the Elman & McClelland results reflect processes which are adequately described as occurring at the phonological level only. We tested whether the statistical structure of the phonological sequence could support restoration without reference to a separate lexical level of representation. Note that we only address the restoration part of the Elman & McClelland data, but this is all that is necessary to disqualify their results. Studying compensation for coarticulation was only the mechanism by which the restoration effect could be proved to employ top-down information flow. If one can show that restoration can be driven from the bottom-up, then the data are invalidated, irrespective of the compensation effect. There is no interest in modelling the compensation part, since this by all accounts follows automatically once restoration has occurred.

If the phonological context is sufficiently informative, it is possible that the network might use this information to restore the degraded word-final phoneme (e.g. /s/ in "Christmas"), without recourse to lexical level representations. To test this we transcribed the six initial words of the word pairs used by Elman & McClelland: "Christmas", "copious", "ridiculous", "foolish", "English" and "Spanish". The segments /s/ and /ò/ differ by only a single feature in our government phonology representation, specifically: the absence or presence, respectively, of the element *palatality*. An intermediate segment was created with 0.5 as the value of this element, and this segment replaced the final fricative in the six words. These transcriptions were then used as input to the model, with the same left sentential context. For each test word, the cosine between the network output for the final fricative and all possible segments was calculated. The network output was interpreted categorically as being the segment which had the largest cosine with the actual output vector.

Each of the six initial words in Elman & McClelland's stimulus materials was appropriately restored at the model's output, either to /s/ or to /ʃ/, according to the cosine best match procedure (see Table 15.1 – note that what appear to be

Table 15.1 Table showing the cosines of network output for final fricative of each item with targets of /s/ and /ʃ/.

Item	Cosine with /s/	Cosine with /ʃ/
Christmas	**0.99**	0.89
foolish	0.96	**0.98**
copious	**0.99**	0.89
English	0.96	**0.98**
ridiculous	**0.99**	0.89
Spanish	0.96	**0.98**

small differences in cosine value translate into large differences in value of the ambiguous feature, e.g. "Christmas" = 0.01, "foolish" = 0.44). This was despite the fact that they had all been identically represented at input as a hybrid vector ambiguous between /s/ and /ʃ/.

We are concerned here only with phoneme restoration at the end of the first word; we take it for granted that the actual compensation effect follows naturally in a model which is trained to respect such a sequential dependency. Recall that Elman & McClelland only use the compensation for the coarticulation effect as proof that phonemic restoration can effect pre-lexical processing. If restoration is itself a pre-lexical phenomenon then the argument disappears, regardless of whether there is an interaction between restoration and compensation, or not.

Elman & McClelland present a further experiment in which larger final portions of a subset of the initial words in the stimulus pairs were made ambiguous. This was intended to meet the objection that phonetic colouring of the final vowel was determining the identity of the final fricative. In the words "Spanish" and "ridiculous" the final VC (vowel–consonant) portions were replaced by ambiguous versions pretested to sound halfway between /ɪʃ/ and /əs/. In addition, the final syllable of the words "foolish" and "ridiculous" was replaced by a syllable pretested to sound halfway between /lɪʃ/ and /ləs/. Listeners again made compensations for co-articulation which were comparable with the veridical stimuli. Elman & McClelland regard these results as compelling evidence against a sublexical explanation of the effect.

We simulated these cases in the same way as before. In the "Spanish"/ "ridiculous" pair, the appropriate restoration effect was observed, with the ambiguous fricative becoming /ʃ/ and /s/, respectively. In only the one instance, the final "foolish"/"ridiculous" pair, did the model not succeed in making the appropriate restoration; the ambiguous fricative was marginally restored to /s/ in both cases. Elman & McClelland report, however, that the stimuli for this final /lɪʃ/–/ləs/ case were "not as natural sounding as the [VC replaced stimuli]".

Thus, the network successfully simulates all but the very final result. Its behaviour partly reflects *n-gram* statistics of consecutive segments, derivable from the initial transcription of the LLC (before the conversion into government phonology elements). Bigram statistics show that $p(\text{ɪ},\int | \int) = 0.22$ and $p(\text{ə, s} | \text{s}) = 0.18$

are preferred to $p(\text{I}, \text{s}\,|\,\text{s}) = 0.19$ and $p(\text{ə},\int|\int) = 0.04$ (we have used probabilities that are conditional on the fricative because /s/ is much more frequent than /∫/. Using conditional probabilities enables one to factor out the effect of this predominance). The probabilities are derived from bigram counts of the whole LLC, reported in Shillcock et al. (1995).

Our results show that the Elman & McClelland study is compatible with a bottom-up architecture, since phoneme restoration, which triggers compensation for coarticulation, may be occurring on the basis of statistics of the phonological sequence, rather than being driven by a lexical level representation. This interpretation of the data is bolstered by recent work in which we successfully simulated (in a double-blind test) further compensation for coarticulation results obtained by McQueen (1994). For a more detailed exposition of the phoneme restoration simulations, see Chater et al. (1994).

Bootstrapping segmentation

Now we move on to discuss the application of our network to modelling the acquisition of lexical segmentation.

Most lexical boundary points in conversational speech are not marked by acoustic discontinuities. Therefore, a neonate must either have a mechanism that will actively hypothesize boundaries in the continuous input, or must initially only process words spoken in isolation (e.g. Suomi 1993). The viability of the latter option is not supported by data from sources such as the CHILDES database (MacWhinney & Snow 1985) which show that only a small proportion (around 15%) of utterances are single words. Furthermore, tokens of words spoken in isolation tend to be phonologically less reduced than other tokens which are embedded in phrases (Jusczyk 1993). Because of these two problems, it is likely that the infant has some mechanism that facilitates the bootstrapping of lexical junctures.

There are several possible information sources that could aid the bootstrapping of lexical segmentation: (a) acoustic/phonetic juncture markers or pauses (Lehiste 1971); (b) prosodic marking that specifies the initial portion of a word, given a pre-syllabified input (Grosjean & Gee 1987, Cutler & Norris 1988, Cutler & Butterfield 1992, Cutler 1993); (c) distributional cues, for example differing probabilities of certain phonological sequences at various points in the speech stream (*phonotactics* – see Harrington et al. 1988).

The dominant bottom-up approach in current psycholinguistic theory is source (b), which has been thoroughly investigated by Cutler and colleagues amongst others. Her *metrical segmentation strategy* (MSS) holds that segmentation depends on the exploitation of prosodic rhythm. Speakers of a particular language are sensitive to its rhythmic pattern, and chunk the input accordingly, initiating lexical access at the beginning of each chunk. In English, the relevant rhythm is the alternation between strong and weak syllables, where weak syllables have /ə/ ("schwa") or another lax vowel as the nucleus, all others being strong. When a strong vowel is heard, a boundary is hypothesized at the beginning of the syllable of which the vowel is nucleus.

The role of information source (c) (phonotactics) has received scant attention in the psycholinguistic literature. Since our model is sensitive to this type of information, we can use it to quantify the role of phonotactics in bootstrapping segmentation – the rest of this section is devoted to this investigation.

Empirical motivation for examining the role of phonotactics in infant segmentation behaviour comes from experiments using the head-turning paradigm which seem to demonstrate a sensitivity to phonotactic structure in infants as young as 9 months (Jusczyk et al. 1993). If infants are sensitive to such information then it seems likely that they would also come to appreciate its power in predicting boundary location.

Network segmentation

It is generally the case that sequences of segments are more constrained within words than across word boundaries. Thus, the sequence /ŋð/ (as in "si*ng the*") is only licensed in English if there is a morpheme boundary between /ŋ/ and /ð/. However, phonotactics do not have to be absolute constraints; probabilistic structure is present too: thus the sequence /nd/ is very common word internally but much less common across word boundaries. Because phonotactic constraints hold within words, but not between, sequences of phonemes will tend to be much more predictable within words (the *entropy* across boundaries is higher, in information theoretic terms). In our network model, if prediction is hard, then error when predicting will tend to be high. Thus, to model segmentation, boundaries are proposed at prediction error peaks.

It is worthwhile emphasizing two salient characteristics of our input corpus:

(a) There is *no* explicit marking of word boundaries. All rules for co-articulation apply equally inter- and intra-lexically.

(b) Between one in four and one in five words occur after a pause, so pausing is a good cue to segmentation. All pause markers were removed so as to factor out their predictive effect, and also to avoid instantiating pause as a phonemic category.

Because of these facts, we consider the data to represent a "worst case" for testing models of segmentation, in that if segmentation is possible with these data, then the inclusion of pauses and some phonetic/acoustic cues can only serve to improve its performance.

The model was tested by providing as input a noise-free 10 000 phoneme (about 2700 words) stretch of corpus, and measuring the *cross-entropy* error on the prediction subgroup of the output units. This yields a variable error signal in which we define a "peak" by placing a cut-off point at varying numbers of standard deviations above the mean. As stated above, high error will tend to indicate increasing likelihood of a juncture point because lack of phonotactic constraints across boundaries makes prediction of word-initial segments difficult. Perform-ance varies according to where the cut-off is placed, but at the cut-off that maxi-mizes the network's performance, 21% of the boundaries are correctly identified with a hits:false-alarms ratio of 1.5:1. A *false alarm* occurs when the model pos-its a boundary when no boundary is actually present. A *hit* is when the model successfully detects a real boundary. To decide when network segmentation was optimal, we measured the *mutual information* for each cut-off, and maximized this value. This allows our estimation of network performance to be independent of any assumptions about the relative desirability/undesirability of hits/false alarms.

In the following sections we evaluate the significance of these results by com-parison with a random segmentation algorithm which was averaged over five different runs. This algorithm was designed to yield a distribution of pseudo-word lengths similar to that of the network. We consider this to be a more strin-gent test of the network's performance than comparison with a random segmentation algorithm that uses a uniform distribution.

Although network performance peaks with correct identification of about one in five boundaries in the test corpus, there is a sizable proportion of false alarms at this cut-off (i.e. cases in which the network predicts a boundary when in fact there is none). It may well be that although the false alarms do not actually correspond to existing boundaries in the test stretch, they are plausible guesses based on the low-level data that are the only information source available to the model. We tested this hypothesis by examining the phonotactic acceptability of the boundaries that the model postulates, defined by the legality of the sequence of segments over the postulated boundary. For example, the sequence /tp#ra/ is a phonotactically malformed boundary postulate, while /pt#ra/ is well formed. We found that false alarm boundaries of the network are much more likely to be phonotactically well formed than those of the random case (for the initial boundaries: $\chi^2_{(1)} = 221.8$, $p < 0.001$, while for the final boundaries $\chi^2_{(1)} = 119.1$, $p < 0.001$).

In summary, phonotactics provide a fairly weak source of information for the bootstrapping of segmentation, but the cumulative effect of such information may well be useful in the initial phases of compiling a lexicon.

Network segmentation and the MSS

In this section we provide a qualitative analysis of network segmentation and present the surprising result that there is a statistical basis for the emergence of the MSS in our purely bottom-up model.

We investigated the performance of the model by counting the instances in which a boundary is correctly postulated before a strong or weak syllable. As an operational definition of *strong* and *weak* we took the lax vowels /ə/, /I/ and /ʌ/ to be weak, and all other monothongs and dipthongs to be strong. Given this

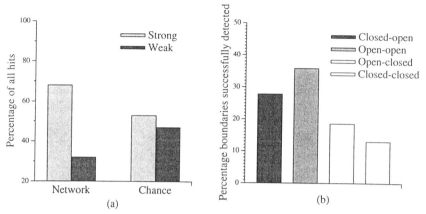

Figure 15.3 Network segmentation performance mimics the MSS. (a) The proportions of correctly inserted boundaries that are placed before strong, as opposed to weak, syllables. (b) The breakdown of boundary insertions for boundaries between open- and closed-class words.

criterion, in the 2700 word test set, 53% of the words are strong-initial. The network performance is proportionally skewed towards successful detection of strong-initial words to a striking degree (Fig. 15.3a, $\chi^2_{(1)} = 77.2, p < 0.001$).

A similar result was obtained when we changed the definition of *weak* to just /ə/ ($\chi^2_{(1)} = 70.4, p < 0.001$). A natural conclusion to draw is that the model is segmenting more before open-class words, and examination of the totals of hits before open-class as opposed to closed-class words shows that this is the case (the majority of open-class content words in English have strong initial syllables). The initial portions of open-class words are much more likely to be detected than beginnings of closed-class items (Fig. 15.3b, $\chi^2_{(1)} = 14.0$, $p < 0.001$). Note also that the boundaries with which the model has most difficulty are the closed–closed boundaries, thus strings of closed-class words such as "up to the" are less likely to be segmented than strings of open-class items.

When we consider the contiguous pairings of correct individual segmentations – the words that correctly emerge from the network – the same pattern is evident. A word count of the LLC revealed that 65% of all items were closed-class, so one would expect that this ratio would hold in correct network output, all other factors being equal. While the network does not segment more whole words from the test stretch than it would by chance (showing that the model does not develop a lexicon), of the correctly extracted tokens 41% are closed-class. This is significantly less than the random segmentation performance: $\chi^2_{(1)} = 19.46, p < 0.001$.

So, our network produces segmentations which broadly mimic the pattern predicted by the MSS, yet the net is not retrodictive in the way that the MSS is: crucially, the nuclear vowel of the initial syllable of a word is not visible to the network when it makes a segmentation decision (at least for syllables with onsets – recall that boundaries are inserted on the basis of the ease of prediction of the first segment in the word). Thus, this model does not need to posit that the strong/weak distinction has *a priori* perceptual salience for the infant. The reason why our network exhibits this pattern of results is simply that the initial segments of strong-initial, open-class words tend to be less predictable than those of closed-class words.

However, we would emphasize that phonotactic information is a relatively weak predictor, and that it is unlikely that a purely phonotactic approach could enable the infant to acquire a lexicon. Rather, we see phonotactic information as being a possible method for bootstrapping of the MSS, which is a more robust and reliable tool for lexical acquisition. This bootstrapping could be mediated by sensitivity to the correlation of the boundaries that phonotactics predict with metrical structure.

Adding categorial knowledge

The results from the previous section were obtained by segmenting with raw scores that were not normalized for phoneme type. This can be seen as simulating the phase of infant development in which phonemic categories, and information about their frequencies, are not available to the infant. However, we know that towards the end of the first year of life the child's phonological space is becoming structured in terms of phonemic categories (e.g. Kuhl 1983, Werker 1993). Therefore, we decided to mimic the effect of this phonemic restructuring in our model, to see if the qualitative pattern of segmentations would remain constant.

We carried out the same segmentation procedure as before, but this time normalizing the network error scores for phoneme type (i.e. dividing each prediction error score by the frequency of the segment being predicted). An intuitive analogy to this process in terms of infant development would be the diminishing of attention directed towards more frequent segments. We found an entirely different pattern of results with respect to strong syllables and word class than before: in general the network no longer mimicked the MSS. Segmentation before strong as opposed to weak syllables was not significantly different from chance: $c^2_{(1)} = 0.387$, not significant. Neither was segmentation before open- as opposed to closed-class items: $c^2_{(1)} = 0.035$, not significant. Furthermore, using phoneme-normalized scores, 78% of correctly extracted word tokens were closed-class, in contrast to the 41% with raw scores. This figure once again differs significantly from the expected distribution: $c^2_{(1)} = 8.07, p < 0.005$, except that now it is the closed-class items that are favoured, rather than the open-class items.

The intuitive explanation of why segmentation behaviour should change in this way when scores are normalized is that closed-class words, because they are most frequent in the language, also contain the most frequent phonemes. Therefore, the network will predict these phonemes more easily than ones which do not occur often in closed-class words. Because predicting these segments is easier, errors are lower. Hence, normalizing for phoneme type will augment the error scores for phonemes that most often occur in closed-class words, and effectively increase the probability of boundaries being proposed before such segments.

Discussion of the segmentation model

We have provided a computational underpinning to the claim that low-level phonotactics could be used by a neonate as a cue for initially breaking up the continuous stream of input speech. Moreover, we have given an account of how the MSS could arise without recourse to positing sensitivity to metrical information as part of a genetic endowment. The network segmentation performance was significantly biased in favour of detecting open-class words that have strong initial syllables. Furthermore, we have shown that our model's mirroring of the MSS disappears when we add knowledge about the frequencies of individual phoneme categories – detection of closed-class words becomes favoured.

We see the overall picture of the role of phonotactics that emerges from these results as follows. Initially, phonotactics could provide tentative segmentations from which the MSS could be induced in the pre-categorial infant. Once the MSS is in place, and the infant's phonological space comes to be structured with the phonemic categories of English, then the MSS would pick out the open-class words, while phonotactics could help in isolating the closed-class items. Also, we would not wish to rule out some form of lexical competition in segmentation once a lexicon is in place.

General summary and conclusions

In the last two sections we have shown how the information that is encoded in our simple network can be used to great effect in diverse psychological modelling: first in countering claims that the Elman & McClelland data are a genuine demonstration of higher-level cognitive influence on phoneme restoration; and secondly in demonstrating that phonotactics can be used to predict segmentation points with no prior lexical knowledge. Neither of these applications has taken the form of our constructing a system which is claimed to directly instantiate the processing of some part of the human cognitive system, as is the case in much psychological modelling work in which neural networks feature. Rather, our approach has been to attempt to abstract, and thus quantify, the utility of a particular type of stimulus information in certain cognitive tasks. The only way of being able to make such quantitative statements is to use input that is truly representative of the input that humans have at their disposal: hence our use of the retranscription of the LLC. Of course, even these input data are not pre-theoretical: our encoding embodies assumptions about phonemes and distinctive features. However, even an account which used the actual speech waveform would still be prey to theoretical bias in terms of how the input was sampled, discretized, quantized and so on.

Concomitant with the use of the corpus has been our ability to quantitatively investigate bottom-up modelling. This allowed us to show that the Elman & McClelland (1988) results were amenable to bottom-up explanation and as such were not unequivocal evidence of interactive processing. Bottom-up modelling can also help to clarify the utility of various information sources in bootstrap learning, as our second study of segmentation has shown. However, as pointed out in the introduction, exploring the utility of bottom-up modelling does not necessarily commit one to a non-interactive view of language processing, since interactive accounts can make use of bottom-up information too. Indeed, there is ample evidence that many parts of the language processor do allow interaction from higher levels. Our goal has been to investigate what can be done using purely bottom-up information, and therefore to clarify the parameters of any more complete model.

One disadvantage of our approach is that in restricting the model to one linguistic domain we are unable to investigate the interaction or integration of various information sources. In the segmentation model, for example, it would have been interesting to see how the addition of information about phonetic juncture clues, or lexical construction, could aid segmentation. A second problem with this approach is that we only posit our model as an abstract encoding of information, with the actual mechanism that we use not being pertinent to the model. Thus, we still need to provide some mechanism by which this information can (a) arise spontaneously in human learning, and (b) be encoded in sufficiently general a form as to allow variant input to match against its codes. A possible answer to (a) is the continual presence of noise in speech stimuli: phonotactics could be used as a cheap and fast method for resolving local phonetic ambiguity. Answering (b) is more problematic, and has been the central thrust of much psycholinguistic work (see Marslen-Wilson & Gaskell 1992).

Notes

1. Note that this usage of phonotactics is different from the standard usage from phonology, which refers to absolute well formedness constraints on surface structures.
2. The training took approximately 7 days CPU time on a Sun SPARCstation 10.

Acknowledgements

Thanks are due to Steve Finch (personal communication) for CHILDES statistics. This work was supported by the UK Economic and Social Research Council (ESRC). Grant number no. R000 23 3649.

References

Cairns, P., R. Shillcock, N. Chater, J. Levy 1994. Bootstrapping word boundaries: a bottom-up corpus-based approach to speech segmentation. *Cognitive Psychology*, submitted.

Chater, N. & P. Conkey 1992. Finding linguistic structure with recurrent neural networks. In *14th Annual Conference of the Cognitive Science Society, Proceedings*, 402–7. Hillsdale, New Jersey: Lawrence Erlbaum.

Chater, N. & P. Conkey 1994. Sequence processing with recurrent neural networks. In *Neurodynamics and psychology*, G. D. A. Brown & M. Oaksford (eds), 269–94. London: Academic Press.

Chater, N., R. Shillcock, P. Cairns, J. Levy 1994. Bottom-up explanation of phoneme restoration. *Journal of Memory and Language*, submitted.

Cleeremans, A., D. Servan-Schreiber, J. L. McClelland 1989. Finite state automata and simple recurrent networks. *Neural Computation* **1**, 372–81.

Cutler, A. 1993. Phonological cues to open- and closed-class words in the processing of spoken sentences. *Journal of Psycholinguistic Research* **22**(2), 109–31.

Cutler, A. & S. Butterfield 1992. Rhythmic cues to speech segmentation – evidence from juncture misperception. *Journal of Memory and Language* **31**(2), 218–36.

Cutler, A. & D. Norris 1988. The role of strong syllables in segmentation for lexical access. *Journal of Experimental Psychology: Human Perception and Performance* **14**, 113–21.

Draper, N. R. & H. Smith 1981. *Applied regression analysis*, 2nd edn. Chichester: John Wiley.

Elman, J. L. 1990. Finding structure in time. *Cognitive Science* **14**, 179–211.

Elman, J. L. 1991. Distributed representations, simple recurrent networks and grammatical structure. *Machine Learning* **7**, 195–225.

Elman, J. L. & J. L. McClelland 1988. Cognitive penetration of the mechanisms of perception: compensation for coarticulation of lexically restored phonemes. *Journal of Memory and Language* **27**, 143–65.

Fodor, J. A. 1983. *Modularity of mind*. Cambridge, Mass.: MIT Press.

Foss, D. J. & M. A. Gernsbacher 1983. Cracking the dual code: toward a unitary model of phoneme identification. *Journal of Verbal Learning and Verbal Behavior* **22**, 609–32.

Ganong, W. F. 1980. Phonetic categorization in auditory word perception. *Journal of Experimental Psychology: Human Perception and Performance* **6**, 110–25.

Grosjean, F. & J. P. Gee 1987. Prosodic structure and spoken word recognition. *Cognition* **25**, 135–56.

Harrington, J., G. Watson, M. Cooper 1988. Word boundary identification from phoneme sequence constraints in automatic continuous speech recognition. In *12th International Conference on Computational Linguistics, Proceedings*, 225–30.

Harris, J. & G. Lindsey 1995. The elements of phonological representation. In *New frontiers in phonology*, J. Durand & F. Katamba (eds). Harlow, England: Longman.

Hinton, G. E. 1989. Connectionist learning procedures. *Artificial Intelligence* **40**, 185–234.

Huang, X. D., Y. Ariki, M. A. Jack 1990. *Hidden Markov models for speech recognition*. Edinburgh: Edinburgh University Press.

Jakimik, J. 1979. Word recognition and the lexicon. In *Speech communication papers presented at the 97th Meeting of the Acoustical Society of America*, J. Wolf & D. Klatt (eds). New York: Acoustical Society of America.

Jusczyk, P. W. 1993. How word recognition may evolve from infant speech perception capacities. In *Cognitive models of speech processing: the Second Sperlonga Meeting*, G. T. M. Altmann & R. Shillcock (eds), 27–55. Hillsdale, New Jersey: Lawrence Erlbaum.

Jusczyk, P. W., A. D. Friederici, J. M. I. Wessels, V. Y. Svenkerud, A. M. Jusczyk 1993. Infants' sensitivity to the sound patterns of native language words. *Journal of Memory and Language* **32**, 402–20.

Kaye, J. D., J. Lowenstramm, J. R. Vergnaud 1985. The internal structure of phonological elements: a theory of charm and government. *Phonology Yearbook* **2**, 305–28.

Kaye, J. D., J. Lowenstramm, J. R. Vergnaud 1990. Constituent structure and government in phonology. *Phonology* **7**(2), 193–231.

Kohonen, T. 1988. The "neural" phonetic typewriter. *Computer Journal* **21**(3), 11–22.

Kuhl, P. K. 1983. Perception of auditory equivalence classes for speech in early infancy. *Infant Behaviour and Development* **6**, 263–85.

Lehiste, I. 1971. The timing of utterances and linguistic boundaries. *Acoustical Society of America, Journal* **51**, 2018–24.

Linsker, R. 1988. Self-organization in a perceptual network. *Computer Journal* **21**(3), 105–17.

McClelland, J. L. & J. L. Elman 1986. Interactive processes in speech perception: The TRACE model. In *Parallel distributed processing: explorations in the microstructure of cognition*, vol. 2. *Psychological and biological models*, J. L. McClelland & D. E. Rumelhart (eds), 58–121. Cambridge, Mass.: MIT Press.

McQueen, J. M. 1994. Lexical or non-lexical effects in compensation for coarticulation. *Cognitive Psychology*, unpublished manuscript.

McQueen, J. M., D. Norris, A. Cutler 1993. Competition in spoken word recognition: spotting words in other words. *Journal of Experimental Psychology: Learning Memory and Cognition* **20**, 621–38.

MacWhinney, B. & C. Snow 1985. The child language data exchange system. *Journal of Child Language* **12**, 271–96.

Mann, V. A. & B. H. Repp 1981. Influence of preceding fricative on stop consonant perception. *Acoustical Society of America, Journal* **69**, 548–58.

Marr, D. 1982. *Vision: a computational investigation in the human representation of visual information*. San Francisco: Freeman.

Marslen-Wilson, W. & G. Gaskell 1992. Match and mismatch in lexical context. *International Journal of Psychology* **27**(3), 61.

Massaro, D. W. 1992. Connectionist models of speech perception. In *Connectionist approaches to natural language processing*, R. G. Reilly & N. E. Sharkey (eds), 351–71. Hillsdale, New Jersey: Lawrence Erlbaum.

Norris, D. 1992. Connectionism: a new breed of bottom-up model? In *Connectionist approaches to natural language processing*, R. G. Reilly & N. E. Sharkey (eds), 351–71. Hillsdale, New Jersey: Lawrence Erlbaum.

Norris, D. G. 1993. Bottom-up connectionist models of "interaction". In *Cognitive models of speech processing: the second Sperlonga meeting*, G. T. M. Altmann & R. Shillcock (eds), 211–34. Hillsdale, New Jersey: Lawrence Erlbaum.

Rumelhart, D. E., G. E. Hinton, R. J. Williams 1986. Learning internal representations by error propagation. In *Parallel distributed processing: explorations in the micro-*

structure of cognition, vol. 1. *Foundations*, D. E. Rumelhart & J. L. McClelland (eds), 318–62. Cambridge, Mass.: MIT Press.

St John, M. F. & J. L. McClelland 1990. Learning and applying contextual constraints in sentence comprehension. *Artificial Intelligence* **46**, 217–57.

Sejnowski, T. J. & C. R. Rosenberg 1987. NETtalk, a parallel network that learns to read aloud. *Complex Systems* **1**, 145–68.

Shillcock, R. C., J. Levy, N. Chater 1991. A connectionist model of word perception in continuous speech. *13th Annual Conference of the Cognitive Science Society, Proceedings*, 340–43. Hillsdale, New Jersey: Lawrence Erlbaum.

Shillcock, R., P. Cairns, J. Levy, N. Chater, G. Lindsey 1995. A statistical analysis of an idealized phonological transcription of the London–Lund corpus. Unpublished paper.

Suomi, K. 1993. An outline of a developmental model of adult phonological organization and behaviour. *Journal of Phonetics* **21**, 29–60.

Svartvik, J. & R. Quirk 1980. *A corpus of English conversation.* Lund: LiberLaromedel Lund.

Tanenhaus, M. K. & M. Lucas 1987. Context effects in lexical processing. In *Spoken word recognition*, U. Frauenfelder & L. Tyler (eds), 213–34. Cambridge, Mass.: MIT Press.

Warren, R. 1970. Perceptual restoration of missing speech sounds. *Science* **167**, 392–3.

Warren, R. M. & G. Sherman 1974. Phonemic restorations based on subsequent context. *Perception and Psychophysics* **16**, 150–56.

Werker, J. F. 1993. Developmental changes in cross-language speech perception: Implications for cognitive models of speech processing. In *Cognitive models of speech processing: the Sperlonga Meeting II*, G. T. M. Altmann & R. C. Shillcock (eds), 23–49. Hillsdale, New Jersey: Lawrence Erlbaum.

Williams, G. & W. Brockhaus 1992. Automatic speech recognition: a principle-based approach. In *Working papers in linguistics and phonetics 2*, A. Goksel & E. Parker (eds), 371–401. London: School of Oriental and African Studies.

CHAPTER 16

Interactive models of lexicalization: some constraints from speech error, picture naming, and neuropsychological data

Trevor A. Harley & Siobhan B. G. MacAndrew

Introduction

Lexicalization is the process in speech production whereby we retrieve the phonological forms of words from a semantic specification. Over the last few years we have developed a connectionist model of lexicalization (Harley 1984, Harley 1988, Harley 1990, Harley & MacAndrew 1992, Harley 1993). This chapter describes the model and shows how it can account for a range of psycholinguistic data. Not only must connectionist models inspire theoretical development, they should not ignore the detailed constraints imposed by human data.

Any model of lexicalization first needs to address two fundamental architectural issues, which can be summarized as how many levels of processing are there, and do they interact?

How many processing levels are there?

There is generally agreement in the speech production literature that lexicalization is a two-stage process (Fig. 16.1). The semantic representation does not access the phonological forms directly, but through an intermediate stage of abstract lexical items called lemmas (Levelt 1992). Data from a number of sources are used to support this claim. First, there are two types of whole word substitution errors found in normal speech (Fay & Cutler 1977): semantic (example (1) below) and phonological (example (2)).

(1) afternoon → morning
(2) clarification → classification

Second, the existence of different types of anomia is suggestive of access in two stages. Patients such as JBR (Warrington & Shallice 1984) and JCU (Howard & Orchard-Lisle 1984) show semantic anomia, while patients such as EST (Kay & Ellis 1987) display phonological anomia. While these data suggest that there

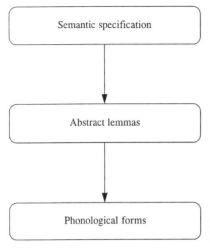

Figure 16.1 A basic two-stage model of lexicalization.

are distinct phases of semantic-to-lexical and lexical-to-phonological access, perhaps the most persuasive evidence comes from a picture-naming experiment by Levelt et al. (1991), which will be discussed in more detail below. Finally, data from experiments on the tip-of-the-tongue state may be neatly explained by this model (Astell & Harley 1993, Harley & Bown 1994). The claim that lexicalization is a two-stage process refers only to the postulated semantic–lexical–phonological form mappings. In addition, other levels of phonological representation may be needed. It might be necessary to distinguish between representations comprising coherent phonological forms (sometimes called the lexemes), discrete phonemes and phonological features (Butterworth 1980). The alternative to the two-stage model is that coherent phonological forms are accessed directly from semantic features.

Do the stages interact?

In particular, can phonological information influence the semantic activation and access of lemmas? Although there is some experimental evidence in support of an interaction between the phonological and lexical levels (see Martin et al. 1989), the primary evidence for interactions between levels comes from speech error analysis. Dell & Reich (1981) showed that there is a lexical bias in sound exchange errors such that sound errors result in word outcomes more often than chance predicts (e.g. example (3)). There are more word substitution errors with both semantic and phonological relationships between the target and intrusion than chance would predict if these factors were independent. These are called mixed errors (Shallice & McGill 1978), such as in example (4). Finally, Harley (1984) showed that high-level intrusion speech errors are more likely to occur if the target and intrusion sound similar, as in example (5). This phenomenon is known as phonological facilitation.

(3) bad cat → cad bat

(4) cushion → curtain

(5) When the telephone man was at work → television man
  (The speaker was looking at a television set at the time.)

A model of lexicalization

Our model is founded upon the constraints that lexicalization is an interactive process and that it takes place in two stages. The model is based upon the interactive activation and competition (IAC) architecture of McClelland & Rumelhart (1981) and Rumelhart & McClelland (1982), as this is well suited to model such constraints. The model also complements previous work and ideas of Dell (1986) and Stemberger (1985). In our model, units are organized into three levels of semantic features, lexical units and phonological units. (See Fig. 16.2 for the

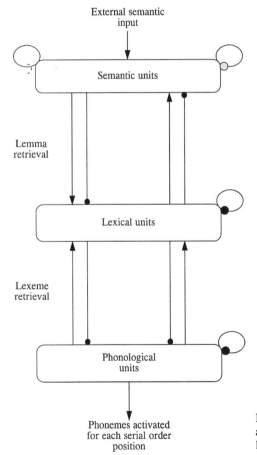

Figure 16.2 The detailed architecture of our IAA model of lexicalization.

detailed architecture of our model.) Each unit is connected to every unit in the following layer; appropriate between-level connections are facilitatory, and inappropriate connections inhibitory. There is within-level lexical and phonological level inhibition, and there are feedback connections between levels. In these simulations the weights of the lexical-to-semantic connections are set to zero; this is discussed further in the final section.

The spread of activation in the model is governed by the usual equations:

$$net_i = \sum_j a_j \cdot w_{ij} \tag{16.1}$$

where a_j is the output of all the units acting upon i, and w_{ij} is the weight of the connection between the j input units and i. In each processing cycle, Δa_i is given by

$$\Delta a_i = \begin{cases} (\mathrm{max} - a_i)net_i - decay(a_i - rest) \text{ if } net_i > 0 \\ (a_i - \mathrm{min})net_i - decay(a_i - rest) \text{ otherwise} \end{cases} \tag{16.2}$$

where max ($=1$) and min (set to -0.2) are the maximum and minimum levels of activation. The activation of units tends to decay to a resting level known as rest. The resting levels of the semantic and phonological units are always set to 0. The resting level of a lexical unit depends upon its frequency. In particular, it depends upon the difference of the natural logarithm of its frequency and the mean logarithm of the frequencies of all the items in the corpus, multiplied by a gain parameter *freqgain*, so that they vary around 0. Units have an activation threshold that they must pass, set to 0 for convenience. The rate at which units decay to their resting levels is governed by the decay parameters *semanticdecay*, *lexicaldecay* and *phonologicaldecay*. Processing cycles serve as a measure of time. In the model, there are 26 semantic features, 70 lexical units and 21 output phonological units. The phonological units are exported into five position-sensitive slots, but the lexical units carry over their activation state from phoneme to phoneme. The semantic level is structured: some units are connected to other units, although, unlike the other within-level connections, connectivity is not complete. Semantic clusters are defined with respect to features that are mutually exclusive; semantic units in such clusters are mutually inhibiting. Some features are connected by excitatory links (e.g. animal and mobile). Every semantic feature is connected to every lexical unit by either an appropriate excitatory or inhibitory connection. We have argued elsewhere (e.g. Harley 1995) that the number of semantic features that are switched "on" or activated corresponds to the imageability of a word. The more semantic features that are switched on, the more quickly activation is fed to the lexical units. This is similar to the account of imageability of Plaut & Shallice (1993). The values of the parameters and weights we use in our model are summarized in Table 16.1.

We do not view as a critical matter for this model the issue of exactly how the access of units occurs. We certainly want any access procedure to make errors,

Table 16.1 Standard values of the parameters and connection strengths in the model. *noofextsemcycles* is the number of processing cycles in which the semantic feature input units are activated. Connection strengths are shown as excitatory or inhibitory, from the originating to the destination levels, e.g. *excitpl* refers to the excitatory feedback connections between the phonological and lexical levels.

semanticdecay	0.20
lexicaldecay	0.20
phonologicaldecay	0.05
freqgain	0.01
noofextsemcycles	2
excitsl	0.06
excitlp	0.30
excitpl	0.15
excitls	0
inhibsl	−0.20
inhiblp	−0.05
inhibpl	0
inhibls	0
excitss	0.01
inhibss	−0.01
inhibll	−0.20
inhibpp	−0.05

and for those errors to reflect the constraints on naturally occurring errors. In the past we have considered four alternative methods of converting the output of the model into errors. First, errors can be inferred from inspection of the relative activation level of competing units at an appropriate time slice (Harley & MacAndrew 1992). Secondly, this can be formalized by converting activation levels into response probabilities. This procedure uses the standard method of the exponential transformation of activation levels in combination with Luce's choice rule (Harley 1993). Thirdly, errors can be caused by adding random noise to the resting levels of units. The limited temporal effect of this makes it unsuitable for generating errors deep in the selection process (Harley & MacAndrew 1992). Finally, errors can be caused by adding noise to connections so that competitors actually become more activated than the targets (Harley & MacAndrew 1992). Eventually we expect the consideration of exactly how errors are generated by any narrow model to be irrelevant within the wider context of a more comprehensive model that maps phonological forms onto ordered phonetic representations.

Three issues confront an interactive model: explaining details of word substitution speech errors, modelling the time-course of picture naming and accounting for relevant neuropsychological data.

315

Issue 1: the challenge of the speech error data

We have seen that a primary motivation for the model is the interactive nature of spontaneously occurring slips of the tongue. In this chapter, we want to examine word substitution errors, or paraphasias, more closely to see if some of their other properties can be assimilated into the model. In particular, we focus upon the relationship between target and intrusion.

The data come from our corpus of word substitution speech errors. There are 1330 word substitutions in 6552 errors in the corpus. We analyzed target–error pairs that met the following criteria:

(a) only completed word substitution errors were considered;
(b) both items involved in the error had to be generally accepted words;
(c) only content words were used in the analysis, as there is some evidence that content and function words form distinct computational vocabularies (Garrett 1980);
(d) errors involving proper nouns were not included;
(e) errors that were derivational, inflectional or involved antonyms produced by affixation were excluded;
(f) compound words were excluded.

This leaves 988 substitutions.

Semantic relatedness scores were obtained from a first year psychology practical class on 865 target–intrusion error pairs and 62 control word pairs. The control pairs were generated by randomly pairing non-target words from other speech error categories in the corpus, matching for proportion of syntactic category. The mean semantic relatedness of these control pairs was 1.31, on a scale of 1 (no relationship) to 4 (very strong semantic relationship). In addition, there were 471 sound error control items to act as controls for specific word substitution comparisons. These were randomly selected targets from sound errors in our speech error corpus.

Semantic errors show a variety of clear semantic relationships between the target (the word intended by the speaker) and the error or intrusion (examples (6)–(9)).

(6) cold → warm (antonym)
(7) toads → turtles (co-ordinates from the same semantic field)
(8) jape → jest (near synonym)
(9) boat → beach (associate)

There are 755 paraphasias that are clearly semantic, confirmed by their very high mean semantic relatedness score (on those pairs with a rating) of 3.23.

Phonological errors show phonological similarity between the two words (sharing the same initial phoneme, or later segment overlap), but no discernible semantic relationship (e.g. examples (10) and (11)).

(10) hospital → holiday
(11) sophisticated → sufficient

There are 233 such errors in the corpus, with a mean semantic relatedness rating of 1.40.

How does the model account for the existence of the two types of word substitution error? Semantic errors could arise from three possible sources. First, there could be an error in activating the precise semantic features used to represent a word. Secondly, they could arise from a transmission failure between the semantic and lexical units so that a semantic relative of the target is activated instead of the target. As both of these involve lexical mechanisms we refer to these hypotheses as intra-lexical origins. Thirdly, they could arise from a failure in generating the semantic representation. We refer to this possibility as an extra-lexical origin. Phonological errors arise when phonological forms corresponding to non-target lemmas receive more activation than the target lemmas, through mechanisms such as noisy transmission between the lexical and phonological levels. However, the presence of feedback connections between phonological and lexical units muddies this neat distinction between transmission failures at two levels in two ways. First, simple sound errors tend towards word outcomes (lexical bias). Secondly, semantic errors are more likely to occur if the target and potential intrusion are phonologically similar (leading to mixed errors). One consequence of this is that we view phonological word substitutions as lying on continua with pure semantic word substitutions on the one hand and pure sound errors on the other. The overall preponderance of semantic over phonological word substitutions (a ratio of 3.24:1) is predicted from simulations using the model by Harley (1993).

Imageability

Harley & MacAndrew (1992) predicted from their simulations that imageability should affect the robustness of lemmas to disruption in aphasic word substitution errors. To test this prediction, we took imageability ratings from the *Oxford Psycholinguistic Database* (Quinlan 1992). As not all items have imageability ratings, we were restricted to 408 semantic paraphasia target–error pairs, and 70 phonological paraphasia pairs. Our findings are shown in Table 16.2.

The difference between the imageability of the target and intrusion in the semantic errors is significant ($t[407] = 2.80, p < 0.01$). Hence, as predicted by Harley & MacAndrew (1992), the target is substituted by a more imageable word. In contrast there is no significant difference in imageability between the target and intrusion in the phonological substitutions (although the sample size is considerably smaller). There is a significant difference in imageability between the targets of the semantic errors and the 302 sound error controls that had imageability ratings ($t[708] = 3.77, p < 0.001$), but not between the phonologi-

Table 16.2 Imageability of target and intrusion.

	Target	Intrusion
Semantic errors	4.98	5.06
Phonological errors	4.56	4.61
Sound error controls	4.72	–

cal targets and sound error controls. It follows that the semantic substitution targets were also higher in imageability than the phonological targets ($t[476] = 3.58$, $p < 0.001$). In summary, semantic substitutions tend to occur on high-imageability words, and result in them being replaced by items even higher in imageability. Phonological errors appear to be unaffected by imageability, suggesting they originate at a different processing level from the semantic substitutions.

Frequency

Harley & MacAndrew (1992) also predicted from their simulations that low-frequency words should be less robust to the effects of disruption than high-frequency items. To test this, we took frequencies from the Francis & Kuçera (1982) norms. Our results are shown in Table 16.3.

There was a significant difference between the frequency of the semantic paraphasia targets and controls ($t[1224] = 3.30$, $p < 0.001$), and between the frequency of the phonological paraphasia targets and controls ($t[702] = 4.96$, $p < 0.001$). Hence, as simulations of the model predicted, paraphasia targets are significantly less frequent than control items. There was a significant difference between the semantic and the phonological substitution target words ($t[986] = 2.40$, $p < 0.02$), with the phonological targets even lower in frequency than semantic targets. Our simulations show that there is reason to suppose that although frequency might affect the initial susceptibility of items to replacement, it should have little subsequent effect. That is, intrusions should not be of higher frequency than targets. Indeed, in the human data we find no significant difference in frequency between the target and intrusion in word substitutions. In both cases target and intrusion frequencies are correlated: for the semantic errors $r_p = +0.38$, $N = 755$, $p < 0.001$; for the phonological errors, $r_p = +0.40$, $N = 233$, $p < 0.001$. The targets of both semantic and, particularly, phonological substitutions tend to be of relatively low frequency, yet all items are substituted by words of similar frequency.

Phonological facilitation

Phonological facilitation means that errors are more likely to occur when the words involved sound similar. To simplify this, our criterion for establishing this is that they must share the same first syllable–initial consonant (e.g. /p/ in "power"). Harley (1984) established empirically that the maximum chance

Table 16.3   Frequency of targets and intrusions.

	Target	Intrusion
Semantic errors	163.4	150.6
Phonological errors	108.2	101.0
Sound error controls	226.4	–

probability of the target and error words having this same first syllable–initial consonant is 0.122. Because they contain feedback connections, only interactive models predict phonological facilitation. Of the semantic paraphasias, 29.7% (224 out of 755) were phonologically facilitated ($\chi^2[1, N = 755] = 188.8$, $p < 0.001$), indicative of strong phonological facilitation. Hardly surprisingly, phonological substitutions display massive facilitation, with 82.4% of items related (192 out of 233; $\chi^2[1, N = 233] = 941.3, p < 0.001$).

Semantic relationship effects

Coltheart (1980) distinguished between shared-feature and associative errors. For example, cases (6), (7) and (8) above are clearly shared-feature errors. Their presumed semantic representations show a great deal of overlap, differing in only a few critical features. Associative errors have a much looser relationship between error and target word, such as example (9) above.

Of the errors, 81.5% are shared-feature substitutions, and 18.5% are associative substitutions. There is only one substitution in the corpus from the superordinate and subordinate types. The rarity of these types of errors was also noted by Hotopf (1980). In dyslexic reading errors a significantly higher proportion of errors are associative.

Antonyms (words opposite in meaning) display minimal contrast. Paradoxically, two words that are exact opposites in meaning are in fact closely related, as they differ by only one semantic feature. This suggests that there should be a large number of antonyms in our corpus. Indeed, there are 128 examples, and this may well be an underestimate because it is sometimes difficult to assign word pairs to precise categories.

There is a significant difference between the imageability of the shared-feature (mean = 5.01) and associative (mean = 4.69) targets ($t[406] = 2.44$, $p < 0.02$). We argue that given the greater degree of semantic overlap, shared-feature errors are conceptually closer to the intended target than associative errors. In fact, all the difference between the imageabilities of the semantic targets versus controls is located in the shared-feature errors; there is no difference in imageability between the associative targets (mean = 4.69, with 49 analyzable items) and the sound error controls. We hypothesize that associative errors do not share a lexical origin with the shared-feature errors. We propose that associative errors reflect a greater disruption of performance. In terms of our earlier distinction, we propose that shared-feature errors have an intra-lexical origin, while associative errors have an extra-lexical origin.

At this early stage we believe that these results should be treated with some caution. The distinction between shared-feature and associative errors is not always clear-cut, and our assignment of error pairs to these categories is necessarily somewhat subjective. It may be that some errors we have classified as associative just have a relatively small amount of featural overlap with the target.

Syntactic category effect

In 99.6% of semantic paraphasias, the intruding item is from the same major syntactic category as the target. The effect in phonological paraphasias is less marked, with 92.3% of target–error pairs from the same category. The difference between the semantic and phonological paraphasias ($\chi^2[1, N = 988] = 46.0, p < 0.001$) is significant. Although our model does not address word ordering and syntactic class, the general nature of interactive models suggests that multiple constraints should operate upon errors. This is consistent with these data.

Word length effects

Longer words are phonologically more complex. Hence we expect more phonologically complex strings to be those that are more prone to replacement by phonologically simpler strings. We therefore predict that the targets in phonological word substitutions would tend to be longer words. This is what we find. Phonological targets are significantly longer than both the semantic targets ($t[986] = 6.77, p < 0.001$) and the sound error controls (mean length = 1.56). There was no word length effect in the semantic word substitutions (mean syllable length of the target = 1.91; intrusion =1.89), nor for the phonological paraphasias (mean target length = 2.43, mean intrusion length = 2.45). The difference in performance of semantic and phonological substitutions is further evidence that they arise at different processing levels.

Discussion

In summary, we have shown that semantic and phonological word substitutions are subject to different constraints. The targets in semantic errors tend to be low-frequency, high-imageability items that are replaced by words of about the same frequency but even higher in imageability. There appear to be two types of semantic error, one based on featural overlap where imageability is important; another based on semantic association where imageability is not so important. Imageability is not a concern in phonological substitutions. These errors tend to occur on particularly low-frequency, phonologically complex strings. Phonological error targets are not replaced by words higher in frequency. How can we account for this pattern of results?

First, we argue that the target words involved in both semantic and phonological paraphasias are aberrant in terms of properties such as imageability, frequency and length, such that these items are particularly prone to disruption. We argue that substitution errors arise because of transient processing difficulties within the lexicon. One exception to this is associative semantic errors, for which there is some evidence for an extra-lexical origin. We also conclude that semantic and phonological substitutions arise because of noise or disruption at different points in the retrieval process. This is supported by many differing constraints upon the error types. Imageability effects suggest that semantic errors occur between the semantic and lexical levels – that is, as a result of difficulty in lemma retrieval. High-imageability words usually have a large number of close semantic neighbours. If they are also low in frequency they will have a low resting level, and will tend to get replaced by more highly activated competing items; these correspond to words that are even more imageable. Phonological complexity effects support the notion that phonological errors arise between the lexical and phonological levels; that is, they are a result of a difficulty in lexeme retrieval. Long, low-frequency items are particularly susceptible to replacement by competitors.

Why should the target in semantic paraphasias be aberrant in imageability and frequency? We propose an explanation in terms of semantic fields. We propose that these items are those that form clusters in semantic space, and have a greater number of neighbours. Low-imageability items are more isolated within a semantic field, and therefore have less opportunity for substitution. This hypothesis is supported by the finding that only shared-feature errors display the imageability effect. Items with close semantic neighbours tend to be highly concrete and low in frequency. Hence, neighbourhood effects may be as important in speech production as they are in speech recognition (e.g. see Bard & Shillcock 1993). For further detail, see Harley & MacAndrew (1994).

Although our model provides mechanisms whereby errors can occur and can accommodate the general constraints on them, these findings are well beyond the current scope of the model. Such neighbourhood effects are only likely to be found in simulations that use a much a larger lexicon.

There are two remaining concerns. First, there is at first sight tension between the claims that semantic and phonological substitutions (or, indeed, featural and associative errors) may lie on a continuum yet arise at different processing stages. The continuum arises because of the presence of feedback connections in our model. Hence "proto-errors" arising at the phonological level are more likely to be translated into fully fledged errors because of these connections. The second concern is that the effect of frequency on phonological substitutions is not what we might expect given the dominance otherwise of frequency in the psycholinguistic literature. The result that low-frequency, phonologically complex targets in phonological substitutions are not replaced by higher-frequency, phonologically simpler intrusions is at first sight most surprising. However, Figure 16.5 below shows that in our model, frequency only affects the initial access of items and not their subsequent time-course, and we argue that phonological substitutions arise relatively late in lexicalization.

Issue 2: the challenge of the time-course of lexicalization

Are the two processing stages of lemma access and phonological form access-independent, or do they overlap? As we have seen, the speech error data strongly

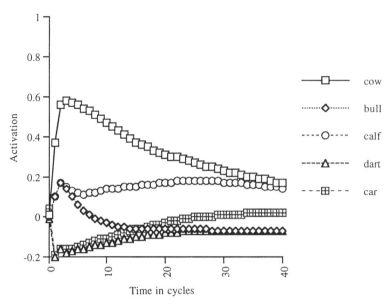

Figure 16.3 The time-course of lexicalization, for the target item lemma (cow), with those of semantic ("bull"), phonological ("car") and mixed ("calf") competitors, with an unrelated item ("dart"). The target is the most activated item, and the unrelated item the least. The high activation level of the mixed competitor is consistent with their frequent occurrence in speech errors.

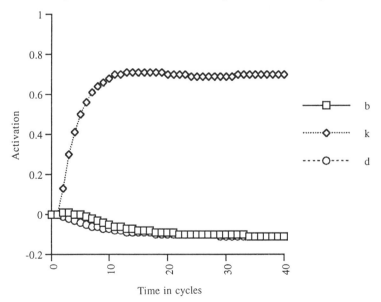

Figure 16.4 The time-course of lexicalization, for the phonological forms (represented as initial phonemes) of the items in Figure 16.3. Only the target form (denoted by K) is highly activated, consistent with the picture naming data of Levelt et al. 1991).

suggest that they are interactive and therefore overlap. Recently there has been a challenge to this position by Levelt et al. (1991), who have argued that important though speech error data may be, models of normal lexicalization must be primarily motivated by data on the normal time-course. Using a cross-modal priming design combining auditory lexical decision with picture naming, they found evidence of an early phase of semantic activation, and a late phase of phonological activation, with very little overlap between the two. There was no mediated priming (such as "goal" priming "goat" phonologically, which then primes "sheep" semantically). However, Harley (1993) showed that it is not difficult to find a range of parameters in our model such that this time-course is displayed. Figure 16.3 shows the time-course of lexicalization for lemmas, and Figure 16.4 for phonological units. Hence our model can account for the time-course of lexicalization (see Harley (1993) for details).

Using these parameters, Harley (1993) shows that given a word substitution error rate of 50/170 000 words and a semantic error rate of 732/937 (from our actual error corpus), we predict a total semantic + mixed substitution error rate of 0.00023. The model gives a value of 0.00025. It does not, however, give quite enough phonological errors. This calculation assumes that both types of error occur in the same time slice, which here is relatively early (= 10). We hypothesize, therefore, that semantic and phonological errors occur at different times. We therefore argue that semantic errors arise relatively early in the lexicalization process, while phonological errors occur late. This would increase the phonological error rate as appropriate. This move also fits in with the Levelt et al. account, and also with our discussion of these error types above.

Issue 3: the challenge from neuropsychology

Connectionist models of normal processing ought to provide an account of abnormal processing. Indeed, the simulated lesioning of connectionist models has led to the emergence of a new field, that of connectionist neuropsychology (Plaut & Shallice 1993). Does our model have anything to say about the neuropsychology of speech production? Harley & MacAndrew (1992) argued that aphasic paraphasias might arise as a consequence of random noise on the semantic-to-lemma connections. Data challenging interactive models of speech production have recently been presented by Nickels (1995). She examined some variables that are predicted to affect the production of different types of error in aphasic picture naming. Her predictions of how connectionist models should behave, and her actual findings, are shown in Table 16.4.

The basis of these predictions is that "the effects of a lesion will spread to affect the network as a whole as activation reverberates between levels". The important finding for our purposes was that she found no imageability effect upon phonological errors. She concluded that stage models provide a better fit to the data than connectionist models. The problem with this conclusion is that it is

Table 16.4 Predictions and findings for discrete stage and connectionist models from Nickels (1995).

	Frequency	Imageability
Stage models		
Semantic errors	Yes	Yes
Phonological errors	No	Yes
Connectionist models		
Semantic errors	Yes	Yes
Phonological errors	Yes	Yes
Actually found		
Semantic errors	Yes	No
Phonological errors	No	Yes

difficult to predict the behaviour of connectionist models without simulation. In particular, Harley (1995) showed that, in principle, just because a model is interactive, it need not follow that variables located at one level (e.g. imageability at the semantic level) must have effects at all levels.

Let us now explore the early and late effects of frequency and imageability in our simulations, within the context of lexical neighbourhoods, in more detail. Figures 16.5–16.10 show the time-courses of different types of item.

First, as we have mentioned above, word frequency has little effect on the time-course when imageability is held constant (Figure 16.5). It merely affects the initial accessibility of items. The speech error and neuropsychological data suggest that frequency is best not instantiated in differences in the resting levels of items. For further discussion, see Harley (1995) and Nickels (1995).

Figure 16.6 shows the effects of imageability upon the time-course of lexicalization of lemmas, with frequency held constant. It has marked early effects, but relatively little late effect. Hence, we would expect imageability to affect errors arising early (hypothesized by us to be semantic substitutions), but to have little effect on errors arising late (hypothesized by us to be phonological substitutions). This fits well with our data. It is also the case that low-imageability items have few semantic neighbours in the simulation, which we assume reflects the properties of the set of words in the language, given that our materials were more or less randomly sampled.

For comparison, the time-courses of high-imageability (Fig. 16.7), low-imageability (Fig. 16.8), high-frequency (Fig. 16.9) and low-frequency (Fig. 16.10) targets are shown with competitors. The precise effects are moderated by their lexical neighbourhoods.

In conclusion, we can say that, at present, connectionist models are not necessarily inconsistent with the neuropsychological data. Further explicit modelling of the data by lesioning connectionist models is necessary, however, before any further claims can be made.

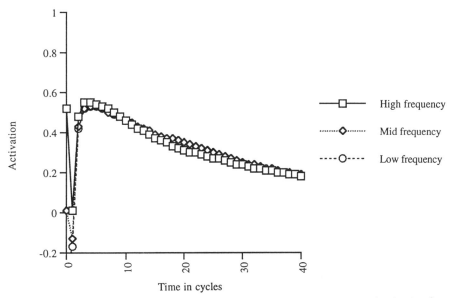

Figure 16.5 The effects of frequency (instantiated as variation around resting levels of lemmas) upon the time-course of lexicalization. The items shown are high frequency (HIF) "man" 3293, mid-frequency (MIF) "mouse" 20 and low frequency (LIF) "mum" 1. All items have eight "on" semantic features. The *freqgain* parameter is amplified to 0.2. Error data show that frequency does not affect which item substitutes for the target.

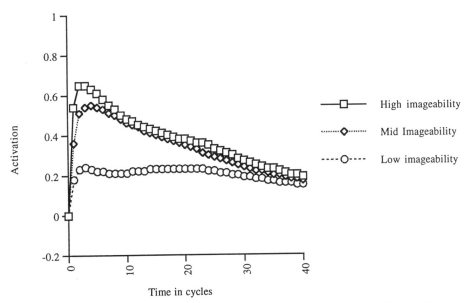

Figure 16.6 The effects of imageability (instantiated as number of activated semantic features) upon the time-course of lexicalization. The items shown are of high-imageability (HII) "bat" (9 "on" semantic features), medium imageability (MIDI) "beer" (6 "on" semantic features) and low imageability (LOI) "bush" (3 "on" semantic features). *freqgain* = 0. Targets are therefore prone to substitution by high imageability competitors, as indicated by the error data.

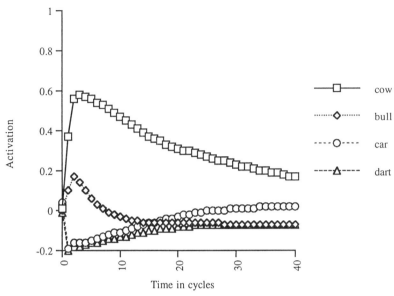

Figure 16.7   The time-course of activation of a high-imageability target, "cow" (high imageability = 9 "on" features), semantic competitor, "bull" (8), phonological competitor, "car" (5) and an unrelated item "dart" (5).

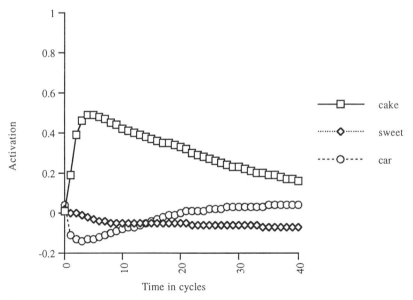

Figure 16.8   For comparison, the time-course of activation of a low-imageability target, "cake" (5 "on" features), semantic competitor, "sweet" (5), and phonological competitor, "car" (5). The principal effect of imageability is early in the time-course and should have greatest effects on errors arising early (semantic errors).

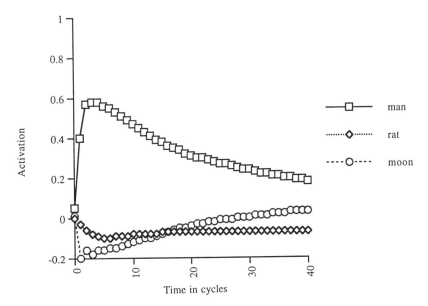

Figure 16.9 The time-course of activation of a high-frequency target, "man" (frequency = 3293, imageability = 8), semantic competitor, "rat" (frequency = 16, imageability = 9) and phonological competitor, "moon" (frequency = 63, imageability = 4).

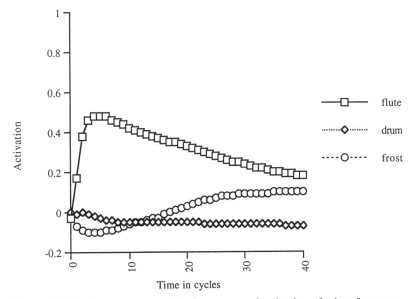

Figure 16.10 For comparison, the time-course of activation of a low-frequency target, "flute" (frequency = 15, imageability = 5), semantic competitor, "drum" (frequency = 26, imageability = 6) and phonological competitor, "frost" (frequency = 8, imageability = 4).

Conclusions

We observed that different constraints operate upon semantic and phonological errors. Hence we propose that semantic errors occur early, while phonological errors occur late. This hypothesis also fits the time-course data. We have argued that there are two types of semantic error, one reflecting intra-lexical featural overlap, the other extra-lexical associations. We argue that lexical neighbourhoods are as important in speech production as they are in speech comprehension, and that much larger simulations will be necessary to accommodate these findings.

A number of other difficulties merit attention. First, we have not addressed the problem of how phonemes become serially ordered within the phonological representation, although there are mechanisms in existence that could accommodate this (e.g. Houghton 1990). Secondly, how are frequency effects to be encoded? The current model provides an adequate account of aspects of the speech error data, but further data from aphasic speakers are problematic. These data suggest that the resting levels of lemmas may not be the ideal location (Harley 1995, Nickels 1995). If this proves to be the case, frequency will have to be instantiated in some other way – candidate solutions include instantiation either in the resting levels of phonological forms, or in connection strengths. A third related issue is the exact way in which processing levels beneath that of phonological forms should be decomposed. For example, locating frequency at the phonological level presupposes some coherence of phonological units into word forms there.

Fourthly, it is worth considering whether imageability is still an appropriate concept to use in this context. It is at least curious to refer to the "imageability" of pictures. Further, the notion that the number of active semantic features is either a direct consequence of imageability or vice versa is rather implausible. Clearly these effects are mediated by other variables: concreteness is one possibility, ease of predication (Jones 1985) another. In all likelihood it is a complex phenomenon. Harley (1995) proposes that the richness of the semantic representation is close to being the underlying explanatory factor for all these variables, and that these variables measure different aspects of it depending upon how it is accessed. Doubtless there is more to semantic richness than the number of semantic features; a simple measure to provide more complexity is that activation levels might well be continuous rather than binary. The current formulation of binary semantic features is, while a working approximation, too simple.

Finally, the presence of inhibition in the model needs re-examination, given that one of its major motivations – inhibition in the tip-of-the-tongue state – is less secure (Meyer & Bock 1992, Perfect & Hanley 1992). It is notoriously difficult, however, to explore the consequences of inhibitory connections (Bard & Shillcock 1993). Indeed, our own simulations suggest that they are unlikely to play a role in the origin of aphasic errors. Finally, although there are active feedback connections between the phonological and lexical units, in this version of the model we set the feedback connections between the lexical and semantic units to zero. This omission is motivated by economy, given that there is a lack of empirical support for such feedback (Levelt 1989). Recent ideas (e.g. Dell & O'Seaghdha 1991) suggest that feedback is motivated by an underlying notion that it embodies processes that are bidirectional. In this case, this means that these networks enable both word production and word recognition within the same system. Given this, we would clearly expect such connections to exist if this model were correct. The search for evidence for these connections therefore merits attention.

References

Astell, A. J. & T. A. Harley 1993. Tip of the tongue in dementing speech. *15th Annual Conference of the Cognitive Science Society, Proceedings*, 209–14. Hillsdale, New Jersey: Lawrence Erlbaum.

Bard, E. G. & R. C. Shillcock 1993. Competitor effects during lexical access: chasing Zipf's tail. In *Cognitive models of speech processing*, G. T. M. Altmann & R. Shillcock (eds), 235–75. Hove: Lawrence Erlbaum.

Butterworth, B. L. 1980. Some constraints on models of language production. In *Language production*, vol. 1. *Speech and talk*, B. L. Butterworth (ed.), 423–59. London: Academic Press.

Coltheart, M. 1980. The semantic error: types and theories. In *Deep dyslexia*, M. Coltheart, K. Patterson, J. C. Marshall (eds), 146–60. London: Routledge & Kegan Paul.

Dell, G. S. 1986. A spreading activation theory of retrieval in speech production. *Psychological Review* **93**, 283–321.

Dell, G. S. & P. G. O'Seaghdha 1991. Mediated and convergent lexical priming in language production: a comment on Levelt et al. 1991. *Psychological Review* **98**, 604–14.

Dell, G. S. & P. A. Reich 1981. Stages in sentence production: an analysis of speech error data. *Journal of Verbal Learning and Verbal Behavior* **20**, 611–29.

Fay, D. & A. Cutler 1977. Malapropisms and the structure of the mental lexicon. *Linguistic Inquiry* **8**, 505–20.

Francis, W. N. & H. Kuçera 1982. *Frequency analysis of English usage*. Boston: Houghton Mifflin.

Garret, M. F. 1980. Levels of processing in sentence production. In *Language Production*, vol. 1. B. L. Butterworth (ed.), ch. 8, 177–220. London: Academic Press.

Harley, T. A. 1984. A critique of top-down independent levels models of speech production: evidence from non-plan-internal speech production. *Cognitive Science* **8**, 191–219.

Harley, T. A. 1988. Automatic and executive processing in semantic and syntactic planning: a dual process model of speech production. In *Advances in natural language generation: an interdisciplinary perspective*, vol.1, M. Zock & G. Sabah (eds), 160–72. London: Pinter.

Harley, T. A. 1990. Paragrammatisms: syntactic disturbance or failure of control? *Cognition* **34**, 85–91.

Harley, T. A. 1993. Phonological activation of semantic competitors during lexical access in speech production. *Language and Cognitive Processes* **8**, 291–309.

Harley, T. A. 1995. Connectionist models of aphasia: a comment on Nickels. *Language and Cognitive Processes* **10**, 47–58.

Harley, T. A. & H. E. Bown 1994. What causes tip-of-the-tongue states? Submitted for publication.

Harley, T. A. & S. B. G. MacAndrew 1992. Modelling paraphasias in normal and aphasic speech. *14th Annual Conference of the Cognitive Science Society, Proceedings,* 378–83. Hillsdale, New Jersey: Lawrence Erlbaum.

Harley, T. A. & S. B. G. MacAndrew 1994. Determinants of normal word substitutions. Unpublished paper.

Hotopf, W. H. N. 1980. Semantic similarity as a factor in whole-word slips of the tongue. In *Errors in linguistic performance*, V. Fromkin (ed.), 97–109. New York: Academic Press.

Houghton, G. 1990. The problem of serial order: a neural network model of sequence learning and recall. In *Current research in natural language generation*, R. Dale, C. Mellish, M. Zock (eds), 287–319. London: Academic Press.

Howard, D. & V. Orchard-Lisle 1984. On the origin of semantic errors in naming: evidence from the case of a global aphasic. *Cognitive Neuropsychology* **1**, 163–90.

Jones, G. V. 1985. Deep dyslexia, imageability, and ease of predication. *Brain and Language* **24**, 1–19.

Kay, J. & A. W. Ellis 1987. A cognitive neuropsychological case study of anomia; implications for psychological models of word retrieval. *Brain* **110**, 613–29.

Levelt, W. J. M. 1989. *Speaking: from intention to articulation.* Cambridge, Mass.: MIT Press.

Levelt, W. J. M. 1992. Accessing words in speech production: stages, processes and representations. *Cognition* **42**, 1–22.

Levelt, W. J. M., H. Schriefers, D. Vorberg, A. S. Meyer, T. Pechmann, J. Havinga 1991. The time course of lexical access in speech production: a study of picture naming. *Psychological Review* **98**, 122–42.

McClelland, J. L. & D. E. Rumelhart 1981. An interactive activation model of context effects in letter perception: part 1. An account of the basic findings. *Psychological Review* **88**, 375–407.

Martin, N., R. W. Weisberg, E. M. Saffran 1989. Variables affecting the occurrence of naming errors: implications for models of lexical retrieval. *Journal of Memory and Language* **28**, 462–85.

Meyer, A. S. & K. Bock 1992. The tip-of-the-tongue phenomenon: blocking or partial activation? *Memory and Cognition* **20**, 715–26.

Nickels, L. 1995. Getting it right? Using aphasic naming errors to evaluate theoretical models of spoken word production. *Language and Cognitive Processes* **10**, 13–45.

Perfect, T. J. & J. R. Hanley 1992. The tip-of-the-tongue phenomenon: do experimenter-presented interlopers have any effect? *Cognition* **45**, 55–75.

Plaut, D. C. & T. Shallice 1993. Deep dyslexia: a case study of connectionist neuropsychology. *Cognitive Neuropsychology* **10**, 377–352.

Quinlan, P. T. 1992. *The Oxford psycholinguistic database.* Oxford: Oxford University Press.

Rumelhart, D. E. & J. L. McClelland 1982. An interactive activation model of context effects in letter perception: part 2. The contextual enhancement effect and some tests and extensions of the model. *Psychological Review* **89**, 60–94.

Shallice, T. & J. McGill 1978. The origins of mixed errors. In *Attention and performance*, vol. 7, J. Requin (ed.), 193–208. Hillsdale, New Jersey: Lawrence Erlbaum.

Stemberger, J. P. 1985. An interactive activation model of language production. In *Progress in the psychology of language*, vol. 1, A. W. Ellis (ed.), 143–86. London: Lawrence Erlbaum.

Warrington, E. K. & T. Shallice 1984. Category-specific semantic impairments. *Brain* **107**, 829–54.

Index

Early Experience and the Development of Competence

William Fowler, *Editor*

NEW DIRECTIONS FOR CHILD DEVELOPMENT
WILLIAM DAMON, *Editor-in-Chief*

Number 32, June 1986

Paperback sourcebooks in
The Jossey-Bass Social and Behavioral Sciences Series

William Fowler (Ed.).
Early Experience and the Development of Competence.
New Directions for Child Development, no. 32.
San Francisco: Jossey-Bass, 1986.

New Directions for Child Development
William Damon, *Editor-in-Chief*

New Directions for Child Development (publication number USPS
494-090) is published quarterly by Jossey-Bass Inc., Publishers.
Second-class postage rates are paid at San Francisco, California, and at
additional mailing offices.

Correspondence:
Subscriptions, single-issue orders, change of address notices,
undelivered copies, and other correspondence should be sent to
Subscriptions, Jossey-Bass Inc., Publishers, 433 California Street,
San Francisco, California 94104.

Editorial correspondence should be sent to the Editor-in-Chief,
William Damon, Department of Psychology, Clark University,
Worcester, Massachusetts 01610.

Library of Congress Catalog Card Number 85-60825

International Standard Serial Number ISSN 0195-2269

International Standard Book Number ISBN 87589-799-1

Cover art by WILLI BAUM

Manufactured in the United States of America